Chemical Microbiology

An Introduction to Microbial Physiology

Dedicated to my wife
Jane

Chemical Microbiology

An Introduction to Microbial Physiology

Third Edition

Anthony H. Rose

School of Biological Sciences
University of Bath, England

PLENUM PRESS · NEW YORK

Published in the U.S.A. by
PLENUM PRESS
a division of
PLENUM PUBLISHING CORPORATION
227 West 17th Street, New York, N.Y. 10011

ISBN 0-306-30888-6

First published by
Butterworth & Co. (Publishers) Ltd.

First Edition 1965
Reprinted 1967
Second Edition 1968
Third Edition 1976

Suggested U.D.C. No. 576.8.098
Library of Congress Catalog Card Number 75-29515

Printed in Great Britain by Page Brothers, Norwich

PREFACE TO THE THIRD EDITION

Biologists have been fascinated by the activities of
micro-organisms ever since they became aware of the
existence of microscopically small creatures. To begin
with, their interests centred mainly on those micro-
organisms that cause diseases in Man and animals, and it
was some time before they came to appreciate the vitally
important roles that micro-organisms play in the natural
cycling of elements. Studies on the chemical activities
of micro-organisms, on which all other activities of these
organisms are based, had to await the birth of the new
science of biochemistry during the latter half of the
nineteenth century. Since then, interest in the chemical
activities of micro-organisms has grown at a tremendous
pace, and chemical microbiology is now recognised as a
subject in its own right.

In the preface to the first edition of this book, which
was published in 1965, I admitted to audacity in having
attempted to describe all aspects of the chemical
activities of microbes in a relatively small volume. The
reception given to the first edition, and also the second
edition published in 1968, somewhat allayed my apprehen-
sions. In preparing this, the third edition, I have
retained the same basic format, and updated all chapters
so that, not surprisingly, the volume is somewhat larger.
I have also removed all references from chapters,
preferring to include important review articles and papers
in a list of suggested reading at the end of each chapter.
This change, I believe, makes for a less interrupted text.

As with previous editions, many friends and colleagues
have helped with advice, unpublished articles, and
micrographs, and to all of them I am deeply grateful.
My thanks are especially due to Dr. Ronald Archibald of
the University of Newcastle-upon-Tyne, with whom I spent
some weeks when we were Visiting Professors at the
Biological Laboratories of the Gulbenkian Science
Foundation, near Lisbon in Portugal. During these weeks,
extensive revisions were made to the second edition of
this text, and I greatly valued Dr. Archibald's advice.
Several colleagues have kindly read chapters, and offered
their criticisms, none more assiduously than Dr. L. Julia
Douglas of the Microbiology Department in the University
of Glasgow. It would not have been possible to prepare
this third edition without the unfailing help and
encouragement of my wife Jane. My son Simon gave timely
and valuable help in preparing the index.

ANTHONY H. ROSE

Microbiology Laboratories
Bath University

vi

CONTENTS

ABBREVIATIONS AND SYMBOLS

NUCLEIC ACIDS AND NUCLEOTIDES

AMP, CMP, GMP, IMP, TMP and UMP	5'-Phosphates of adenosine, cytidine, guanosine, inosine, thymidine and uridine
ADP, CDP, GDP, IDP, TDP and UDP	5'-Pyrophosphates of adenosine, cytidine, guanosine, inosine, thymidine and uridine
ATP, CTP, GTP, ITP, TTP and UTP	5'-Triphosphates of adenosine, cytidine, guanosine, inosine, thymidine and uridine Deoxyribonucleotides are distinguished by the prefix d; e.g. dATP for the 5'-triphosphate of deoxyadenosine
DNA	Deoxyribonucleic acid
mDNA	Mitochondrial deoxyribonucleic acid
RNA	Ribonucleic acid
mRNA	Messenger ribonucleic acid
rRNA	Ribosomal ribonucleic acid
tRNA	Transfer ribonucleic acid

AMINO ACIDS

Ala	Alanine
Arg	Arginine
Asp	Aspartic acid
DAP	Diaminopimelic acid
Glu	Glutamic acid

Gly	Glycine
Ile	Isoleucine
Leu	Leucine
Lys	Lysine
Met	Methionine
Phe	Phenylalanine
Pro	Proline
Ser	Serine
Thr	Threonine
Val	Valine

SUGARS

Ara	Arabinose
Abe	Abequose
Col	Colitose
Gal	Galactose
Glc	Glucose
GlcNAc	N-Acetylglucosamine
Man	Mannose
MurNAc	N-Acetylmuramic acid
PRPP	Phosphoribosyl pyrophosphate
Rha	Rhamnose

VITAMINS AND COENZYMES

FAD, $FADH_2$	Flavin adenine dinucleotide, oxidised and reduced forms
NAD, $NADH_2$	Nicotinamide adenine dinucleotide, oxidised and reduced forms
NADP, $NADPH_2$	Nicotinamide adenine dinucleotide phosphate, oxidised and reduced forms
CoA.SH	Coenzyme A
TPP	Thiamine pyrophosphate

OTHER ABBREVIATIONS

Pi	Inorganic phosphate
PPi	Inorganic pyrophosphate

ENZYMES

Most of the enzymes mentioned in this book are referred
to by the trivial names recommended by the Commission on
Enzymes of the International Union of Biochemistry

TEMPERATURES

All temperatures recorded in this book are in degrees
Centigrade

INTRODUCTION

The aim of the microbial physiologist in studying the chemical activities of microbes is to explain in molecular terms the living processes of these organisms. Information on the chemical activities of micro-organisms has been steadily accumulating ever since the late seventeenth century when Antonie van Leeuwenhoek first observed micro-organisms. The early microbiologists had no option but to study mixed populations of micro-organisms. Following the work of Lord Joseph Lister and Robert Koch, it became possible to study organisms in pure culture, but although this greatly simplified the task of observing and classifying organisms, it meant that they were studied under conditions that were vastly different from those that obtain in natural environments.

The first experiments on the chemical activities of micro-organisms were mainly concerned with the ability of growing organisms to change the chemical composition of their environment by removing some compounds and excreting others. Many famous names were associated with this early period in chemical microbiology, including Louis Pasteur, Sergius Winogradsky and Martinus Willem Beijerinck. A major turning point in the history of the subject came in 1897 with the discovery by the Buchner brothers that fermentation of sucrose could be carried out by cell-free extracts of yeast, an observation which is often said to have given birth to the science of biochemistry. The Buchners' discovery sparked off a series of studies into the metabolic pathways used by micro-organisms beginning with the process of alcoholic fermentation by yeast. It showed, too, the tremendous advantages that can be gained by using micro-organisms for studying the chemical activities of living cells, and in the years that followed research in many areas of general biochemistry came to depend more and more on the Microbial Kingdom for experimental material. Today, most of the major metabolic pathways used by micro-organisms have been charted at least in outline, and the comprehensive metabolic maps which adorn the walls of biochemical laboratories provide eloquent testimony to the success that the biochemist has achieved in this venture.

During the past decade, microbiologists have become increasingly interested in the physiological aspects of microbial activity, aspects that were temporarily overshadowed by the spectacular success of the enzymologist in charting metabolic pathways. Great strides have been made in studies on regulation of microbial metabolism. This has been accompanied by an expanding interest in the relationship between structure and function in microbial organelles, particularly of the cell wall, plasma membrane, mitochondria and chloroplasts, and therefore in the physiology of the intact micro-organism. Differentiation processes in micro-organisms, though restricted compared with those carried out by higher organisms, are also attracting an ever-increasing number of workers, mainly because of the advantages which they offer as systems for elucidating the basic molecular processes of differentiation.

The chemical activities of microbes are studied by many different groups of biologists and, as a result, the literature on the subject is rather widely scattered. The main task in writing this book has therefore been to condense the vast quantity of published data on the various branches of microbial physiology. Inevitably, there have been occasions when it has been necessary to omit material that would have found a place in a more lengthy treatise, but I hope that these omissions have not been too arbitrary. The book is intended for readers who already have a basic knowledge of microbiology and biochemistry, and who wish to combine and extend this knowledge to make a study of the chemical activities of algae, bacteria, fungi, protozoa and yeasts. The tempo of research on this subject is feverish, and it has to be admitted that many sections of this book will rapidly become out of date. Readers who wish to keep abreast of developments in the subject are recommended to peruse the following review organs which regularly furnish up-to-date and authoritative reviews on a variety of aspects of the chemical activities of micro-organisms. In this way they will be able to share in the excitement as the molecular secrets of Leeuwenhoek's animalcules continue to be revealed.

Advances in Microbial Physiology, published biannually by Academic Press in London.

Annual Review of Biochemistry, published annually by Annual Reviews Inc., Palo Alto, California, U.S.A.

Annual Review of Microbiology, published annually by Annual Reviews Inc., Palo Alto, California, U.S.A.

Bacteriological Reviews, published quarterly by the American Society for Microbiology, Washington, D.C., U.S.A.

Critical Reviews in Microbiology, published by the Chemical Rubber Company Press Inc., Cleveland, Ohio, U.S.A.

2

MOLECULAR ARCHITECTURE

The chemical composition of micro-organisms has been analysed ever since techniques were devised for growing large quantities of microbes in pure culture. Inevitably, the results of these analyses reflected the precision of the analytical methods employed. Today, a large number of extremely sensitive micro-analytical techniques are available for studying the chemical composition of micro-organisms, and these methods permit the detection and separation of molecules that are present in microbes in very small amounts.

However, data on the overall chemical composition of micro-organisms tell us nothing about the ways in which the component molecules are located in the microbial cell. Fortunately, at the same time that progress was being made in the development of analytical methods, important advances were taking place in the subject of microscopy. The limit of resolution of the light microscope (about 0.2 µm) severely restricts its use in studying intracellular structures in microbes. By using the electron microscope, which has a limit of resolution nearer 0.001 µm, it is possible to examine intracellular structures in even the smallest microbe. The link between analyses of the chemical composition of micro-organisms on the one hand, and investigations into their fine structure on the other, came from studies in which intracellular structures and organelles in microbes were separated, and then analysed chemically or examined in the electron microscope.

5

2.1   PROKARYOTES AND EUKARYOTES

The detailed information on the subcellular organisation
of micro-organisms which emerged from studies with the
electron microscope confirmed what had, in fact, been
suspected for some time, namely that a major morphological
discontinuity exists between bacteria and blue-green
algae, and other micro-organisms. That bacteria and
blue-green algae are atypical organisms was suspected as
long ago as 1875 by the botanist Ferdinand Cohn, and
since the availability of fine-structure data,
microbiologists have appreciated that bacteria and blue-
green algae are fundamentally different from other micro-
organisms in that they are always much less differentiated
intracellularly. The writings of Roger Y. Stanier and
Cornelius van Niel of the University of California have
contributed significantly to an understanding of this
basic division in the Microbial Kingdom, and it was
largely following their suggestion that bacteria and
blue-green algae are now known as *prokaryotes*, and other
micro-organisms as *eukaryotes*. The basic distinction
between prokaryotic and eukaryotic micro-organisms is
that the former do not possess a nuclear membrane,
a structure which envelopes the genome in all eukaryotic
organisms. Other differences, such as the absence of
mitochondria, chloroplasts and endoplasmic reticulum from
prokaryotes, also exist and these, together with differences
in the molecular architecture of other organelles, are
discussed in later sections of this chapter.

2.2   METHODS USED IN STUDYING THE MOLECULAR
ARCHITECTURE OF MICRO-ORGANISMS

In order to study the molecular architecture of micro-
organisms it is necessary to isolate subcellular
structures from the organisms in an intact condition and
free from contamination with other subcellular components.
This is a two-stage process. In the first, micro-organisms
are disintegrated to free subcellular structures from one
another; in the second, these structures are separated by
various types of centrifugation regime. The isolated
structures and organelles can then be subjected to
chemical analysis or microscopic examination.

2.2.1 DISINTEGRATION OF MICRO-ORGANISMS

Methods for disrupting organisms have undergone extensive
development since 1897 when the Buchners first obtained a
cell-free yeast juice by grinding yeast with a mixture of
kieselguhr and sand. The number of different methods that
are used today is legion, and it seems that each
laboratory has developed its own techniques for preparing
those subcellular structures and organelles in which its
workers are principally interested.

Basically, there are two classes of method for
disintegrating micro-organisms. With the first, mechanical
forces which are often quite large are applied to a
suspension of micro-organisms, and this stress causes the
disintegration of some and frequently all of the organisms
in the suspension. Other, gentler methods do not employ a
mechanical force and do not cause as much damage to
subcellular structures and organelles.

Mechanical methods

These methods can be subdivided into those which employ a
solid shear and those which involve application of a
liquid or hydraulic shear. Because of the high tensile
strength of their walls, hydrodynamic gradients of up to
10^8 s^{-1} are required to disrupt micro-organisms; it is
important to note that subcellular organelles are often
damaged by gradients as low as 10^4 s^{-1}.

Some microbial physiologists still prefer to use an
up-dated version of the Buchners' original method which
creates a *solid shear* in the suspension of organisms.
Grinding a paste of microbes with glass powder, washed
alumina or sand in a chilled pestle and mortar is a very
effective disruption procedure, even with organisms that
have very tough walls. Another effective method for
disrupting micro-organisms exploits the solid shear
created by compression of ice crystals. In these methods,
a frozen suspension or paste of organisms is forced through
a small orifice into a receiving chamber at very high
pressures of about 5.5×10^8 Pa (or 80 000 psi). Two
widely used pieces of apparatus which employ this method
are the Hughes and 'X' presses.

Liquid-shear methods are of three main types.
Ultrasound (10-15 kHz) can be used to create a liquid
shear. This causes gaseous cavities to form in the
suspension which, in turn, lead to acoustic or micro-

streaming of liquid round each bubble, and it is thought
that the acceleration due to this streaming generates a
sufficiently large force to disrupt the organisms.
Several pieces of equipment are available for ultrasonic
disruption of microbes, including those marketed by
M.S.E.-Mullard and Bronwill Scientific. Another way of
imposing a liquid shear on microbes is to force a chilled
suspension through a small orifice, using a pressure of
$0.69-2.07 \times 10^8$ Pa. The French pressure cell and the
Chaikoff press employ this method, and both pieces of
equipment are very effective in causing disruption of
microbes. A third type of method imposes a liquid shear
on the microbes by rapidly shaking a chilled suspension
with small glass beads (such as Ballotini beads) in what
is essentially a vibration mill. Several manufacturers
market pieces of equipment for this purpose, notably
B. Braun of Melsungen in West Germany, all using high-
amplitude shaking. Considerable heat is generated in
these machines during the disruption process and it is
customary to cool the suspension with liquid carbon
dioxide. *Figure 2.1* shows electron micrographs of walls
obtained after disrupting bacteria by shaking with glass
beads. They show how the rigid wall structure has been
ruptured by the liquid shear imposed on the bacteria.

Non-mechanical methods

On the whole, non-mechanical methods are less commonly
used than those which involve application of a mechanical
force. Drying or desiccating a population of micro-
organisms is a long-established way of preserving enzyme
activity in cells, but clearly it does not permit
isolation of subcellular structures and organelles. The
most elegant way of isolating subcellular structures and
organelles by a non-mechanical method is first to convert
the organisms into protoplasts or sphaeroplasts, and then
to lyse or disrupt these osmotically sensitive structures
under gentle conditions. As a result, subcellular
structures and organelles are released from the protoplast
or sphaeroplast, and these can then be isolated and
examined.

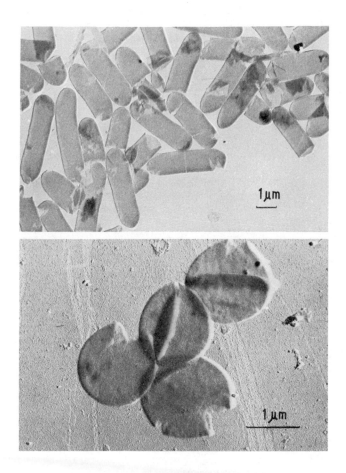

Figure 2.1 Electron micrographs of isolated walls of
Bacillus polymyxa *(top)* and Staphylococcus aureus *(bottom)*
showing how the wall was ruptured by a liquid shear
imposed on the bacteria. (By courtesy of E.H. Boult)

2.2.2 PROTOPLASTS AND SPHAEROPLASTS

By definition, *protoplasts* are the structures derived
from vegetative cells of micro-organisms by removal of
the entire cell wall; when part of the cell wall remains
attached to the structures, they are then termed
sphaeroplasts. Protoplasts and sphaeroplasts are
osmotically sensitive and can be formed only in solutions
of osmotic pressure comparable with that of the contents
of the protoplast or sphaeroplast. They are capable of
carrying out many of the metabolic reactions of the
intact organisms, including sometimes the ability to
regenerate cell-wall material.

A large number of different methods have been devised
for preparing protoplasts and sphaeroplasts. The basis
of most of these techniques is to remove enzymically, or
to prevent the biosynthesis of, those components of the
cell wall that are responsible for its rigidity. The
most successful technique so far devised for preparing
bacterial protoplasts involves treating a thick suspension
of certain Gram-positive bacteria in a stabilising
solution (e.g. 20% sucrose) with lysozyme (*see* page 33)
which fragments the polysaccharide backbone in the cell-
wall peptidoglycan. However, the walls of only a limited
number of bacteria, including *Bacillus megaterium,*
Micrococcus lysodeikticus and *Sarcina lutea,* are completely
susceptible to lysozyme action, although with some other
strains protoplasts can be formed by the combined action
of ethylenediamine tetra-acetic acid (EDTA) and lysozyme.
Other enzymes, including some from streptomycetes and
phages, have been used to prepare bacterial protoplasts.
Methods based on preventing the formation of cell-wall
peptidoglycan include treatment with penicillin, and
growing bacteria that are auxotrophic for peptidoglycan
constituents (e.g. lysine, diaminopimelic acid) in media
containing suboptimum concentrations of these nutrients.
Techniques have also been described for preparing
protoplasts of certain yeasts and filamentous fungi.
A commercial snail-juice preparation frequently used for
making yeast protoplasts contains a large number of
hydrolytic enzymes the combined action of which dissolves
the walls of certain yeasts and moulds. Unfortunately,
commercial snail juice contains phospholipase activity
which could damage the membrane that surrounds the yeast
protoplast. For this reason increasing preference is now
being given to purer preparations of β-glucanase, such as
that elaborated by Basidiomycete QM 806. Snail juice,

often in combination with other lytic enzymes such as
cellulase and chitinase, is used also to prepare fungal
protoplasts.

2.2.3 SEPARATION OF CELLULAR STRUCTURES AND ORGANELLES

The indispensible piece of equipment for separating
mixtures of microbial cellular structures and organelles
is the high-speed centrifuge. Without one the microbial
physiologist is virtually unable to isolate structures
of a quality that justify further examination. The
centrifuge offers two ways of separating subcellular
structures and organelles, namely on the basis of
differences in *sedimentation rate* and in *buoyant density*.
   The oldest and still the simplest way of separating
these structures and organelles in a preparation of
disrupted microbes is to subject the preparation to a
series of increasingly fast centrifugations, and to
remove the pelleted material after each centrifugation.
This procedure is referred to as *differential centrifugation*
and is based on differences in sedimentation rate. *Figure
2.2* shows an idealised scheme for separation of subcellular

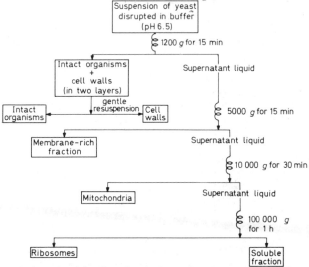

*Figure 2.2 Idealised scheme for the separation of cell
structures and organelles from a suspension of disrupted*
Saccharomyces cerevisiae. *All operations are carried out at
0-2°C. For separation of mitochondria, the yeast must be
grown aerobically and the disruption performed in 0.25 M
sucrose*

structures of *Saccharomyces cerevisiae* by differential
centrifugation. The technique does not permit isolation
of structures and organelles that are completely free
from contaminating material, except possibly for cell
walls, and is most useful as a means for preparing
fractions enriched in certain subcellular structures,
fractions which can then be freed from contaminating
material by other techniques.

The resolving power of differential centrifugation can
be increased by layering the preparation of disrupted
microbes onto a solution of higher density. On
centrifuging, structures and organelles are pelleted
through the solution under conditions that minimise
contamination. Alternatively, the disrupted microbes
can be layered onto the top of a solution which provides
a shallow gradient of densities. The solution is then
centrifuged briefly to bring the structures and
organelles into the gradient in the form of zones in
positions which reflect the different sedimentation rates
of the structures. This technique is known as *zonal
centrifugation*. Although for use on a large scale it
requires a special centrifuge rotor, the technique is
now recognised as one of the most powerful methods for
separating microbial subcellular structures and
organelles.

If a subcellular structure or organelle is centrifuged
through a gradient of densities to a point where its
density equals that of the surrounding liquid, its
centrifugal motion ceases. This forms the basis of the
equilibrium or *isopycnic density-gradient* method for
separating mixtures of structures and organelles. The
mixture of organelles can be loaded onto a discontinuous
or a continuous gradient of densities. Centrifugation
must be powerful enough to allow the distribution of
structures on the gradient to reach equilibrium. This
usually requires about 10^8 *g*.-min. *Figure 2.3* depicts
an idealised separation of the major subcellular
structures and organelles from lysed sphaeroplasts of
Sacch. cerevisiae using isopycnic density-gradient
centrifugation.

After analysing the chemical composition of the
subcellular structures and organelles obtained from an
organism, the microbiologist can begin to visualise the
overall distribution and arrangement of molecules within
the microbe. Certain generalisations regarding the
molecular architecture of micro-organisms have emerged
from these studies. It is clear for instance that,

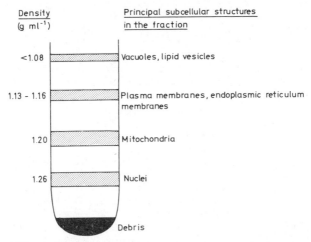

Figure 2.3 An idealised separation by isopycnic density-gradient centrifugation of the major subcellular structures from a preparation of lysed sphaeroplasts of aerobically grown Saccharomyces cerevisiae. *See text for details*

although micro-organisms vary tremendously in size and shape and in nutritional requirements, they nevertheless resemble plant and animal cells in containing a high proportion of water (around 80%) and also the same types of molecules (nucleic acids, proteins, lipids, poly-saccharides). Moreover, certain subcellular structures, including plasma membranes and ribosomes, appear to be similar in size and molecular architecture in all types of cell. Other subcellular organelles, such as mitochondria, are found only in larger micro-organisms and are absent from prokaryotic organisms.

This chapter deals with the molecular architecture of the major subcellular structures in micro-organisms. For convenience, the outermost structures are considered first, followed by those which lie more deeply inside the micro-organism. It should be realised, however, that many of the terms in which microbial physiologists currently express their ideas of cell structure - capsule, cell wall, plasma membrane - are essentially operational. Some of these layers may well be nothing more than arbitrary dislocations in structure with interrelationships that are not yet amenable to study.

2.3   SURFACE APPENDAGES

Some micro-organisms have structures protruding from
the cell surface, the commonest of which are cilia,
flagella and pili. Other types of appendage, such as
the large threads that occur on the iron bacterium
Gallionella ferruginea, are also known but these are not
dealt with in this account.

2.3.1   CILIA AND FLAGELLA

Cilia and flagella are thread-like structures found on
certain algae, bacteria, fungi and protozoa. Their
normal function is the movement of liquid relative to the
point of attachment of the organelle. If the micro-
organism can move, this results in locomotion of the
organism; if on the other hand the organism is incapable
of moving, then it serves to move liquid over the surface
of the micro-organism.

Fortunately for the microbial physiologist, cilia and
flagella can usually be removed from micro-organisms by
treating a suspension of organisms in a Waring blender.
The detached appendages are obtained in a clean condition
by differential centrifugation followed by repeated
washing, and are then subjected to analysis and
examination.

Eukaryotic cilia and flagella

In algae and protozoa, the appendages are usually
150-300 nm in diameter and 20-2000 μm long. Each one is
attached to a basal body within the organism. Electron
microscope studies show that algal and protozoal cilia
and flagella are similar in fine structure. Each
appendage is surrounded by a unit membrane (about 8 nm
wide) which is continuous with the plasma membrane.
Internally, thin sections of the appendage show a ring
of nine pairs of microtubules (each 20-24 nm in diameter)
surrounding a further two microtubules in the centre.
This arrangement is referred to as the '9 + 2' structure.

Cilia and flagella from a few selected eukaryotic
microbes have been the subjects of a more intensive
examination. Notable among these are the flagella from
the unicellular alga *Chlamydomonas reinhardii* which has
been favoured by Sir John Randall and his colleagues of

King's College, University of London. Flagella can be
detached from *Ch. reinhardii* simply by exposing the algae
to calcium ions. Flagella isolated from this alga have
a somewhat more intricate anatomy than that indicated
above, although it may be one which is shared with other
eukaryotic flagella. In addition to the tubules which
go to make up the '9 + 2' structure, flagella from
Ch. reinhardii have several subsidiary structures
(*Figure 2.4*). Firstly, there are radial spokes attached

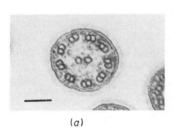

(a)

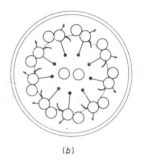

(b)

Figure 2.4 Fine structure of the flagellum of
Chlamydomonas reinhardii. *(a) An electron micrograph of
a thin section through a flagellum; the bar indicates a
length of 0.1 μm. (b) A drawing showing the typical
'9 + 2' structure of eukaryotic flagella as well as the
subsidiary structures found in flagella of* Ch. reinhardii.
(The micrograph is reproduced by courtesy of J.M. Hopkins)

to one of each of the pairs of outer tubules and extending
into the centre of the flagellum. However, these radial
spokes do not reach the centre tubules. At the free end
of each radial-spoke fibre is a hammer-head shaped
structure whose axis is parallel to the long axis of the
flagellum. Also attached to one of the tubules in each
outer pair are 'side arms' which are hooked at their ends,
and at right angles to the radial spokes. Microtubules
of the type detected in algal and protozoal flagella have
recently been found as intracellular structures in other
eukaryotic microbes, often associated with membranes and
nuclear spindles.
     Very little indeed is known about the chemical
composition of eukaryotic flagella. The King's College

group have, however, published preliminary data on the
composition of flagella from *Ch. reinhardii*.   The proteins
in the various tubules would seem to be different.   A low
ionic-strength buffer containing EDTA extracts protein
from just one of the centre pair of tubules and part of
one of each of the fibres associated with the outer nine
tubules.   Moreover, the outer nine tubules are not
dissolved by dialysis against 0.6 M potassium chloride,
but are soluble in ionic detergents.   Obviously, much
remains to be discovered about the molecular anatomy of
these eukaryotic appendages.

Bacterial flagella

Bacterial flagella (*Figure 2.5*), which can account for
up to 2% of the dry weight of a bacterium, differ in many
respects from their counterparts on eukaryotic microbes.
They measure only one-tenth of the diameter of eukaryotic
flagella (10-30 nm in diameter) and consist almost
entirely of protein.

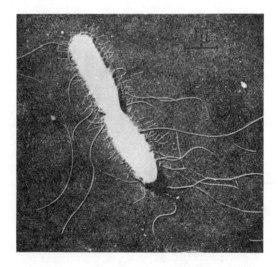

Figure 2.5   Electron micrograph of the bacterium
Salmonella typhi NCTC 6029 *showing flagella and pili.*
(By courtesy of J.P. Duguid and J.F. Wilkinson)

Anatomically, a bacterial flagellum consists of three parts: a basal structure, a hook and the main filament. The basal structure, the composition of which is not known, is often spherical (20-50 nm in diameter) and is very fragile. It anchors the flagellum in the plasma membrane and cell wall of the bacterium. A hook connects the filament of a flagellum to the basal structure, and can account for up to 1% of the weight of the flagellum. Reasonably pure preparations of hooks, which consist of protein antigenically different from that of the filament, have been obtained from several bacteria.

The filament consists almost exclusively of protein and is the site of the 'H' (Hausch) antigen. A few filaments, such as those from *Bdellovibrio bacteriovorans,* are sheathed. If a suspension of filaments is brought to a pH value of 3-4, or is treated with detergents, the filaments disintegrate to yield a soluble protein termed *flagellin.* The molecular weights of flagellins vary with the species of bacterium from which they are obtained, but are usually in the range of 30 000-50 000 daltons. It has been suggested that with some filaments, such as those from *Bacillus pumilus,* there may be two different species of protein.

Individual flagellins differ in their amino-acid composition, but all are free from, or very low in, residues of cysteine, histidine, proline, tryptophan and tyrosine. Flagellins from *Salmonella typhimurium* and *Spirillum serpens* contain the unusual amino-acid residue ε-N-methyllysine, while those from *Bacillus stearothermophilus* and *Spirillum serpens* are glycoproteins.

One of the most puzzling problems associated with bacterial flagella is the arrangement of the protein subunits in the organelle. Electron microscopy shows flagellin molecules to be globular or ovoid, but limitations in the microscopic and physical techniques used to examine isolated filaments are largely responsible for the lack of detailed knowledge of the ways in which the flagellin subunits are stacked. It is clear that the packing arrangement varies in flagella from different bacteria. Many filaments appear to be hollow cylinders with the subunits arranged in a helical fashion (*Figure 2.6*). With subunits that have an ovoid or elongated shape, the filament may have an appearance rather like that of a pine cone.

Flagellin subunits have the remarkable capacity to reaggregate. When a solution of flagellin is adjusted to pH 6.0, the subunits reaggregate spontaneously to form helical fibrils which, although larger, look like normal filaments in the electron microscope.

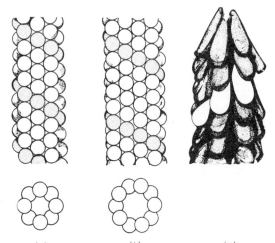

(a)   (b)   (c)

Figure 2.6 *Side and end-on views of three models*
showing various arrangements for flagellin molecules in
bacterial flagella. The models were constructed from
data obtained by electron-microscope and X-ray
crystallographic examination of flagella. (a) Arrangement
of molecules in flagella from Salmonella typhimurium, *and*
in (b) that in flagella from Pseudomonas fluorescens. *The*
end-on views below the side views in (a) and (b) are
intended only to show the arrangement of flagellin
molecules at the periphery. It has yet to be demonstrated
conclusively that the flagella are hollow. (c) Side view
of the pine-cone model for arrangement of molecules in
flagella from Proteus mirabilis

Pili

Pili (which are also known in Europe as *fimbriae*) are
filamentous appendages which occur on many strains of
Gram-negative bacteria especially on enteric bacteria
and some pseudomonads. These hair-like structures were
discovered in the 1950s, independently by Charles Brinton
in Pittsburgh, U.S.A. and by James Duguid working in
Edinburgh, Scotland. They are generally narrower than
bacterial flagella, and are only about 5-10 nm wide.
Pili have been classified into a number of different
types on the basis of their width and length and on the
nature of the bacteria on which they occur.
   Clean preparations of pili can be obtained simply by
shaking a suspension of non-flagellated bacteria that

possess pili. So far, the chemical composition of only
a few of these has been studied in any detail. Type I
pili, which are found on *Escherichia coli* K-12, consist
almost entirely of protein known as *pilin*. Pilin, which
has a molecular weight of about 16 600 daltons, contains
a high proportion of amino-acid residues with hydrocarbon
side chains, and a low proportion of basic amino-acid
residues. The proportion of free carboxyl groups is also
low. The preponderance of non-polar side chains in the
amino-acid residues of pilin may account for the
haemagglutinating properties of Type I pili.

The presence of the sex factor, or of certain drug-
resistance factors, in bacteria permits the organism to
form another type of pilus, called an *F-pilus*. F-Pili
can be distinguished from other types of pili in that they
have a terminal knob (*Figure 2.7*), and are often wider and

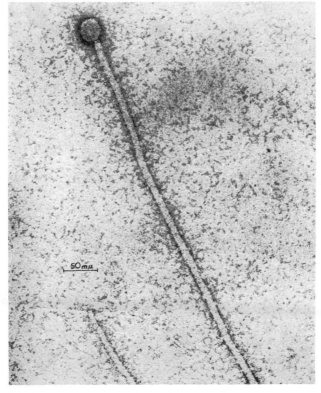

Figure 2.7   Electron micrograph of an F-pilus from
Escherichia coli K-12.   *(By courtesy of A.M. Lawn)*

longer. Recently, reports have appeared on the
composition of F-pili, so much so that more is now known
about these pili than about any other type of pilus.
This work has been aided by the isolation of mutant strains
of bacteria that produce abnormally large numbers of
F-pili, and by the discovery that these organelles are
released into the culture liquid and can therefore be
isolated in large quantities without the need for
mechanical stress. F-Pilin, the protein of F-pili, has
a molecular weight of 11 800 daltons and is of interest
in that it is devoid of arginine, cysteine, histidine
and proline residues, in addition to being low in
tryptophan and tyrosine residues. Moreover, each
polypeptide chain of F-pilin contains two phosphate
residues and a D-glucose residue, although the function
of these residues is not understood. It has also been
established that the terminal knob on the F-pilus is not
chemically different from the rest of the filament, but
is simply a vesicle formed from F-pilin. Like bacterial
flagella, F-pili can be separated into their components
and reassembled. Electron micrographs of sections through
F-pili suggest that the pilus is built up of two parallel
rods of protein, each consisting of an assembly of
monomers of F-pilin. The groove between the two rods
forms the channel along which DNA is transferred from the
donor to the recipient bacterium during conjugation.

2.4 CAPSULES AND SLIME LAYERS

Many micro-organisms are surrounded by layers of material
that lie external to the rigid cell wall; these are the
capsules and slime layers (*Figure 2.8*). Like the surface
appendages, these extramural layers can usually be
removed from the micro-organism without affecting its
viability. Anatomically they can be divided into the
following three categories: (a) *macrocapsules* which are
demonstrable by light microscopy (and so are at least
0.2 μm thick) and have a definite external surface.
Since the materials making up these capsules are not
easily stained, the layers are usually detected by
negative staining using Indian ink, or more specifically
using homologous antibody; (b) *microcapsules,* which are
capsular layers less than 0.2 μm thick and so cannot be
demonstrated using the light microscope; they can, however,
be detected immunologically. It is a moot point whether
or not certain of these microcapsule materials should be

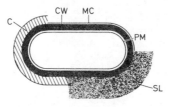

Figure 2.8 A diagrammatic representation of the anatomical relationship between the outer layers of a micro-organism, showing the plasma membrane (PM), cell wall (CW), slime layer (SL), capsule (C) and microcapsule (MC)

considered as cell-wall components; (c) *slime layers*, which accumulate at the surface of the micro-organism and have little definite anatomical significance. Micro-organisms which produce capsules often form a layer of slime similar if not identical in composition to the capsular material. The material making up the slime layer is often found in the culture fluid.

The extramural material produced by a micro-organism may appear in any one of these anatomical forms, depending upon the strain of micro-organism and the composition of the environment. Capsulated micro-organisms can also mutate to non-capsulated varieties and, since the former give mucoid or smooth (S) colonies on plate cultures whereas the non-capsulated strains appear as rough (R) colonies, this is known as the S → R transformation.

Several different techniques have been used for extracting capsule and slime materials from micro-organisms. Slime can often be separated simply by centrifuging the culture. Capsule material can occasionally be removed from the organisms by shaking the culture, but it is usually necessary to extract with dilute acid or alkali. Reagents such as trichloroacetic acid, phenol-water or diethylene glycol are required to release the components of some microcapsules, but use of these more drastic reagents should be avoided if chemical degradation of the capsule material is to be kept to a minimum.

Several functions have been suggested for microbial capsules and slime layers. The most plausible suggestion is that these extramural layers protect the organism against desiccation as a result of the ability of the layers to retain considerable quantities of water. The numerous carboxyl groups in these extramural polymers

could also bind cations and so provide a role for the
layers in cation absorption.   It has also been suggested
that capsules and slime layers protect micro-organisms
against phagocytosis, possibly by inhibiting action of
the enzyme lysozyme.

2.4.1   STRUCTURE

Water is the principal component of capsules and slime
layers.   Certain species of *Bacillus* have capsules in
which the main organic component is polyglutamic acid
either alone or in association with polysaccharide.
Production of polyglutamate capsules appears to be
confined to these bacteria.   The glutamate residues in
the polymer are γ-linked and may be in the D- or L-form.
Capsule materials which yield both D- and L-glutamate on
hydrolysis probably consist of two polypeptides each
composed of a single monomer.   Certain streptococci
possess a microcapsule containing protein, and this has been
shown to be the M antigen in these bacteria.   Nevertheless,
the commonest organic constituents of capsules and slime
layers are polysaccharides.

Homopolysaccharides

Large numbers of micro-organisms produce extracellular
homopolysaccharides, which are polysaccharides containing
only one type of sugar residue.   Among the best studied
of these polysaccharides are glucans which are formed,
often in copious amounts, in cultures of species of
Leuconostoc and *Pediococcus*.   These glucans, which are
branched molecules with predominantly α-1,6 but also
some α-1,3 and α-1,4 linkages, have been used commercially
as blood-plasma extenders.   Some species of *Acetobacter*
produce extracellular cellulose which can form a tough
pellicle on cultures of these bacteria.   Levans are
secreted by many plant-pathogenic bacteria belonging to
the genera *Pseudomonas* and *Xanthomonas* as well as by some
bacilli and *Streptococcus salivarius*.   Levans are
polyfructoses, often with a molecular weight in excess of
one million daltons, in which the predominant linkage is
β-D-fructosyl-2,6-D-fructose.   Extracellular homopolysaccharides are also often produced by fungi.   *Pullularia
pullulans,* for example, produces a polysaccharide called
pullulan, which is a linear polymer of α-D-glucopyranose

residues with 1,4 and 1,6 linkages. A few micro-
organisms produce phosphorylated homopolysaccharides.
The yeasts *Hansenula holsti* and *H. capsulata* secrete
phosphomannans which have phosphodiester linkages between
the mannose residues. Acetylated homopolysaccharides
have also been detected in microbial capsules. The Vi
antigen, which is found overlying the O-antigen in
virulent strains of *Salmonella typhi,* is made up of
N-acetyl-galactosaminuronic acid residues. Some micro-
organisms secrete two or more separable homopolysaccharides.
Penicillium charlesii, for example, produces a galactan
and a mannan. The galactan is of more than passing
interest since it appears to contain a large proportion
of D-galactofuranosyl residues.

Heteropolysaccharides

Most microbial extracellular polysaccharides contain more
than one type of sugar residue, that is, they are
heteropolysaccharides. Representatives from all of the
major groups of micro-organisms produce extracellular
heteropolysaccharides. The bacterium *Pseudomonas
aeruginosa,* for example, secretes an acidic
polysaccharide which contains D-glucose, D-galactose,
D-mannose, L-rhamnose and D-glucuronic acid residues.
Yeasts of the genus *Cryptococcus* characteristically form
capsules; that produced by *C. laurentii* contains
D-mannose, D-xylose and D-glucuronic acid residues.
Algal mucilages also frequently contain heteropolysaccharides. In the polysaccharide secreted by
Chlamydomonas ulvaensis, glucose and xylose are the main
constituent sugars.
   Data on the detailed structure of microbial capsular
heteropolysaccharides are most extensive for certain
medically important bacteria. Because of the need in
medical microbiology for rapid means of identifying
pathogenic bacteria, a considerable amount of information
has accumulated on the serological properties of these
bacteria, and schemes have been drawn up for subdividing
genera and species into serologically distinguishable
groups and subgroups. The availability of these
serological classifications has prompted chemical
microbiologists to study the fine structure of the
antigenically active components in capsulated bacteria,
and to attempt to relate chemical structure with antigenic specificity.

Strains of *Diplococcus pneumoniae* (Pneumococcus) were among the first pathogenic bacteria to be classified into serotypes. The number of recognised types and subtypes of *D. pneumoniae* is now well over 75. Only a few of these are of clinical importance, and representatives of Types 1-8 account for about three-quarters of all cases of lobar pneumonia in human adults. Each type of *D. pneumoniae* possesses a macrocapsule which differs from other types in chemical structure. *Figure 2.9* shows the structures of two pneumococcal polysaccharides. Type III polysaccharide has one of the simplest chemical structures known in microbial extramural polysaccharides. Knowledge of the chemical structures of many of the pneumococcal capsular polysaccharides is sufficiently detailed in that some workers, notably Michael Heidelberger working in Columbia University in New York, U.S.A., have used interactions with type-specific pneumococcal antisera to predict the structures of unknown polysaccharides.

Another group of bacteria that has been intensively studied serologically and biochemically is that which comprises the species *Escherichia coli*. These strains

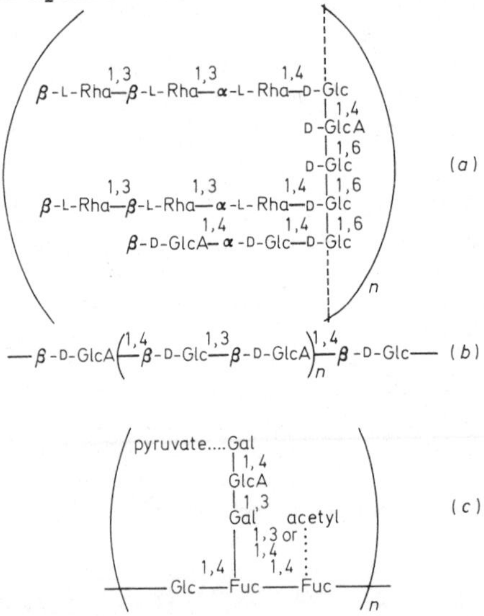

Figure 2.9 Structures of the repeating units in capsular polysaccharides produced by some bacteria. (a) Type II capsular polysaccharide formed by Diplococcus pneumoniae. *(b) Type III capsular polysaccharide formed by* Diplococcus pneumoniae. *(c) Colanic acid produced by* Salmonella typhimurium. *See page x for an explanation of the abbreviations used for names of sugar residues*

are subdivided into three groups, referred to as A, B and L. Of these, the B and L types produce microcapsules whereas the A types form macrocapsules. In addition, a large number of strains of *E. coli* together with strains of most *Salmonella* species and *Aerobacter cloacae* are capable, under suitable conditions, of excreting an extracellular slime containing a polysaccharide known as *colanic acid*. The structure of the colanic acid in the slime produced by *Salmonella typhimurium* is shown in *Figure 2.9*. The hexasaccharide repeating unit is common to colanic acids formed by a variety of strains of enteric bacteria. Variations do, however, occur in the non-carbohydrate residues (acetate and pyruvate in the colanic acid from *S. typhimurium*) in other colanic acids, both as regards their nature and their position on the sugar residues.

Little has been reported on the ultrastructure of microbial capsules. It has however been established that, when stained with ruthenium red and osmium tetroxide, the capsule of *Diplococcus pneumoniae* has the appearance in the electron microscope of a tightly woven mat, while that surrounding *Klebsiella pneumoniae* has a fibrous appearance.

The nature of the association between capsular and slime-layer components and the underlying cell wall is largely unknown. Many workers believe that ionic bonding is largely responsible for 'trapping' the capsule around the micro-organism. There is also evidence that, in certain micro-organisms, covalent bonding may exist between capsular and cell-wall components, as shown by the discovery that a polysaccharide isolated from capsules on *Bacillus anthracis* contained a covalently bound peptidoglycan. It is not unreasonable to wonder whether this capsular polysaccharide may not have been retained on the surface of the bacillus by covalent bonding to the cell-wall peptidoglycan.

2.5   CELL WALLS

The main structural component in most micro-organisms, with the exception of mycoplasmas, L-forms and certain protozoa, is the rigid cell wall. This is the envelope which surrounds the protoplast of the micro-organism and gives the microbial cell its characteristic morphological form, e.g. rod or coccus. Actively metabolising micro-organisms take up low molecular-weight compounds (amino acids and ions) from the environment and concentrate these within the cell, with the result that a high osmotic pressure difference is set up across the surface

layers of the micro-organisms. The plasma membrane,
which is the osmotic barrier in the micro-organism, is
unable to withstand any appreciable osmotic pressure
difference and, unless protected in some way, would
quickly rupture as water flowed into the organism. This
protection is afforded by the rigid cell wall.

The cell wall can account for between 10 and 50% of
the dry weight of a micro-organism. The amount of cell-
wall material in a micro-organism may vary during the
growth cycle, and usually increases as the organism ages.
The thickness of the wall may be determined to some
extent by the chemical composition of the environment.

Until about 20 years ago, knowledge of the chemical
composition of microbial cell walls had come largely from
histochemical studies and was on the whole rather meagre.
During the past 20 years, techniques have been developed for
isolating preparations of microbial cell walls (*see* page
7) which, after being separated from the suspension of
disrupted organisms, can be washed repeatedly to remove
contaminating material before being analysed.

The walls of most micro-organisms are made up of two
main types of component: a more or less organised
network of *microfibrils* which gives the wall its rigidity,
and a *matrix* in which the microfibrils are embedded and
which can often be removed chemically without altering
the shape of the wall. As such, the microbial cell wall
can be compared to pre-stressed concrete which is made up
of a network of steel rods in a concrete matrix.

2.5.1 ALGAE

In the majority of algae, the structural microfibrils are
made of cellulose. The fibrils are usually about 10-20 nm
thick, although in *Chlorella pyrenoidosa,* which has a very
thin cell wall, they are narrower (3-5 nm). Evidence from
X-ray diffraction studies has shown that, at least in
some algal walls, the cellulose microfibrils occur in
layers. Other polysaccharides such as xylans and mannans
replace cellulose in the structural microfibrils of some
algae. The microfibrils usually account for 50-80% of
the dry weight of the algal cell wall.

A number of different polymers have been reported to
make up the matrix of the algal cell wall. Many of these
are heteropolysaccharides, although protein and lipid may
also be present. In members of the Phaeophyta, alginic
acid, a polysaccharide containing D-mannuronic acid and

L-glucuronic acid residues, occurs as a cell-wall matrix
constituent. Fucoidin, a sulphated polysaccharide made
up of L-fucose residues, is found in the walls of brown
seaweeds. Information on the cell-wall matrix
polysaccharides in *Chlorella pyrenoidosa* is more complete.
There are two polysaccharides, one a β-linked galacto-
rhamnan which accounts for 80% of the matrix
polysaccharide, the other being a related but chemically
more complex polysaccharide. These two matrix
polysaccharides are concentrated partly at the outer
and inner surfaces of the cell wall.

Certain algae are rather exceptional in that their
cell walls become impregnated with deposits of silica
or calcium carbonate. Silicification is particularly
common in the diatoms in which over 50% of the dry weight
of the organism may be silica, and in the Chrysophyta
which form silica scales on the cell surface.

2.5.2 FUNGI

The walls of most of the fungi that have been examined
consist mainly of polysaccharide together with small
amounts of protein and lipid. Little is known of the
properties of the minor components. Polysaccharides
form both the microfibril and matrix components, the
chemical nature of which varies with the taxonomic
grouping of the fungi. These variations are shown in
Table 2.1. A close relationship between taxonomy and
wall composition is, of course, not surprising since the
morphological form of a fungus, which is a reflection of
the chemical composition of the wall, is a character
used in fungal taxonomy. In the lower fungi (e.g. the
Acrasiales), the wall microfibrils are composed of
cellulose which, in the higher fungi, is replaced by
chitin. But when fungi exist in the single-cell or
yeast state, glucan usually takes over as the structural
component of the wall.

Only a few fungi have been examined in detail for wall
composition, and these are principally ascomycetes,
basidiomycetes and deuteromycetes, mainly because these
classes accommodate the majority of fungal species. One
of the better understood fungal walls is that from the
deuteromycete *Aspergillus niger.* The polysaccharides in
walls of this fungus are, in addition to chitin, glucans
which mainly have α-1,3 linkages between the residues.
Three different glucans, of which two are soluble in

Table 2.1 Relationships between wall composition and taxonomic grouping in fungi

Principal polymers in the wall	Taxonomic class	Examples of organisms which have been analysed
Cellulose, glycogen	Acrasiales	Dictyostelium discoideum
Cellulose, glucan	Oomycetes	Pithium debaryanum
Cellulose, chitin	Hyphochytridiomycetes	Rhizidiomyces sp.
Chitin, chitosan	Zygomycetes	Mucor rouxii
Chitin, glucan	Chytridiomycetes	Allomyces macrogynus
	Ascomycetes (mycelial forms)	Neurospora crassa
	Basidiomycetes (mycelial forms)	Schizophyllum commune
	Deuteromycetes	Aspergillus niger
Glucan, mannan	Ascomycetes (yeast forms)	Saccharomyces cerevisiae
	Deuteromycetes (yeast forms)	Candida utilis
Chitin, mannan	Basidiomycetes (yeast forms)	Sporobolomyces rosei
Galactan, polygalactosamine	Trichomycetes	Amoebidium parasiticum

alkali, have been separated from hyphal walls of
A. niger. One of the alkali-soluble fractions, known
as nigeran, is also soluble in water and is thought to
have alternating α-1,3 and α-1,4 linkages between the
glucose residues. Less is known about the alkali-
insoluble glucan, but it probably has β-1,3 linkages
between the residues. An ascomycete fungus whose walls
have been the subject of more detailed examination is
Neurospora crassa. In this fungus, two-thirds of the
hyphal wall is accounted for by chitin and glucan. While
it is generally agreed that chitin is the main
microfibrillar structural component, it is possible that
a β-linked glucan may share this role.

Yeasts, which are single-celled fungi, have been the
subjects of biochemical study for over a century. One
might reasonably expect, therefore, that knowledge about
the yeast cell wall would be fairly extensive. Sadly,
this is not so, a situation which is explained largely
by the chemical and structural complexity of the yeast
cell wall.

In *Saccharomyces cerevisiae,* the yeast species that
has been most intensively studied, the wall contains
approximately equal quantities of glucan and mannan, which
together account for about 85% of the dry weight of the
wall. The remainder of the wall consists of protein,
glucosamine, phosphate and lipid. When walls of *Sacch.
cerevisiae* are treated with hot dilute sodium hydroxide
solution, the mannan and some of the glucan are extracted,
leaving a shell which retains the shape of the intact
wall. It is presumed, therefore, that glucan constitutes
the structural component of the wall in this yeast.
Fibrils of glucan have, indeed, been observed on the
surfaces of protoplasts of *Sacch. cerevisiae* that have
been incubated under conditions which allow a partial
regeneration of the cell wall.

Although it has been a popular subject of study among
carbohydrate biochemists for many years, the structure of
the glucan from walls of *Sacch. cerevisiae* is still not
completely understood. Current opinion favours the view
that most of the yeast cell-wall glucans that are
examined are heterogeneous and consist of two separable
components. The major component, which can account for
as much as 85% of the total glucan, is a branched
β-1,3 molecule, with a molecular weight of around
240 000 daltons and containing a small percentage of
β-1,6 glucosidic interchain linkages. The minor component
is a highly branched β-1,6 glucan with some β-1,3 interchain

linkages. The mannan in walls of *Sacch. cerevisiae* is
also a branched polymer, with a backbone of α-1,6-linked
residues in which many, and possibly all, of the residues
bear a side chain of 2-5 mannose residues joined by
α-1,2 or α-1,3 linkages. About 10% of the mannan side
chains have at their branch points two mannose residues
joined by a phosphodiester linkage.

The main problems which face the physiologist who
wishes to research on the walls of *Sacch. cerevisiae* are
the complex ways in which the wall polymers are joined
together. Here, the cell-wall protein holds the key,
since it is linked to both the glucan and the mannan, as
shown by the isolation of glucan-protein and mannan-
protein complexes following fractionation of isolated
walls. Protein may be linked to mannan through serine
or threonine residues in the polypeptide. Alternatively,
these polymers may be linked through an N-acetylglucosamine
residue, which is joined to C-3 or C-4 of a mannose
residue and by a nitrogen-glycosyl bond to an aspartamide
residue in a polypeptide, as illustrated. Enzymic
activities, including invertase and melibiase, have been
associated with mannan-proteins in the wall of *Sacch.
cerevisiae*. In addition to acting as a link between
mannan and protein, some N-acetylglucosamine in the wall
of *Sacch. cerevisiae* is thought to exist in a polymerised
form as chitin, which is concentrated to some extent in
the bud scars.

What is thought to be the molecular anatomy of the wall
of *Sacch. cerevisiae* is shown in *Figure 2.10*. Evidence
for the surface location of the mannan came from the
discovery that the antigenic determinants in the yeast
wall are mannans. The net negative charge or *zeta*

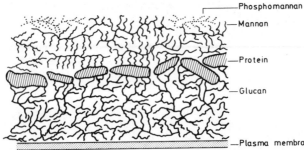

Figure 2.10. Schematic structure of the cell wall of
Saccharomyces cerevisiae. *The proteins shown in the mannan
layer, which are covalently bound to mannan, represent
glycoprotein enzymes, such as invertase and melibiase, that
are located in the wall of this yeast. The wall is about
70 nm thick*

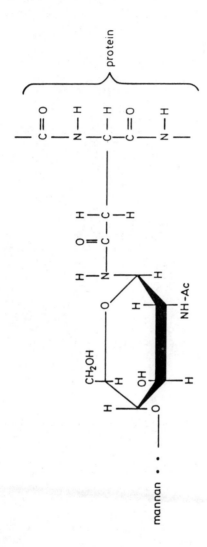

potential (ζ) on cells of *Sacch. cerevisiae* is thought
to be mainly attributable to the phosphodiester linkages
between mannose residues which lie on or near the cell
surface.
   Information on the cell-wall composition in other yeast
species is fragmentary. While the walls of most of these
organisms contain a glucan, this is not always
associated with mannan. Walls of *Saccharomycopsis
guttulata* and *Endomycopsis capsularis* contain only small
amounts of mannan, while those in *Nadsonia fulvescens* and
in the fission yeast *Schizosaccharomyces* spp. are devoid
of this polymer. With the exception of *Schizosaccharomyc*
spp., all mannan-deficient yeasts contain increased
amounts of chitin in their walls.

2.5.3   BACTERIA

Few microbiologists would deny that the work of the past
15 years on the biochemistry of the bacterial cell wall
represents one of the most exciting developments in
contemporary chemical microbiology. Not only have these
studies allowed the biochemist to visualise the molecula:
architecture of the bacterial cell wall, but they have
also led to the discovery of two hitherto unknown classe:
of heteropolymers, peptidoglycans and teichoic acids, as
well as to knowledge of the mode of action of certain
antibiotics and a deeper insight into the nature of
bacterial group-specific antigens.

Peptidoglycans

The structural components in almost all of the bacterial
walls so far examined are peptidoglycans, which have als
been found in the walls of blue-green algae and ricketts
The amount of peptidoglycan in the bacterial wall varies
from as much as 95% in some Gram-positive bacteria to as
little as 5-10% in Gram-negative species.
   Peptidoglycans (they are also called mucopeptides,
glycopeptides, glycosaminopeptides and mureins) form the
residue when isolated walls of preferably Gram-positive
bacteria are treated with trypsin and extracted with
trichloroacetic acid or hot formamide. On acid hydrol₃
all cell-wall peptidoglycans yield glucosamine and a
small number of amino acids, some of which have the
D-configuration.

All of the bacterial cell-wall peptidoglycans so far examined are made up of a polysaccharide backbone which usually consists of alternating residues of N-acetyl-glucosamine and N-acetylmuramic acid (the 3-O-D-lactyl ether of N-acetylglucosamine) linked by β-1,4 linkages (*Figure 2.11*). The average chain length of the cell-wall peptidoglycan varies between 20 and 140 hexosamine residues depending upon the species. A few bacteria (*Staphylococcus aureus* and a few strains of *Lactobacillus acidophilus* and *Micrococcus lysodeikticus*) have O-acetyl groups on C-6 of some of the N-acetylmuramic acid residues. This substitution renders the peptidoglycan resistant to the action of lysozyme, an enzyme obtained from egg white and which splits the glycosidic linkage between C-1 on the N-acetylmuramic acid residue and C-4 on the adjacent N-acetylglucosamine residue (*Figure 2.11*). Another variation in the structure of the polysaccharide backbone has been observed in peptidoglycans from *Mycobacterium smegmatis* in which N-acetylmuramic acid residues are replaced by N-glycolylmuramic acid residues.

Some (as in *Micrococcus lysodeikticus*) or all (*Staph. aureus*) of the N-acetylmuramic acid residues in the backbone are linked to tetrapeptide chains through the D-lactic acid residues (*Figure 2.11*). The tetrapeptide units have the general sequence:

$$R_1 - \gamma\text{-D-glutamyl} - R_3 - \text{D-alanine}$$

The R_1 residue is frequently L-alanine, or occasionally L-serine or glycine. Linked to the R_1 residue is a γ-D-glutamyl residue, the α-carboxyl group of which may be substituted with an amide group (as in *Staph. epidermidis*) or linked by a peptide bond to a single glycine residue (as in *Micrococcus lysodeikticus*). The R_3 residue can be a neutral amino acid such as homoserine, but it is more frequently a diamino acid such as L-ornithine, L-lysine, L,L-diaminopimelic acid or *meso*-diaminopimelic acid. The carboxyl groups of L,L- and *meso*-diaminopimelic acid which are not engaged in peptide linkage may be substituted with an amide group.

The tetrapeptide chains are joined to peptide chains in another stretch of peptidoglycan by a linkage which involves the terminal D-alanine residue of one chain and either the free amino group of the diamino acid or the α-carboxyl group of the D-glutamic acid residue of another peptide chain. The cross-linking between two tetrapeptide chains constitutes the most variable element in peptidoglycan structure. It may consist of:

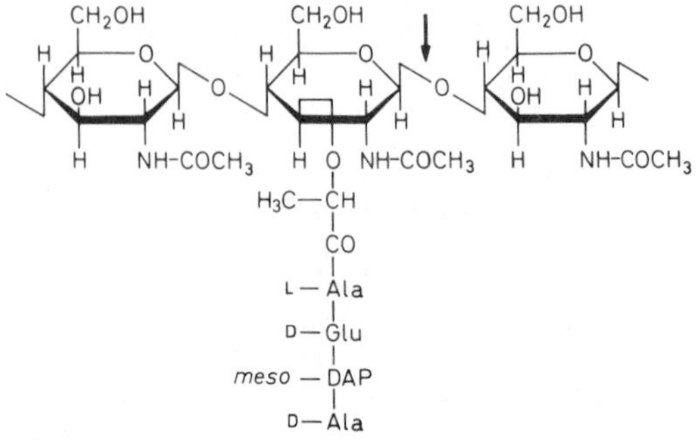

Figure 2.11   Structure of part of the cell-wall peptidoglycan in Escherichia coli. *The arrow indicates the lysozyme-sensitive glycosidic bond.   The abbreviations are explained on page (ix)*

(a) a direct bond, such as D-alanyl-(D)-*meso*-diamino-pimelic acid, as in *Escherichia coli*; (b) a single additional amino acid, such as glycine, a neutral L-amino acid or a D-isoasparagine residue; (c) an intervening peptide chain composed of glycine and/or neutral amino-acid residues, a cross-linking which occurs in the peptidoglycan from *Staph. aureus*; (d) one or several peptides each having the same amino-acid sequence as the tetrapeptide attached to the polysaccharide back-bone; peptidoglycans with this structure have been detected in walls of *Micrococcus lysodeikticus*; (e) a diamino acid residue, such as D-ornithine or D-lysine, extending between the α-carboxyl group of the glutamic acid residue and the free carboxyl group of the D-alanine residue; this structure is found in peptidoglycans from some plant-pathogenic corynebacteria.   Examples of these structures are shown in *Figure 2.12*.

Peptidoglycans are really bag-shaped molecules which, with their extensive cross-linking, are uniquely designed to accommodate a limited amount of bond breakage which is required during cell-wall growth (*see* page 367) without destroying the overall structure of the wall.   They have been compared to a three-dimensional rope ladder, with relatively rigid polysaccharide rungs and flexible polypeptide ropes.

A variety of compounds are associated with bacterial cell-wall peptidoglycans.   These matrix materials are most conveniently discussed separately for walls of Gram-positive and Gram-negative organisms.

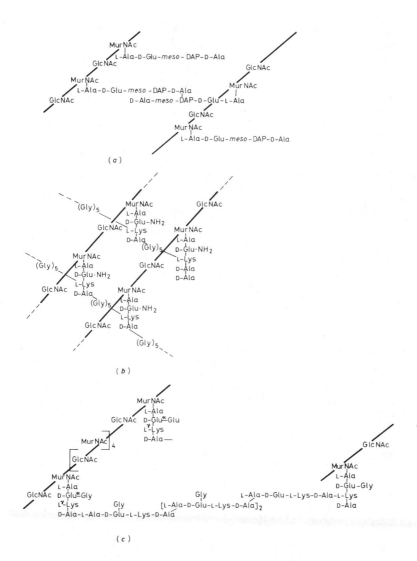

*Figure 2.12 Structures of peptidoglycans from walls of
(a)* Escherichia coli, *(b)* Staphylococcus aureus, *and
(c)* Micrococcus lysodeikticus. *The abbreviations are
explained on page (ix)*

Walls of Gram-positive bacteria

The matrix materials in the walls of Gram-positive
bacteria can be polysaccharides. They include the
immunologically active C-polysaccharide of *Streptococcus
haemolyticus,* which is made up of a backbone of rhamnose
residues with N-acetylglucosamine side-chains, and the
glucose-N-acetylmannosaminuronic acid polymer present in
the walls of *Micrococcus lysodeikticus.*

However, the most interesting and the most widespread
group of matrix polymers in the walls of Gram-positive
bacteria are the *teichoic acids.* These polymers, which
can be extracted from isolated walls with cold
trichloroacetic acid, dilute alkali, or a dilute solution
of N,N-dimethylhydrazine, can account for as much as 50%
of the dry weight of the wall. So far, teichoic acids
have been found only in walls of Gram-positive bacteria.
The structures of many of these teichoic acids are now
known, largely as a result of the efforts of James Baddiley
and his colleagues in the University of Newcastle-upon-
Tyne, England.

The first teichoic acids to be examined were polymers
of ribitol phosphate or glycerol phosphate, and these
have come to be referred to as *ribitol teichoic acids* and
glycerol teichoic acids, respectively.

In ribitol teichoic acids, a phosphodiester linkage
joins C-1 and C-5 of adjacent residues (*Figure 2.13*),
while most of the ribitol residues have a D-alanine residue
in labile ester linkage on C-2. Various sugar substituents
may be present on C-4 and C-3. In *Staph. aureus,* for
example, the wall ribitol teichoic acid is substituted at
C-4 by an N-acetylglucosaminyl residue in either α or β
linkage (*Figure 2.13*). Other sugar substituents which
have been detected in ribitol teichoic acids include
β-glucopyranosyl (*Bacillus subtilis*) and α-glucopyranosyl
(*Lactobacillus plantarum*), both on C-4. In the
lactobacillus, additional α-glucopyranosyl residues occur
on C-3. Interesting variations on this basic structural
theme are seen in the structures of ribitol teichoic
acids from walls of *Actinomyces streptomycini* in which
succinate residues, instead of D-alanine residues, occur
at C-2, and from walls of *Actinomyces violaceus* which has
acetate groups at C-2. Yet another structural variation
is seen in the ribitol teichoic acid from *Streptomyces
griseus* which has a glycerol phosphate residue linked to
the hydroxyl group at C-3 of the ribitol residues.

Glycerol teichoic acids are more widespread than

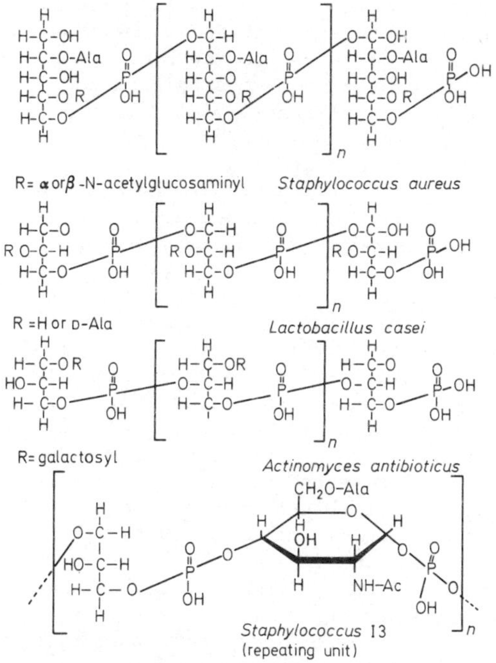

Figure 2.13 Structural formulae of different types of teichoic acids from walls of representative microorganisms. Ala indicates a D-alanine residue

ribitol teichoic acids and, as we shall see (page 58) are also found in bacterial membranes. In most glycerol teichoic acids, C-1 and C-3 of adjacent glycerol residues are joined by a phosphodiester linkage (*Figure 2.13*). The hydroxyl group at C-2 may be substituted with a D-alanine ester residue, or with a glycosyl residue which includes α-D-glucosyl (as in *Lactobacillus buchneri*) and N-acetylglucosamine (*Staph. albus*). Another variation in the structure of the wall glycerol teichoic acid is seen in the polymer from walls of *Actinomyces antibioticus* in which the glycerol phosphate residues are linked through C-2 and C-3, while a galactosyl residue is attached at C-1 (*Figure 2.13*).

The walls of some Gram-positive bacteria contain polymers which, while resembling ribitol teichoic acids and glycerol teichoic acids in structure, differ in that sugar residues occur in the polymer chain. The first of

these polymers to be identified was from *Staph. lactis*
13, in which the repeating unit in the backbone consists
of a glycerol residue joined to the hydroxyl group at
C-4 of N-acetylglucosamine 1-phosphate through a
phosphodiester linkage (*Figure 2.13*). Several polymers
of similar structure have since been isolated from walls
of other Gram-positive bacteria.

The wall of any one strain of Gram-positive bacterium
contains just one type of teichoic acid. The only known
exception to this generalisation is the wall of
Micrococcus sp. 24 which contains a glycerol teichoic acid
in which N-acetylglucosamine residues form part of the
polymer chain, in addition to small amounts of a
poly(ribitol phosphate) teichoic acid.

Although the essential details of teichoic acid
structure are known, there is a lack of information on
the fine structure of these wall polymers. Little is
known, for example, of the frequency with which the
backbone residues are substituted with D-alanine and
sugars, or of the possible occurrence in the walls of any
one bacterium of teichoic acid molecules with different
degrees of substitution in the backbone. In view of the
role of teichoic acids as immunological determinants,
these clearly are data which are urgently required.

Teichoic acids are held in the cell wall by covalent
attachment to peptidoglycan. It is thought that the
linkage involves a terminal phosphodiester bond on the
teichoic-acid chain and an N-acetylmuramic-acid residue
in the peptidoglycan. However, a direct demonstration
of this linkage has been established only with the wall
teichoic acid of *Staph. lactis* 13. In this bacterium
about 40% of the peptidoglycan chains are linked to a
teichoic acid; each glycan chain carries only one
teichoic acid molecule.

Several pieces of evidence indicate that at least
some of the teichoic acid is located on the surface of
many Gram-positive bacteria. Thus, teichoic acids form
part of the specific receptor sites for bacteriophages,
while both walls and intact cells are agglutinated by
lectins and antibodies which are specific for the
teichoic-acid component of the wall. The arrangement of
the peptidoglycan chains - and therefore the teichoic
acid molecules - in the wall has been the subject of
considerable speculation. The two most likely arrangements of the polymers in the walls of *Staph. lactis* 13
are shown in *Figure 2.14*. In one of these arrangements
the peptidoglycan chains are arranged radially, while in

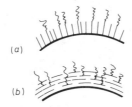

*Figure 2.14 Diagram showing two possible arrangements
for the peptidoglycan chains and teichoic acid molecules
in the walls of* Staphylococcus lactis 13. *(a) This shows
the rigid peptidoglycan chains arranged radially with the
teichoic acid molecules lying externally; (b) shows
peptidoglycan lying parallel with the surface of the
bacterium*

the other they lie parallel to the surface of the
membrane. In any consideration of the arrangement of
teichoic acids in walls, it must be remembered that
teichoic acids are fairly long and flexible molecules
which are held covalently at only one end.

Teichoic acids are negatively charged polymers, and
it is believed that this acidic character permits the
bacteria to maintain a high concentration of bivalent
ions, especially of magnesium, in the region of the plasma
membrane. Magnesium ions are needed to maintain the
integrity of the plasma membrane (*see* page 45) and are
also required in fairly high concentrations for the activity
of enzymes concerned in synthesis of bacterial cell-wall
components. These polymers are probably largely
responsible for the net negative charge (zeta potential)
on most Gram-positive bacteria.

In addition, teichoic acids may be involved in the
action of autolytic enzymes that are present in bacterial
walls. Autolytic enzymes are thought to play an important
role during cell growth and division (*see* page 367), and to
be involved in turnover of wall material and in the
acquisition of cell competence. Teichoic acids bind
strongly to certain autolytic enzymes, and it has been
suggested that they may be involved in the localisation
of autolytic enzymes in the wall and in the modulation of
their activity. Teichoic acids have also been shown to
be responsible for the resistance of certain bacteria to
lysins. Finally, wall teichoic acids participate in the
binding of phages by several species of Gram-positive
bacteria.

Walls of Gram-negative bacteria

The walls of Gram-negative bacteria have long been known
to be chemically more complex than those of Gram-positive
strains. They contain less peptidoglycan than walls of
Gram-positive strains and, on hydrolysis, yield lipids
and a full range of amino acids which indicate the
presence of protein.

Electron micrographs of thin sections through walls
of Gram-negative bacteria show them to be made up
usually of two layers. Overlying the plasma membrane
there is a dense layer that is itself covered by an outer
membrane that cytologically has the appearance of a unit
membrane. The intermediate layer is 3.0-8.0 nm, and the
outer layer 6.0-10.0 nm thick.

The dense layer outside the plasma membrane is made up
of peptidoglycan. The peptidoglycans in walls of Gram-
negative bacteria have not been so intensively studied
as their counterparts in the walls of Gram-positive
bacteria, but the available data suggest that the
composition of the polymer is more uniform than that in
walls of Gram-positive strains.

The outermost layer in the walls of Gram-negative
bacteria consists of a lipid-protein membrane and a
lipopolysaccharide. In *E. coli,* the peptidoglycan in
the middle layer is covalently linked to an acidic
lipoprotein in the outer membrane. The bond joins a
terminal amino-acid residue in the protein to a diaminopimelic acid residue in the peptidoglycan. Little is
known about the composition of the lipoprotein, or of
other components of the outer membrane. In the outer
membrane of the wall in *E. coli* it has, however, been
shown that the principal protein has a molecular weight
of about 44 000 daltons.

Rather more has been published on the composition of
the lipopolysaccharide which lies on top of the outer
membrane in the walls of Gram-negative bacteria. These
lipopolysaccharides can be extracted from whole cells
or wall preparations with a warm aqueous solution of
phenol or with cold trichloroacetic acid. Some of these
extracts contain lipopolysaccharide linked to protein,
and it is thought that the protein may be a component of
the underlying lipid-protein membrane. Cell-wall
lipopolysaccharides from Gram-negative bacteria carry the
O-antigen specificities of the bacteria, and it is largely
on this account that much of the work on these wall polymers
has been done. Microbiologists have devised procedures

for rapid identification of many groups of Gram-negative
bacteria, mainly because of the actual or potential
pathogenic nature of these bacteria. Many of these
diagnostic schemes are based on immunological
specificities which prompted microbial biochemists to
relate chemical structure to antigen specificity. Much
of this work has been done with strains of *Salmonella*
which have been subdivided into about 1000 serotypes
based on the different properties of the thermolabile
flagellar H-antigens and the thermostable somatic or
O-antigens.

Bacterial cell-wall lipopolysaccharides consist of
complex heteropolysaccharides covalently linked to a
glucosamine-containing lipid known as *lipid A*. The
structures of several of these lipopolysaccharides have
been examined, some in considerable detail. Lipid A
contains glucosamine, phosphate and fatty acids which
include lauric, myristic and especially β-hydroxymyristic
acids. The basic structure is thought to be a
disaccharide, glucosaminyl-glucosamine, which is fully
substituted by ester- and amide-linked fatty acids and
by phosphate. Lipid A is responsible for the endotoxicity
of the lipopolysaccharide.

The polysaccharide portion of the cell-wall lipopolysaccharide contains between five and ten different types
of sugar residue, depending on the chemotype. These
residues include hexoses, hexosamines, and the much
rarer 3,6-dideoxyhexoses including abequose (3,6-dideoxy-
D-galactose), colitose (3,6-dideoxy-L-galactose),
tyvelose (3,6-dideoxy-D-mannose), ascarylose (3,6-dideoxy-
L-mannose) and paratose (3,6-dideoxy-D-glucose). In
addition, they contain two sugars which are apparently
unique to bacterial cell-wall lipopolysaccharides,
namely, L-glycero-D-mannoheptose (sometimes abbreviated
to Hep) and 2-keto-3-deoxyoctonate (KDO).

Structures have been published for several *Salmonella*
lipopolysaccharides and for polymers from some strains of
E. coli. Two of these structures are shown in *Figure 2.15.*
The inner core polysaccharide, which is thought to be the
same in all strains of *Salmonella*, consists of a short
polysaccharide chain linked covalently to lipid A. The
core chains are linked to each other by phosphodiester
linkages. Lipopolysaccharides in rough strains of Gram-
negative bacteria consist only of the inner and outer
core structures. In smooth strains, the outer core
polysaccharide is linked to an O-antigen chain made up
of repeating trisaccharide, tetrasaccharide or penta-

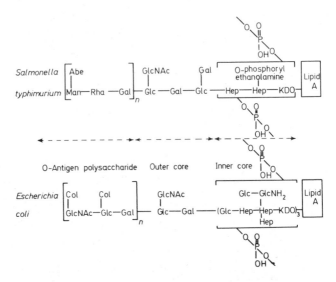

Figure 2.15 Structures of the lipopolysaccharides from walls of Salmonella typhimurium *and* Escherichia coli. Abe *indicates a residue of abequose;* Col *of colitose;* Hep *of L-glycero-D-mannoheptose; and KDO of 2-keto-3-deoxyoctonate. Other abbreviations are explained on page (x). See text for an explanation of the structures*

saccharide units. Semi-rough mutant strains of some Gram-negative bacteria have been isolated, and these have walls with lipopolysaccharides that contain part but not all of the O-specific polysaccharide.

A schematic representation of the wall structure in *E. coli* is shown in *Figure 2.16*. The figure depicts not only the covalent bonding between the peptidoglycan in

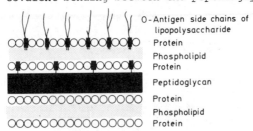

Figure 2.16 The location of components of the cell envelope of Escherichia coli, *and their interrelationships. See text for details*

the middle layer with a lipoprotein in the outer membrane, but also the way in which the lipopolysaccharide is thought to be anchored into the outer membrane through lipid A.

The Gram stain

The differences in molecular architecture between the walls of Gram-positive and Gram-negative bacteria are thought to form the basis of the staining technique by which bacteria are divided into two groups. The Gram stain, devised by Christian Gram in 1884, is the most widely used staining procedure in bacteriology. It is based upon the ability of some bacteria (Gram-positive) to take up *p*-rosaniline dyes such as crystal violet (CV) and, after mordanting with iodine (I), to retain the CVI complex following extraction with ethanol or acetone. Bacteria which lose the CVI complex after ethanol treatment are Gram-negative. Most of the early work on the biochemical basis of the Gram stain was aimed at isolating specific compounds responsible for retention of the CVI complex in the Gram-positive bacterium. It is now clear that a wide range of compounds, including lipids, polysaccharides and RNA, react with the complex to some extent, which suggests that the Gram-positive character cannot be correlated with the presence of a specific compound in the bacterium.

Recent work suggests that the basis of the Gram stain lies in permeability differences between the two groups of bacteria. In Gram-positive bacteria, the CVI complex appears to be trapped in the wall following ethanol treatment, which presumably causes a diminution in the diameter of the pores in the wall peptidoglycan. Walls of Gram-negative bacteria contain a much smaller proportion of peptidoglycan, and this is less extensively cross-linked than that in the walls of Gram-positive bacteria (*see* page 40). The pores in the peptidoglycan in walls of Gram-negative bacteria are thought to be sufficiently large, even after ethanol treatment, to allow the CVI complex to be extracted.

Walls of acid-fast bacteria

Certain Gram-positive bacteria possess a property known as *acid fastness*. Acid fastness is determined by a procedure known as the Ziehl-Neelson stain. Fixed cells are treated with a hot dilute solution of phenol containing a basic red dye, fuchsin. After washing, the stained cells are treated with a 20% solution of hydrochloric acid or sulphuric acid. Cells of acid-fast bacteria retain the fuchsin dye after acid extraction, while other bacteria are decolourised.

The group of acid-fast bacteria which has attracted most attention are the mycobacteria. The walls of mycobacteria, like those of other bacteria, contain peptidoglycan, but in addition are rich in waxes which include a group of compounds known as *mycolic acids*. Extracted mycolic acids are acid fast, and it is generally agreed that the acid-fast property of intact mycobacteria is attributable to the presence of mycolic acids in the walls. Although a detailed explanation of the role of mycolic acids in conferring the acid-fast character on mycobacteria has yet to be forthcoming, it is suggested that they react with fuchsin and that the mycolic acid-fuchsin complex, or simply the mycolic acid, acts as a permeability barrier and impedes penetration of the mineral acid.

Mycolic acids are β-hydroxy acids which are substituted in the α-position with a moderately long aliphatic chain. The mycolic acid from walls of *Mycobacterium smegmatis* has the following structure:

$$CH_3-(CH_2)_{\overline{17}}CH=CH-(CH_2)_{13}-CH=CH-\overset{\overset{\displaystyle CH_3}{|}}{CH}-(CH_2)_{17}-\overset{\overset{\displaystyle OH}{|}}{CH}-\underset{\underset{\displaystyle C_{22}H_{45}}{|}}{CH}-COOH$$

The β-hydroxyl group is thought to be essential for the mycolic acid to confer acid fastness on the bacterium.

Despite the remarkable progress of the past two decades, much clearly remains to be discovered about the molecular architecture of the microbial cell wall. The number of species so far examined for cell-wall structure is quite small and the molecular architecture of the walls in other micro-organisms may well be found to differ from the anatomical pattern observed in the few species so far studied. Already an exception to the generalisation that

all microbial walls possess structural microfibrils has
been found with certain extremely halophilic bacteria,
the walls of which do not contain peptidoglycan. This
is understandable since the plasma membranes in these
bacteria are not subject to the differences in osmotic
pressure that are imposed on the membranes of non-
halophilic bacteria. However, one of the most fascinating
problems in cell-wall biochemistry still remains virtually
unexplored. This is the relationship between the
macromolecular structure of the walls and their
morphological form (rod-shaped, coccoid). It promises to
offer a formidable challenge to the molecular biologist.

2.6 PLASMA MEMBRANES

The vitally important structures and organelles of a
micro-organism are located within the cytoplasm, which is
itself bounded by a membrane usually referred to as the
plasma membrane or occasionally by the older name of
cytoplasmic membrane. Because there is normally a large
difference in osmotic pressure between the cytoplasmic
contents and the aqueous environment, this membrane is
distended and, in thin sections of organisms, is seen to
lie directly beneath the rigid cell wall. However, the
cytoplasm and the plasma membrane can, under certain
conditions, retract away from the inside of the cell wall
to leave a gap between the wall and the membrane. This
gap is referred to as the *periplasmic space,* and it is
the location for several hydrolytic enzymes. In many
enteric bacteria, the enzymes acid phosphatase, cyclic
phosphodiesterase and 5'-nucleotidase are located in the
periplasmic space, as is an esterase in *Saccharomyces
cerevisiae.*

   Some micro-organisms, such as bacterial L forms,
mycoplasmas and certain protozoa, are devoid of a cell
wall and, in these organisms, the plasma membrane forms
the outermost layer of the organism. In some protozoa,
the plasma membrane is covered with a glycoprotein
(*Figure 2.17*). The outer membrane in protozoa is usually
referred to as the *plasmalemma.*

   Although the existence of plasma membranes was
recognised over half a century ago, knowledge about their
composition, structure and function has only recently
become available as methods have been devised for
preparing isolated membranes. The plasma membrane has
four main functions in a micro-organism: (a) to act as

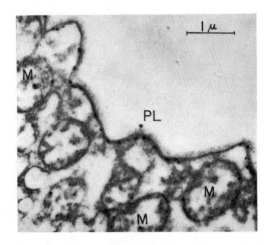

Figure 2.17   Electron micrograph through a thin section of the protozoon Amoeba proteus, *showing the plasmalemma* (PL) *and mitochondria* (M). *(By courtesy of D.G. Pappas)*

the ultimate containing organelle for a micro-organism. Microbes can exist without extramural structures, capsules, and even walls.   But if the plasma membrane is destroyed, a micro-organism ceases to exist; (b) to act as an organelle which concentrates nutrients within the cell and excretes waste products; (c) to act as a site for biosynthesis of certain cell constituents, particularly cell-wall and capsule components; and (d) to locate certain enzymes and cell organelles, such as ribosomes.   This diversity of function makes a study of the plasma membrane particularly fascinating.   At the same time, it is a study which bristles with problems, largely because these functions are to a great extent dependent upon the spatial arrangement of this organelle within the micro-organism.

2.6.1   COMPOSITION

In electron micrographs of thin sections of micro-organisms, the plasma membrane is seen as a continuous layer some 7.5 nm wide, made up of an electron-transparent layer sandwiched between two darker staining layers each 2-3 nm wide.   Observations on artificially prepared lipid-protein mixtures suggest that the inner layer is of lipid

and the outer layers of protein. The spacing between
the layers, which is similar in almost all types of
membrane so far examined, agrees with that expected from
two fatty-acyl chains each 15-18 carbon atoms long. This
arrangement is essentially that proposed by Danielli and
Davson for the typical cell membrane as far back as 1935,
and now usually referred to as the *unit membrane (see
Figure 2.20).*

Analyses have been reported for plasma membranes from
only a small number of micro-organisms. Most of these
membranes have been obtained from organisms, particularly
bacteria, which can readily be converted into protoplasts
or sphaeroplasts. On the whole, the analyses so far
reported confirm the suggestion regarding membrane
composition deduced from electron micrographs, namely
that they are composed principally of lipid and protein.
The proportion of protein in the membrane varies from
about 50 to 65%. Some of the protein and lipid may be
present as lipoprotein, but it is generally believed
that the bulk of these components are not covalently
linked. Some microbial plasma membranes contain, in
addition, small amounts of carbohydrate, RNA and DNA.
Certain of these minor components probably represent
contaminating material from the cytoplasm, although some
may well be genuine membrane components.

Lipid

Much more is known about the lipid composition of
microbial plasma membranes than about their protein
composition. Since the lipid composition of plasma
membranes in prokaryotic micro-organisms differs in many
respects from that of the eukaryotic plasma membrane, the
two are dealt with separately in the following account.

Prokaryotic micro-organisms. Although most prokaryotic
micro-organisms have only one membrane - the plasma
membrane - it cannot always safely be assumed that the
lipid extracted from cells of prokaryotic microbes is
identical in composition with that of the plasma membrane.
We have already noted (page 40) that Gram-negative
bacteria have lipid in their cell walls, while other
bacteria, including some Gram-positive strains, contain
intracellular membraneous structures (page 60). Not all
of the lipids described in the account that follows have,

therefore, unequivocally been shown to be present in the
plasma membrane. However, reference is made only to
those lipids which are in all likelihood true plasma-
membrane components.

The bulk of the lipids in the plasma membranes of the
majority of prokaryotic micro-organisms are polar lipids.
The commonest of these are *glycerophospholipids*; these
are derivatives of *sn*-glycerol-3-phosphate in which C-1
and C-2 are esterified with long-chain fatty-acyl residues,
and the phosphate group on C-3 with one of a range of
organic bases, amino acids or alcohols. Structural
formulae of representative glycerophospholipids that
occur in microbial plasma membranes are given in *Figure
2.18*. One of the most abundant of these is phosphatidylethanolamine, particularly in bacteria. The N-methylated
derivative of phosphatidylethanolamine, phosphatidylcholine,
is more often found in fungal than in bacterial membranes.
Phosphatidylserine is an intermediate in the synthesis of
phosphatidylethanolamine and phosphatidylcholine, and is
found in only trace amounts in most bacterial plasma
membranes. Phosphatidylglycerol, on the other hand, is
found in membranes of both Gram-positive and Gram-negative
bacteria, as is cardiolipin or diphosphatidylglycerol.
Amino-acylphosphatidylglycerols are, apparently, unique
to bacterial membranes. One of the most common of this
class of membrane lipids is lysylphosphatidylglycerol
which occurs in the plasma membrane in *Bacillus subtilis*
and *Streptococcus faecalis*. Phosphatidylinositol, which
is an important constituent of the yeast plasma membrane,
is found in trace amounts in just a few bacteria.

*Figure 2.18 Structural formulae of representative
glycerophospholipids and of a sterol that occur in
microbial plasma membranes. (a) A phosphatidylethanolamine with a palmitic acid residue on C-1 and a lactobacillic acid residue on C-2; (b) a phosphatidylglycerol
with a stearic acid residue on C-1 and a branched-chain
C_{17} residue on C-2; (c) a phosphatidylethanolamine
plasmalogen with an α,β-unsaturated C_{16} acid linked to
C-1 by an ether linkage, and an oleic acid residue on
C-2; (d) a phosphatidylcholine with a palmitic acid
residue on C-1 and an oleic acid residue on C-2; (e)
a phosphatidylinositol with a stearic acid residue on
C-1 and a linoleic acid residue on C-2; (f) ergosterol.
Saturated fatty-acyl chains ($-CH_2-CH_2-CH_2-$) are
represented by* ⋀⋀

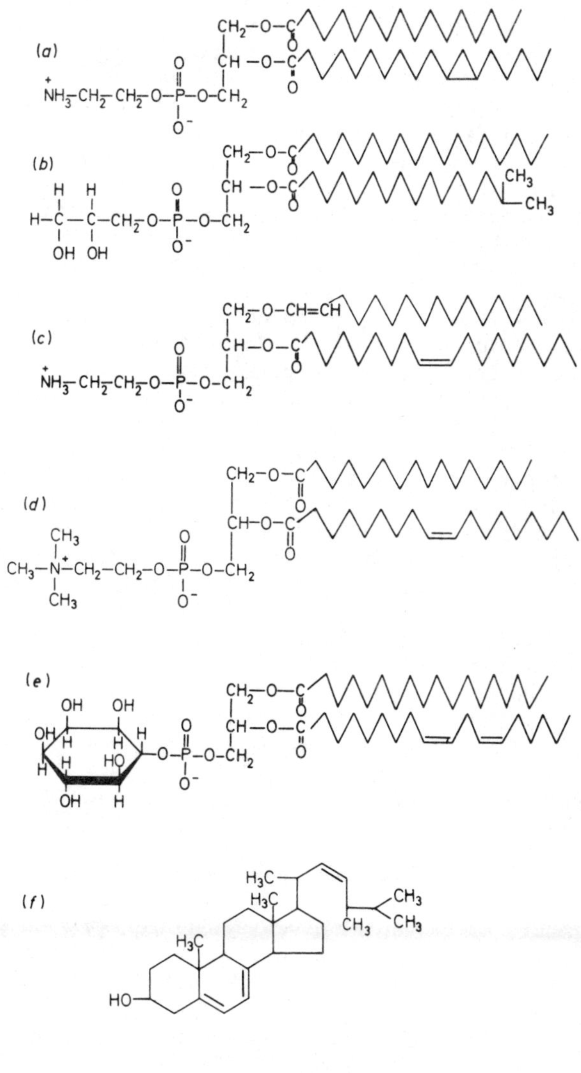

Figure 2.18 See facing page for details

Bacterial glycerophospholipids contain a variety of different fatty-acyl chains on C-1 and C-2. Generally, these acyl chains are 10-20 carbon atoms in length, with $C_{15}-C_{19}$ chains predominating. There are four main types of fatty-acyl chain on bacterial glycerophospholipids, namely straight-chain saturated, straight-chain unsaturated, branched chain particularly *iso* and *anteiso*, and cyclopropane fatty-acyl residues. Glycerophospholipids with straight-chain fatty-acyl substituents are commonest, with an unsaturated residue if present usually linked to C-2. Cyclopropane fatty-acyl residues are frequently found in glycerophospholipids of lactobacilli, streptococci and clostridia. One of the most abundant of these acids is *cis*-11,12-methylene octadecanoic acid which is usually known as lactobacillic acid.

Many strictly anaerobic bacteria contain glycerophospholipids with an unsaturated ether substituted on C-1. These lipids are known as *plasmalogens*. As yet, little has been reported on the polar groups of bacterial plasmalogens. It has been established, however, that over half of the phosphatidylethanolamine in *Clostridium butyricum* is of the plasmalogen type (*Figure 2.18*).

Another variation in the type of acyl substituent on C-1 and C-2 of bacterial glycerophospholipids, although admittedly a rare one, is found in the extremely halophilic bacterium *Halobacterium cutirubrum* (see page 113). This bacterium has little if any fatty-acyl residues in its lipids. Instead it synthesises phosphoglycerol phosphates with long-chain alcohols substituted on C-1 and C-2 through ether linkages. The major alcohol substituent is dihydrophytyl alcohol.

The lipids in many Gram-positive and in some Gram-negative bacteria contain small amounts of *glycolipids*. Bacterial glycolipids can be grouped into two categories. The first are *glycosyldiacylglycerols* in which carbohydrate residues are linked glycosidically to the 3-position of an *sn*-1,2-diacylglycerol. The glycosyl grouping may consist of up to four sugar residues, which are usually hexoses but can include uronic acids (*Figure 2.19*). The second category of bacterial glycolipids are *acylated sugars*, in which one or more of the hydroxyl groups on a sugar are esterified with long-chain fatty acids. An example is given in *Figure 2.19*.

It has recently been established that in certain bacteria, some of the plasma-membrane diglycosyldiacylglycerol is linked covalently to a glycerol teichoic acid to form a *lipoteichoic acid*. In one strain of *Streptococcus faecalis,* the lipoteichoic acid is thought

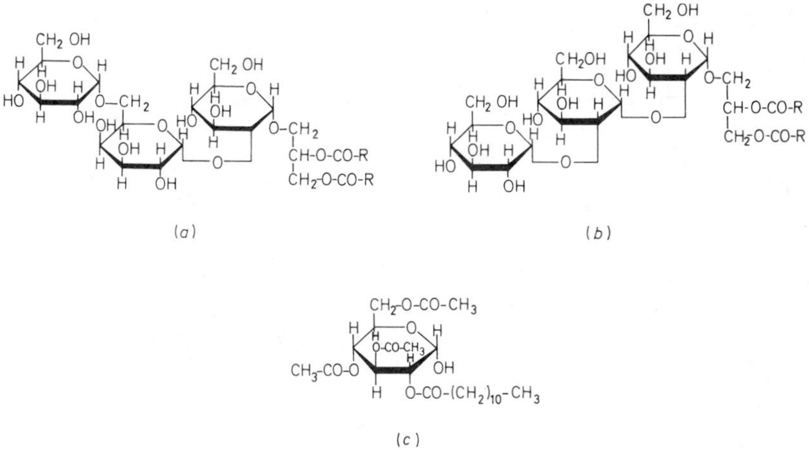

Figure 2.19 Structural formulae of some bacterial glycolipids. (a) Glucosylgalatosylglucosyl diglyceride from Lactobacillus casei; *(b) triglucosyl diglyceride from* Streptococcus haemolyticus; *(c) acylated glucose from* Streptococcus faecalis

to consist of a glycerol-phosphate chain linked covalently, presumably through its terminal phosphate group, to a diglucosyldiacylglycerol which is also found unsubstituted in the plasma membrane of this bacterium. The discovery of lipoteichoic acids is of particular interest because it had for many years been thought that intracellular or 'membrane' teichoic acids were located on the outside of the plasma membrane, probably in the periplasmic space, and that they were not covalently linked to a membrane component.

A few bacteria synthesise phospholipids based not on glycerol but on sphingosine as the alcohol moiety; these are known as *sphingolipids.* In *Bacteroides melaninogenicus,* for example, over half of the total extractable lipids are sphingolipids. One of the main sphingolipids in this bacterium is ceramide phosphorylethanolamine.

Numerous minor but nevertheless physiologically important lipids are also present in bacterial plasma membranes. These include the quinone coenzymes, coenzymes Q and vitamin K, as well as carotenoids. Bacterial plasma membranes also contain small amounts of polyprenols which function as lipid intermediates in the synthesis of various cell-wall and extramural polymers (*see* page 299).

The plasma membrane in *mycoplasmas* has a strange lipid composition, quite unlike that found in other classes of prokaryotic micro-organism. The principal phospholipid is phosphatidylglycerol, sometimes accompanied by small amounts of diphosphatidylglycerol. Other major constituents of the membrane in mycoplasmas are glycolipids, which include glycosyldiacylglycerols as well as acylated sugars. This lipid composition is hardly unusual. However, most mycoplasmas require cholesterol or some other sterol for growth, and the sterol is found in the plasma membrane, both free and as sterol glycosides. In the plasma membrane from a strain of *Mycoplasma gallinarum*, cholesteryl-β-D-glucopyranose accounts for about half of the total glycolipid, and about 10% of the total lipid. Another unusual glycolipid has been found in the membrane of *Acholeplasma laidlawii*. This glycolipid, the structure of which is given below, is essentially a phosphoglycolipid, a class of lipids that also occurs in some lactic-acid bacteria. As indicated in this structure, some uncertainty exists as to the exact nature of the linkage between the glycerol phosphate and sugar residues.

Eukaryotic micro-organisms. Although numerous analyses have been made of the overall lipid composition of eukaryotic micro-organisms, only of one organism, namely *Saccharomyces cerevisiae*, has the plasma membrane been isolated in a reasonably uncontaminated state and analysed for lipid composition. The plasma membrane in this yeast contains two main classes of lipid: glycerophospholipids and sterols. Two other classes of lipid, namely triacylglycerols (otherwise known as triglycerides) and sterol esters, which account for a large proportion of the lipid extracted from *Sacch. cerevisiae,* are present in isolated plasma membranes in only very small amounts. As we shall see later (page 80), triacylglycerols and sterol esters are located mainly in intracellular structures in *Sacch. cerevisiae.*

The main glycerophospholipids in the plasma membrane of *Sacch. cerevisiae* are phosphatidylethanolamine, phosphatidylinositol and phosphatidylcholine, with small amounts of phosphatidylserine. Phosphatidylinositol is phosphorylated by a kinase present in the yeast plasma membrane to give diphosphatidylinositol. However, diphosphatidylinositol is rapidly dephosphorylated to yield phosphatidylinositol, which explains why the amount of diphosphatidylinositol in the yeast plasma membrane is quite small. The commonest sterols in the plasma membrane of *Sacch. cerevisiae* are ergosterol, zymosterol and a tetraethenoid sterol (dehydro-ergosterol) which is a biosynthetic precursor of ergosterol (*see* page 316). The molar ratio of phospholipid to sterol in the plasma membrane is around 5:1, which is substantially greater than the ratio in many mammalian plasma membranes.

Protein

Despite the fact that the principal metabolic functions of the proteins that occur in the microbial plasma membrane are understood, at least in outline, little is as yet known about the properties of these proteins. As membrane components, they are undeniably less tractable than lipids.

From what is known about membrane function, it is possible to list the major types of protein that occur in microbial plasma membranes. Firstly, there are transport proteins that effect movement of solute molecules across the membrane. The activities of these proteins can only be demonstrated in intact organisms, in protoplasts and sphaeroplasts, or vesicles. There is also a wide range of enzymes associated with the microbial plasma membrane. In all microbes, these probably include enzymes which catalyse certain of the reactions that lead to synthesis of membrane components (lipid and protein) and of new cell-wall and extramural polymers. Also, almost all microbial plasma membranes so far examined possess adenosine triphosphatase (ATPase) activities. These

ATPase activities differ in their properties, particularly
with regard to their requirements for cations (Na^+, K^+,
Mg^{2+}, Ca^{2+}). Plasma-membrane ATPases probably have a
role in solute transport across the membrane, a role that
is assessed on page 173. Other enzyme activities that
are commonly detected in isolated microbial plasma
membranes include phospholipases, proteases and peptidases.
Less is known about their metabolic roles, although some
are most likely to be involved in turnover of membrane
lipids and proteins.

In prokaryotic micro-organisms, the variety of proteins
associated with the plasma membrane is much wider than in
eukaryotes. In addition to those proteins already
referred to, the bacterial plasma membrane contains
proteins involved in electron transport (*see* page 203)
and in DNA replication and RNA synthesis (page 281).
Whether all plasma-membrane protein can be accounted for
by enzymes and transport proteins is far from clear.
Some researchers hold that the microbial plasma membrane
contains catalytically inactive structural proteins;
however, as one sees fewer and fewer references to
membrane structural proteins, it must be assumed that
the microbial physiologist is at least sceptical of their
existence.

The various types of protein which have been detected
in microbial plasma membranes differ in the ease with
which they can be extracted from isolated membranes. Many
transport proteins are released quite easily; even from
intact organisms they can be 'squeezed' out of the
membrane into the suspending liquid using a technique
known as *cold osmotic shock* (page 166).

Adenosine triphosphatases also come off microbial plasma
membranes with a minimum of perturbation. Several
ATPases from bacterial plasma membranes have been
isolated and purified to homogeneity. The purified
enzymes do not appear to require a lipid specifically
for activity although, when bound in the membrane, their
activities may be modulated by interaction with membrane
lipids. By contrast, many of the plasma-membrane enzymes
that catalyse synthesis of wall and extramural polymers
are extracted only with difficulty. With these enzymes,
it is often necessary to use special reagents, called
chaotropic agents, to effect extraction. These reagents
include sodium trichloroacetate and sodium perchlorate.

Structure and function in plasma membranes

It has long been accepted that the amphipathic character
of phospholipids (that is the presence of both
hydrophilic and hydrophobic groupings in the molecule)
makes them ideal molecules to form a barrier between two
hydrophilic regions, and at the same time accommodate
protein molecules in the barrier layer. What has not
been accepted, and indeed still forms the basis for
controversy, is the manner in which phospholipid and
protein molecules are arranged in membranes. Studies on
the erythrocyte membrane by the Dutch workers E. Gorter
and F. Grendel in the 1920s, followed by those of
J.F. Danielli and H. Davson in Britain a decade later,
led physiologists to believe that membranes exist as
triple-layered structures, with protein molecules
arranged on either side of a double layer of phospholipid
molecules (*Figure 2.20*). This view of membrane structure
was supported by observations made on electron micrographs
of thin sections through membranes, and it was dubbed the
unit membrane by the American microscopist, J.D. Robertson.
  Recently, several lines of investigation have suggested
that membranes may exist not as triple-layered structures
but rather as a continuous layer of phospholipid molecules
in which are embedded globular proteins that may extend
across the width of the membrane. This alternative
structure, which is currently receiving a great deal of
support from physiologists, is known as the *fluid mosaic
model* (*Figure 2.20*). The immense difficulties involved
in studying membrane structure suggest that an understanding of the true structure of biological membranes
may be a long way off. Meanwhile, it is worth noting
that the application of physical methods of examination
to biological membranes suggests that phospholipid and
protein molecules may be present in both the triple-
layered and fluid mosaic arrangements in any one type of
membrane.
  Leaving aside this somewhat vexed question of membrane
structure, let us turn to the functions of the vast array
of different lipids in microbial membranes. The various
models for membrane structure imply that all phospholipid
molecules have an equivalent function in membranes.
However, there are reasons to believe that this is not
true in many if not all membranes. It has been shown,
for example, that certain bacterial membrane-bound enzymes
require the presence of a particular phospholipid for
optimum activity. In *Staphylococcus aureus,* the membrane-

C

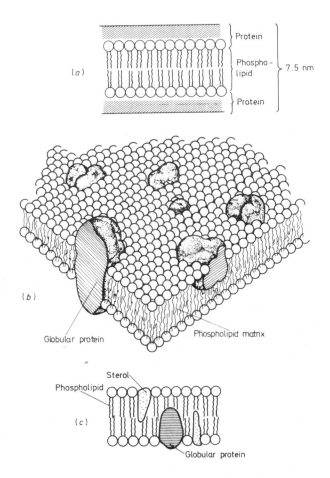

Figure 2.20 *Proposed structures for biological membranes.*
(a) Triple-layered structure of the unit membrane;
(b) schematic three-dimensional model of a bacterial
membrane as a fluid mosaic showing globular proteins
embedded in a phospholipid matrix; (c) schematic cross-
sectional view through a eukaryotic membrane in the form
of a fluid mosaic, showing globular proteins and also
sterols intercalated with phospholipids

bound enzyme undecaprenyl alcohol phosphokinase, which catalyses phosphorylation of a C_{55}-undecaprenyl alcohol, an intermediate in peptidoglycan biosynthesis (see page 299), requires phosphatidylglycerol for optimum activity. Phosphatidylglycerol is also required for activity of phosphotransferases that are involved in transport of some sugars across the plasma membrane in *E. coli*. It needs to be remembered, too, that names such as phosphatidyl-ethanolamine and phosphatidylcholine are generic, since there exist in membranes several representatives of each class of phospholipid which differ in the nature of the fatty-acyl residues on C-1 and C-2. Introduction of one, or two, double bonds into the fatty-acyl chains of phospholipids perturbs the close packing of the chains, and lowers the temperature at which the transition from the gel to the liquid-crystalline phase takes place. This explains, in part, why the presence of unsaturated fatty-acyl residues in membrane lipids helps to maintain the activity of membrane-bound proteins at low temperatures (see page 134). It is worth noting, too, that the activity of the undecaprenyl alcohol phosphokinase in the plasma membrane of *Staph. aureus* is influenced by the degree of unsaturation of the associated phosphatidyl-glycerol, the activity being greater the lower the proportion of unsaturated fatty-acyl residues in the phospholipid.

Although the occurrence of sterols in membranes of eukaryotic micro-organisms has been recognised for many years, the role of these compounds in membrane function is still far from clear. Force-area studies on phospholipid monolayers show that the inclusion of a sterol into a monolayer causes the phospholipid molecules to be arranged in a more compact fashion. How this effect is brought about is not understood, but it has led to the supposition that the presence of sterols in membranes serves to strengthen the membrane. Nutritional studies with certain mycoplasmas, which require a sterol for growth, and with anaerobically grown *Sacch. cerevisiae* which is also auxotrophic for a sterol, show that the structural requirements for a steroid to be incorporated into a cellular membrane are firstly that the steroid nucleus must be planar; secondly that the steroid must have a free hydroxyl group in the 3-β position (in other words it must be a sterol); and finally that there must be a long alkyl chain on C-17 of the molecule. It is likely that the sterol molecules lie alongside phospholipids in the membrane, with the C-3 hydroxyl

group near the polar head group on the phospholipid,
and the alkyl chain on C-17 extending into the hydrophobic
portion of the bilayer alongside the end of the fatty-acyl
chains (*Figure 2.20*). Almost nothing, however, is known
about the way in which variations in sterol structure
influence the behaviour of the molecule in a membrane,
although it has been established that the structure of
the alkyl side chain on C-17 is important in determining
the extent to which a membrane can stretch.

Glycolipids, although they are quantitatively minor
components in many bacterial plasma membranes, may have
an important role in pore formation in the membrane.
Because the conformation of bacterial glycolipids is such
that all of the hydroxyl groups lie on one side of the
molecule and all of the lipophilic groupings (fatty-acyl
chains and glycosidic oxygen atoms) on the other, they
are ideally structured to act as molecules which line
hydrophilic channels through the membrane. However, this
role for glycolipids in membranes is far from proven and
indeed is just one – albeit the most plausible – of
several possible roles for these lipids.

Lipoteichoic acids are large structures which are
envisaged as being anchored into the plasma membrane by
the fatty-acyl residues on the glycolipid moiety, with
the hydrophilic teichoic acid moiety extending into the
cell wall. The role of lipoteichoic acids in plasma
membranes is probably to provide a localised concentration
of magnesium ions which are required for optimal activity
of several membrane-bound enzymes.

The physiological function of several of the other minor
lipid components of the plasma membrane is, ironically,
much better understood. In the bacterial membrane many
of them, including quinones and coenzyme Q, form parts of
the electron-transport chain. Isoprenoid alcohols, which
have been detected in bacterial plasma membranes and which
almost certainly are present too in the plasma membranes
of eukaryotic micro-organisms, play an important role in
the biosynthesis of wall and extramural polymers (*see*
page 298).

Mesosomes. Many Gram-positive bacteria, as well as a few
Gram-negative strains, produce an invagination of the
plasma membrane filled with clusters of vesicles and
tubules or membraneous whorls, or both (*Figure 2.21*).
These structures have been called *mesosomes* following the
suggestion of Philip Fitz-James of the University of

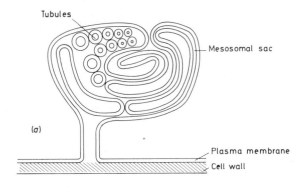

Tubules

Mesosomal sac

(a)

Plasma membrane

Cell wall

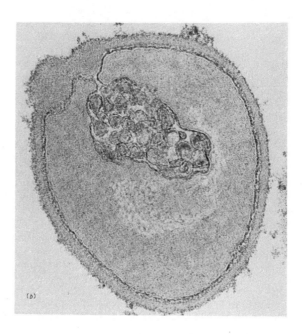

(b)

Figure 2.21 Molecular anatomy of mesosomes. (a)
A schematic interpretation of mesosome structure, showing
the relationship with the plasma membrane, and the
mesosomal sac and tubules. (b) An electron micrograph
of a thin section through Staphylococcus aureus;
magnification x 82 500. (The micrograph was kindly
provided by David Ellar)

Western Ontario, Canada, who first examined these
structures in detail. Bacilli produce particularly
prominent mesosomes and many of the data on these
structures have come from experiments on strains of
Bacillus.

Mesosomes can be isolated by converting bacilli into
protoplasts when they are released as vesicles into the
supernatant liquid, from which they can be isolated on
sucrose density-gradients. Mesosomes have a lipid
composition similar to that of the plasma membrane.
However, the proteins in mesosomes differ to some extent
from those in the plasma membrane, as judged by gel
electrophoretic patterns of the extracted proteins and
by cytochrome assays.

The physiological function of mesosomes remains
something of a mystery although there has been no
shortage of suggestions. The most plausible suggested
role is that they are concerned in septum formation,
mainly because they are usually seen near the developing
septum in electron micrographs of thin sections through
bacteria. It is conceivable that these membraneous
structures are involved in synthesis of wall polymers or
in control of the action of autolytic enzymes necessary
for wall formation.

2.7 INTRACELLULAR MEMBRANES

Microbiologists are now fairly well acquainted with the
highly complex fine structure of the microbial cytoplasm.
By definition, prokaryotic microbes differ fundamentally
from eukaryotes. The cytoplasm in bacteria and blue-
green algae contains many different structures and
assemblies of macromolecules, but none of these is
enveloped in a membrane. However, microbial
physiologists do not envisage these structures and
macromolecules in the prokaryotic cell floating freely
in the cytoplasm. More likely, they are bound, either
covalently or by secondary bonds, to the plasma membrane.
With some of these structures, including the bacterial
genome, there is firm evidence to show that they are linked
to the plasma membrane. Nevertheless, bacteria and blue-
green algae are not entirely devoid of internal membranes.
Some bacteria, especially those that grow autotrophically
or can use gaseous nutrients such as methane and
dinitrogen, have fairly well developed internal membrane
systems (*see Figure 2.22*). Often in the past, these

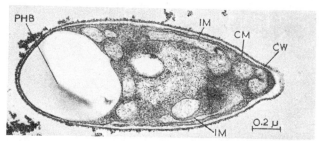

Figure 2.22   Electron micrograph of a thin section through the budding bacterium Hyphomicrobium *strain H-526, showing a well developed internal membrane system (IM), plasma membrane (PM), cell wall (CW) and a granule of poly-β-hydroxybutyrate (PHB). (By courtesy of S.F. Conti)*

intracellular membranes have not been detected in electron micrographs of thin sections through the bacteria because they are frequently heavily encrusted with ribosomes.

In eukaryotic micro-organisms, by contrast, the nucleus as well as other internal organelles such as mitochondria and chloroplasts are enclosed in membranes (*Figures 2.23,*

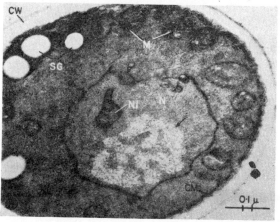

Figure 2.23   Electron micrograph of a thin section through the yeast Saccharomyces cerevisiae *showing the cell wall (CW), plasma membrane (PM), mitochondria (M), nucleus (N) and nucleolus (NI). Areas of different density are apparent within the nucleus and are indicated by arrows. (By courtesy of A.W. Linnane)*

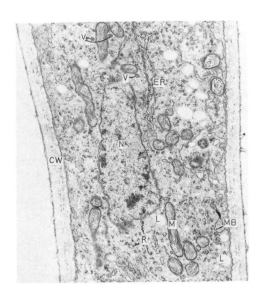

Figure 2.24 *Longitudinal section through part of an aerial hypha of the fungus* Pyronema domesticum *showing a nucleus (N), mitochondria (M), rough endoplasmic reticulum (ER), cytoplasmic ribosomes (R), vesicles (V), lipid droplets (L), a microbody (MB) and a layered cell wall (CW); magnification x18 700. (The micrograph was kindly provided by Alan Beckett)*

Figure 2.25 *Electron micrograph of a cross-fractured cell of* Saccharomyces cerevisiae *showing a plasma membrane (Pl), endoplasmic reticulum (ER), the nucleus (N), nuclear membrane (Nm) and a mitochondrion (M). (The micrograph was kindly provided by H. Moor)*

2.24 and *2.25*). Membrane-bound cytoplasmic organelles in eukaryotes are often connected by membranes to form a simple *endoplasmic reticulum,* although not usually one as extensive as the reticula that are found in cells of higher organisms (*Figures 2.24* and *2.25*). Occasionally, the membranes of the endoplasmic reticulum in some eukaryotic micro-organisms become stacked to form structures which resemble *dictyosomes* or *Golgi bodies* that are detectable in cells of higher organisms. Unfortunately nothing is known of the composition or structure of these eukaryotic membraneous structures.

2.8 GENOMES

The information for the running of a cell – that is information which directs synthesis of specific proteins with or without enzymic activity – is encoded in molecules of deoxyribonucleic acid (DNA). Deoxyribonucleic acid consists of a linear unbranched chain of deoxyribonucleo-side residues joined by phosphodiester bonds between the 3' and 5' positions in the 2'-deoxyribose residues. Essentially, it is a deoxyribose phosphate backbone from which bases (adenine, cytosine, guanine and thymine) project in a linear array (*Figure 2.26*). The DNA in micro-organisms can easily be extracted as a viscous solution, the viscosity being attributable to the highly organised secondary structure of the molecule, in which a pair of rigid deoxyribose phosphate backbones is stabilised by base-pairing between strictly complementary nucleotides. The brilliant discoveries of James D. Watson and Francis Crick in the 1950s established the importance of the specific base-pairing in the DNA molecule, and its significance in the replication and expression of genetic information. Two polydeoxyribonucleotide chains are wound round one another to form a *double-stranded helix* (*Figure 2.26*). The planar bases lie inside the helix, perpendicular to the long axis, and so resemble steps in a circular staircase. The molecule is about 0.2 nm in diameter and has approximately ten nucleotide residues in each turn of the helix. The bases on the polynucleotide chains are paired, an adenine residue on one chain with a thymine on the other, and a guanine with a cytosine residue. It follows therefore that the molar ratios of adenine:thymine and guanine:cytosine in DNA both equal

*Figure 2.26 Structure of DNA. The deoxyribose
phosphate backbone is represented by -S-P-S;
A=T represents the adenine-thymine pairing, and G≡C the
guanine-cytosine pairing*

unity, although the proportions of each pair of bases
vary widely in DNAs from different organisms. The paired
bases are linked by hydrogen bonding, and the chains can
be separated by heat (page 332) or by reagents (e.g. 8 M
urea) that break hydrogen bonds. Each of the polynucleotide chains in the DNA molecule is a complement of the
other, and this suggested a possible mechanism for gene
replication. After the chains have unfolded, each could
act as a template for the formation on itself of the
companion chain, thus providing for duplication of the
entire molecule.

Approximately 2-3% of the dry weight of a micro-organism
is DNA. The exact amount in a microbial cell depends on
the rate at which the cells in a population are dividing,
the faster the growth rate the greater the DNA content.
The DNA content of a micro-organism is easily assayed
using the diphenylamine method, which depends on a
reaction between diphenylamine and 2-deoxy-D-ribose
residues in DNA.

The location of DNA in micro-organisms can be revealed
using a modification of the Feulgen staining reaction.
This reaction is based on the ability of DNA to stain
deep purple when treated with Schiff's reagent (fuchsin

decolourised with sulphurous acid) due to the formation
of free aldehyde groups in the DNA. The major obstacle
in staining DNA, especially in bacteria, is the presence
of large amounts of RNA. This RNA can be hydrolysed by
warming a preparation of fixed organisms with N-HCl for
7-8 min at 60°C; RNA hydrolysis products do not react
with the Feulgen stain. Many modifications of the
Feulgen reaction have been reported, most of them using
other dyes, particularly compound dyes of the Romanowsky
series (e.g. Giemsa's stain). Ribonuclease has been used
instead of acid for hydrolysing the RNA.

2.8.1 GENOMES IN PROKARYOTES

The DNA in prokaryotic microbes is associated with the
plasma membrane, although it is not itself bounded by an
intracellular membrane. In several bacteria, the genome
has been shown to be a single circular loop of DNA, a
structure usually referred to as a *chromosome*. A very
elegant demonstration of the circularity of the bacterial
chromosome has come from John Cairns. He labelled DNA
in growing *E. coli* with ^{3}H-thymidine at high specific
activity. After very carefully extracting the DNA from
the bacteria, he spread it on a membrane under conditions
which minimised fragmentation and caused it to unravel.
The membrane was then covered with photographic emulsion,
and the structure of the chromosome inferred from the
distribution of silver grains in the exposed emulsion.
When fully extended, the chromosome from *E. coli* has a
circumference of 1.1-1.4 mm (*Figure 2.27*).
   E. coli can code for about 2000 different enzymes. If
one allows for additional proteins, such as structural
proteins in membranes and ribosomes, and assumes that
repressor proteins can account for a reasonable proportion
of the cellular protein, it is reasonable to assume that
there are about 5000 genes on the genome in *E. coli*. If
each gene has an average of 600 base pairs, then the total
DNA will be 3×10^6 base pairs with a molecular weight of
2×10^9 daltons, and a length of 1000 µm (1 mm). As such,
the DNA content of *E. coli* (and a similar calculation can
be made for other bacteria) is roughly what is to be
expected.
   It has been known for some time that the DNA in bacteria
often contains, in addition to adenine, cytosine, guanine
and thymine residues, small amounts (as little as 0.05 to
1.0 mole per cent) of methylated bases which include

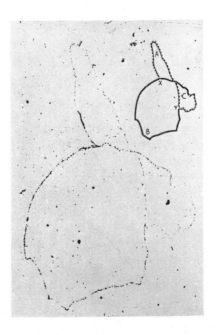

Figure 2.27   An autoradiograph of a chromosome of
Escherichia coli K-12, *labelled with [³H] thymidine for*
two generations.   The small inset shows the same structure
diagrammatically and divided into three sections (A, B
and C) that arise at the two forks (X and Y).   (By
courtesy of John Cairns)

6-methylaminopurine and 5-methylcytosine, but the
significance of these rare bases has only recently been
realised.   There are several means by which DNA can be
transferred from one bacterium to another which include
conjugation, transformation and transduction, and which
are discussed on page 335.   Once inside a bacterium, a
DNA molecule may survive, possibly to replicate or to
combine with the host genome, or it may be degraded by a
process known as *restriction*.   Whether a DNA molecule
survives or is subject to the action of restriction
enzymes depends upon, firstly, the properties of the
restriction enzymes, and secondly the ability of the cell
from which the DNA came to carry out certain structural
modifications to the DNA molecule.   *Modification* results
from the methylation of DNA, presumably at or near the

nucleotides in the DNA chain where restriction nucleo-
tidases cleave the chain. Modification therefore explains
the occurrence in bacterial DNA of a small proportion of
methylated bases. Information on those parts of the DNA
chain that are subject to modification is confined
largely to *Haemophilus influenzae* and *E. coli.*

Whereas the DNA in higher eukaryotic organisms is
associated with appreciable amounts of basic protein,
known as *histone,* there is no evidence for the presence
of large amounts of histone on the bacterial genome. At
the same time, it is known that repressor proteins are
attached to the bacterial genome (*see* page 347).

Not all of the DNA in bacteria is in the form of a
single large circular chromosome; a small amount -
sometimes as little as 0.1 to 0.2% - of the cellular DNA
is present extrachromosomally in the form of *plasmids* or
episomes. These subsidiary DNA structures resemble
chromosomes in that they are capable of autonomous
self-replication in the bacterial cell, in other words
they are *replicons.* Also, plasmids are attached to the
bacterial plasma membrane and, like chromosomes, they
exist as closed circular duplex molecules. However,
plasmids differ from chromosomes in that they are not
essential for the existence of a bacterium and can be
removed from a bacterial cell by a process of *curing*
which involves treating cells with reagents that include
acriflavine and other acridines, rifampicin, ethidium
bromide, or cobaltous ions. Several types of plasmid
have been detected in bacteria, and they differ in the
nature of the genetic information which they carry.
Among the best studied are *resistance factors* which carry
information that directs synthesis of enzymes which confer
resistance by the bacterium to certain antibiotics such
as penicillin, streptomycin, chloramphenicol and
tetracyclines (*see* page 138). The ability of some enteric
bacteria to produce highly specific extracellular
antibiotic proteins, called colicins, is also carried on
plasmids, known as *colicinogenic factors.* The *sex factor*
is also a plasmid, one which carries genetic information
for mediating conjugation including the production of
F-pili (*see* page 335). Not all bacterial plasmids have
been characterised. For example, *Micrococcus lysodeikticus*
can carry a plasmid with a molecular weight of 63×10^6
daltons, the function of which is as yet not understood;
these are referred to as *cryptic plasmids.*

Although they are capable of autonomous replication in
the bacterial cell, plasmids do not always exist in this

state. Part or whole of the plasmid DNA can be
incorporated into the chromosomal replicon, when
replication and distribution of the plasmid genes become
associated with chromosomal replication and distribution.
As well as integrating with the chromosomal replicon,
plasmids may merge, either transiently or permanently,
with other plasmids.

2.8.2 GENOMES IN EUKARYOTES

In eukaryotic micro-organisms, the DNA is compacted into
an organelle known as the *nucleus* (*Figures 2.23, 2.24* and
2.25). Electron micrographs of thin sections through
eukaryotic microbes show that each nucleus is bounded by
a double-unit membrane, the outer of which often appears
to be continuous with the endoplasmic reticulum. Gaps
or pores can often be seen in the nuclear membrane,
although in some organisms these are few in number. The
pores in the nuclear membrane are frequently quite wide.
Those in the nuclear membrane in amoebae, for example,
allow the passage of particles as wide as 10 nm. Inside
the nucleus, many eukaryotic micro-organisms have one
or more nucleoli, apparently also membrane-bound
(*Figure 2.23*).
  The DNA in eukaryotic nuclei exists in the form of
chromosomes. By contrast with the prokaryotic cell,
there is no evidence that the eukaryotic chromosomes are
circular. Moreover, there is not just one but several
chromosomes in the eukaryotic microbe. Information on
the number of chromosomes in various eukaryotic microbes
is restricted mainly to those organisms in which the
genome can be mapped. *Sacch. cerevisiae,* for example,
has at least 16 identifiable chromosomes, and each must
be similar in size and information content to the
bacterial chromosome. Very little indeed is known of
the way in which chromosomes are arranged in the nucleus,
although it is generally assumed that they are intimately
associated with the nuclear membrane. In higher
eukaryotic organisms (plants and animals), the chromosomes
are associated with appreciable amounts of histones.
While evidence so far available tends to be somewhat
equivocal, there is doubt as to the existence of large
amounts of histones in the nuclei of eukaryotic microbes.
  Eukaryotic micro-organisms contain about ten times as
much DNA as prokaryotes, and this corresponds to 50 000–
70 000 genes. Eukaryotes have a more complex cellular

structure than prokaryotes, and in addition they often
have the capacity to differentiate. Nevertheless, it is
difficult to believe that this additional genetic
information, associated with the eukaryotic way of life,
would require eukaryotic micro-organisms to possess quite
so much DNA. If it is assumed that there is excess DNA
in eukaryotes, then there is the need to explain the
nature of the seemingly gratuitous genetic information.
Three possible solutions have been suggested to this
riddle. The first postulates that some of the genes in
eukaryotes are duplicated longitudinally to form a tandem
sequence of copies within a single-stranded chromosome.
It has further been suggested that one of the gene copies -
the 'master' - has control over the remaining copies (the
'slaves') in that, once during each cell cycle, the
slaves are matched against the master. A second solution
proposes that there is lateral multiplicity of genetic
information, in that each chromosome contains not just
one copy of the genetic material but, instead, a number
of parallel strands each containing the full complement
of genetic information. A third possibility is that only
part of the DNA in eukaryotic micro-organisms has a
genetic function, and that the remainder is redundant or
simply junk.

2.9 RIBOSOMES

The bulk (70-90%) of the RNA in micro-organisms is
associated with proteins in the form of small particles
which sediment at around 100 000 g. These are the
ribosomes, which contain 50-70% RNA and 30-50% protein.
Ribosomal composition and structure differ in prokaryotes
and eukaryotes, and the organelles from each of these
classes of microbe are considered separately.

2.9.1 RIBOSOMES FROM PROKARYOTES

There are about 10 000 ribosomes in a bacterium. The
number varies with the rate at which the bacterium is
grown; the greater the rate of growth the higher the
ribosomal content. Ribosomes isolated from bacteria and
blue-green algae have a sedimentation coefficient of
70S. A 70S ribosome is made up of two dissimilar subunits,
the properties of which are listed in *Table 2.2*. The
smaller subunit, which has a sedimentation coefficient
of 30S, is made up of one 16S RNA molecule and 21

Table 2.2 Properties of the small and large subunits from ribosomes of prokaryotic micro-organisms

	Sedimentation coefficient (S)	Dimensions (nm)			Particle mass (daltons)	Sedimentation coefficient of RNA (S)	Molecular weight of RNA (daltons)	Number of separable proteins
Small subunit	30	5.6	22.4	22.4	0.8×10^6	16	0.55×10^6	21
Large subunit	50	13.0	17.5	26.0	1.8×10^6	23	1.2×10^6	about 35
						5	4.0×10^4	

different and separable proteins. Although few data are
available on the properties of the 16S RNA molecule, it
is thought that it contains hydrogen-bonded regions and
hairpin loops. Ribosomal proteins from both the small
and larger subunits of bacterial ribosomes have been
separated by gel electrophoresis. Some are acidic or
neutral proteins, but the majority are basic in nature.
The 21 proteins found in the small subunit range in
molecular weight from 10 900 to 65 000 daltons.

When isolated 30S subunits are centrifuged in 5 M
caesium chloride, about a third of the proteins in the
particle are released. When mixed with the denuded
particle, the released proteins recombine to form a
functional 30S subunit. From this type of experiment,
it has been shown that not all of the 21 proteins in
the 30S subunit are essential for ribosomal activity;
some are apparently needed only to effect assembly of
the subunit.

Far less is known about the properties of the 50S
subunit of the ribosomes from bacteria. It has been
established, however, that the subunit contains one
molecule each of a 23S RNA molecule and a 5S RNA molecule,
but almost nothing has been reported on the properties of
the proteins that go to make up the 50S subunit.

Aggregation of a 30S with a 50S subunit to form a 70S
ribosome is an integral part of the process of polypeptide synthesis. The aggregation process is regulated,
in part at least, by the concentration of magnesium ions,
which are always found associated with ribosomal
particles. The polyamines spermine and spermidine are
also present in ribosomes, and these compounds probably
also have a stabilising role in the 70S ribosome.

Information on the three-dimensional arrangement of
RNA and proteins in the 30S and 50S subunits is lacking.
The reason is that the ribosome is too large and
complicated in structure to apply X-ray crystallography.
It has recently been suggested that neutron diffraction
might prove a more suitable probe for unravelling the
three-dimensional structure of ribosomes. In addition,
electron-microscope examination of ribosomes has, so far,
proved to be of limited value as a technique for
examining the structure of these organelles. It is
hardly surprising, therefore, that it is not yet clear
whether the proteins are located on the inside or on the
outside of ribosomal particles.

Ribosomes are the centres of protein synthesis (page
286) in which they act, not in the form of isolated

particles, but as clusters of particles known as
polyribosomes or *polysomes*. The functional particles in
the *E. coli* polysome are 70S ribosomes and, during
protein synthesis, each of the 30S particles in the 70S
ribosome makes contact with the strand of mRNA.
Messenger-RNA, which conveys the genetic coding from the
DNA to the polysome, accounts for between 2 and 4% of the
total cell RNA in *E. coli*. As one would expect, messenger-
RNA is very heterogeneous with respect to molecular weight
since it is coding for proteins of different sizes.   In
micro-organisms, mRNA has a high turnover rate.

2.9.2 RIBOSOMES FROM EUKARYOTES

Ribosomes from eukaryotic microbes have not been studied
nearly as intensively as their prokaryotic counterparts.
We do know however that, like bacterial ribosomes, those
from eukaryotic microbes are made up of two separable
units.   In *Sacch. cerevisiae*, the smaller ribosomal
subunit has a sedimentation coefficient of 40S and contains
one 17S RNA molecule (molecular weight 0.65×10^6 daltons),
while the larger 60S subunit contains a 26S RNA molecule
which has a molecular weight of 1.2×10^6 daltons,
together with one 5S RNA molecule. The 40S and 60S
subunits combine to form a ribosomal particle with a
sedimentation coefficient of 80S. As yet, very little
is known about the numbers and properties of the proteins
that go to make up the subunits in eukaryotic microbes.
**Eukaryotes do, however, contain 70S ribosomes in certain
intracellular organelles.**
2.10   MITOCHONDRIA

The enzymes and carriers which are involved in electron
transport and oxidative phosphorylation in prokaryotes
are located in the plasma membrane. But, in aerobically
grown eukaryotic microbes, electron microscopy reveals
the presence of mitochondria similar in structure to
those found in animal and plant cells. Mitochondria are
the 'power houses' of the cell, and they contain all the
enzymes and carriers involved in electron transport and
oxidative phosphorylation. Their function is to release
energy by oxidation of substrates, and to conserve this
energy in the form of ATP (*see* page 181). Microbial
mitochondria come in a wide range of sizes. In some
protozoa (such as *Chaos chaos*), they are about the same
size as animal and plant mitochondria (around $0.2-0.5 \times$

0.7-4.0 µm), while in other micro-organisms they are
smaller. In the electron micrograph through a thin
section of the fungus *Rhizopus sexualis* shown in *Figure
2.28,* the smaller mitochondrion measures 0.4 × 1.2 µm;
by comparison, mitochondria in *Sacch. cerevisiae* are only
about 0.6 µm long. Small eukaryotic microbes have as few
as 10-20 mitochondria in each cell, but there are more in
larger cells.

*Figure 2.28 Longitudinal section through parts of two
mitochondria in an aerial hypha of the fungus* Rhizopus
sexualis. *Mitochondria are often lobed, with deep
invaginations which in certain planes give the organelles
a ring-like appearance. Note the numerous tubular cristae;
magnification ×40 000. (The micrograph was kindly
provided by Alan Beckett)*

   Mitochondria are bounded by a double-unit membrane, the
inner of which is folded inwards to form *cristae* (*Figures
2.28* and *2.29*). Cristae appear in many shapes and sizes
in microbial mitochondria and it has been suggested that
the extent to which the inner membrane folds to form
cristae is related to the energy requirements of the
cell.
   It is possible to separate the inner and outer membranes
of mitochondria, and to analyse these membranes as well as

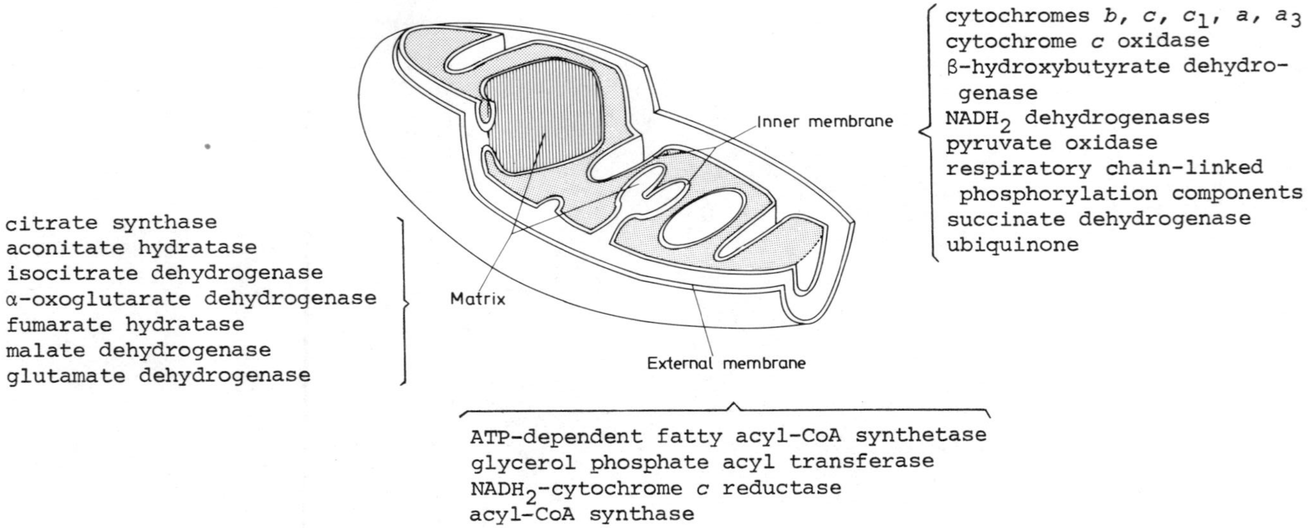

citrate synthase
aconitate hydratase
isocitrate dehydrogenase
α-oxoglutarate dehydrogenase
fumarate hydratase
malate dehydrogenase
glutamate dehydrogenase

cytochromes *b*, *c*, *c*$_1$, *a*, *a*$_3$
cytochrome *c* oxidase
β-hydroxybutyrate dehydrogenase
NADH$_2$ dehydrogenases
pyruvate oxidase
respiratory chain-linked
phosphorylation components
succinate dehydrogenase
ubiquinone

Inner membrane

External membrane

Matrix

ATP-dependent fatty acyl-CoA synthetase
glycerol phosphate acyl transferase
NADH$_2$-cytochrome *c* reductase
acyl-CoA synthase

*Figure 2.29 Diagram illustrating the main structural features of a mitochondrion, and the distri-
bution of the principal enzymes and carriers in the organelle. Many of the locations indicated are
from data on mammalian mitochondria and have not been demonstrated unequivocally in microbial
mitochondria*

the matrix material for proteins and lipids. The location of the various respiratory enzymes and carriers in the mitochondrion is shown in *Figure 2.29*. Many of the enzymes of the TCA cycle occur in the mitochondrial matrix, while the inner membrane, which is analogous to the bacterial plasma membrane, contains the cytochromes and other carriers that go to make up the respiratory chain. These carriers are located in the inner membrane in the sequence in which they act, with each functional unit being known as a *respiratory assembly*. Some carriers are also located in the outer membrane.

Data on the lipid composition of mitochondrial membranes are more extensive for mammalian organelles. It has been found that membranes of animal mitochondria differ to some extent in phospholipid composition and in the molar ratios of sterol to phospholipid. Information on microbial mitochondria is, by comparison, rather meagre, and is confined very largely to organelles from *Sacch. cerevisiae* and *Neurospora crassa*. With yeast mitochondria it has been established that unsaturated fatty-acyl residues in phospholipids are essential for oxidative phosphorylation, and that cardiolipin or diphosphatidylglycerol is confined to the inner mitochondrial membrane.

Mitochondria also contain DNA which confers on them a degree of autonomy since mitochondrial DNA (known as MtDNA) directs synthesis of many, although not all, of the enzymes and carriers involved in respiration and oxidative phosphorylation. In haploid *Sacch. cerevisiae,* as much as 20% of the cellular DNA is accounted for by MtDNA. The mitochondrial DNA in this yeast has an unusual secondary structure; its estimated molecular weight is around 50×10^6 daltons.

2.11 PHOTOSYNTHETIC APPARATUS

In photosynthetically active micro-organisms, the photosynthetic and, when they occur, accessory pigments are located in subcellular structures or organelles. These structures and organelles come in many different shapes and sizes, and with different compositions that depend in large part on whether the organism is a prokaryote or a eukaryote. However, the basic unit of the photosynthetic apparatus in all organisms is an enclosed membraneous sac known as a *thylakoid*. Photosynthetic microbes differ in the manner in which the thylakoids are arranged in the cell.

2.11.1 THE PHOTOSYNTHETIC APPARATUS IN PROKARYOTES

Bacteria

Photosynthetic bacteria fall into three fairly distinct
groups. Members of the **Chlorobacteriaceae** (the green
sulphur bacteria such as species of *Chlorobium*) have a
photosynthetic apparatus that consists of a series of
membrane-bound vesicles (measuring 100-150 × 30-40 μm)
which are not continuous with the plasma membrane. In
the purple sulphur bacteria or **Thiorhodaceae** (which
include *Chromatium* spp.), the photosynthetic apparatus
is a membraneous extension of the plasma membrane and
appears in thin sections as vesicles, tubular membranes
or as a lamellar system all made of thylakoids. An
attachment with the plasma membrane is also a characteristic of the photosynthetic apparatus in the **Athiorhodaceae**
or purple non-sulphur bacteria (which includes species
of *Rhodospirillum* and *Rhodopseudomonas*). Again, the
apparatus can appear in any one of several morphological
forms.

   Subcellular structures that form the photosynthetic
apparatus have been isolated from many photosynthetic
bacteria and have been referred to as *chromatophores*.
Not unexpectedly in view of their morphology, they are
composed of a mixture of protein and lipid, together
with a range of photosynthetic pigments. In the green
sulphur bacteria, the basic or *reaction-centre*
photosynthetic pigments are the *Chlorobium chlorophylls
650 and 660* (a nomenclature based on their absorption
maxima in ether), which are sometimes referred to as
bacteriochlorophylls c and *d* (*Figure 2.30*). Green
sulphur bacteria also contain small amounts of
bacteriochlorophyll *a*, although the Chlorobium
chlorophylls are responsible for the bulk of the light-
harvesting in these bacteria. In the purple photosynthetic bacteria, the basic pigment is bacteriochlorophyll *a* (*Figure 2.30*) which has an absorption maximum
between 800 and 1000 nm, depending on the species although
these bacteria often also contain appreciable amounts of
bacteriochlorophyll *b*. Experiments with photosynthetic
structures isolated from *Rhodopseudomonas spheroides*
indicated that bacteriochlorophyll absorbing maximally
at 850 nm is mainly responsible for harvesting light.
The complex which absorbs at 870 nm, and which accounts
for only 5% of the total pigment in the bacterium,
constitutes the photosynthetic reaction centre (*see* page 234).

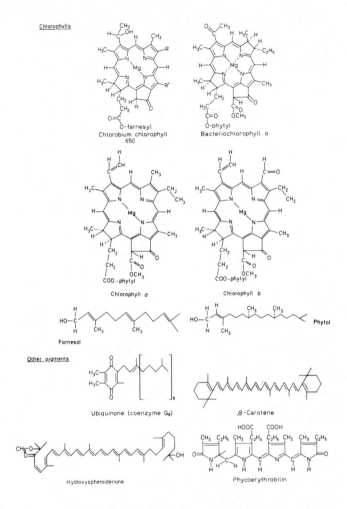

Figure 2.30 *Structural formulae of chlorophylls and other pigments that occur in the photosynthetic apparatus in prokaryotes and eukaryotes*

The lipids in isolated photosynthetic structures from
bacteria are mainly phospholipids. In chromatophores
from a *Chromatium* spp., for instance, the main lipid is
phosphatidylethanolamine; quinones also occur in these
structures. In the green sulphur bacteria this is
usually menaquinone, and in the purple sulphur bacteria,
ubiquinone (*Figure 2.30*). Carotenoids are also found,
but unlike those in chloroplasts, they are often
aliphatic carotenoids, such as hydroxyspheroidenone which
occurs in members of the *Athiorhodaceae* (*Figure 2.30*).
Finally, there are the enzymes and carriers involved in
the electron-transport chain, such as cytochromes and
ferredoxins (*see* page 203) for more information on the
nature of these proteins).

Blue-green algae

In the blue-green algae, or Cyanophyceae, the thylakoids
tend to lie parallel to one another rather than in
closely packed stacks as in many photosynthetic bacteria.
The photosynthetic apparatus of blue-green algae
resembles that of eukaryotic microbes in composition,
the similarity extending to the constituent cytochromes
and ferredoxin. The principal pigment is chlorophyll *a*
which however differs structurally from bacteriochlorophyll
a (*Figure 2.30*). Blue-green algae also have the accessory
pigment phycocyanin in their photosynthetic apparatus and
in this respect they differ from green algae which have
other accessory pigments. Photosynthetic bacteria,
however, do not possess accessory photosynthetic pigments.

2.11.2   THE PHOTOSYNTHETIC APPARATUS IN EUKARYOTES

The characteristic feature of the photosynthetic
apparatus in eukaryotes is that it is contained within
a discrete organelle, the *chloroplast* (*Figure 2.31*).
Although a great deal of the research done on the molecular
architecture of chloroplasts has been on organelles from
higher plant cells, there are reasons for believing that
chloroplasts in algae are basically similar in structure
to those from higher plants. The chloroplast is enveloped
in a double membrane, inside which is a matrix known as
the *stroma* which is traversed by the thylakoids. In
electron micrographs through thin sections of algae,
chloroplasts show much greater diversity in shape, size
and number compared with those in plant cells. Some

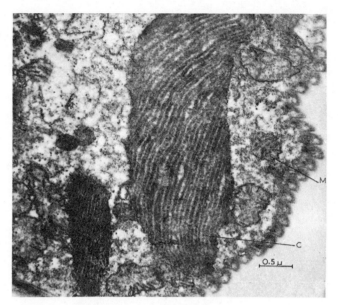

Figure 2.31 Electron micrograph of a thin section through Euglena gracilis *var.* bacillaris, *showing a chloroplast (C) with structure, and mitochondria (M) with cristae. (The micrograph was kindly provided by Sarah P. Gibbs)*

algae, such as *Micromonas* **spp.**, contain only one chloroplast, while others such as *Euglena* spp. may contain hundreds. In general, algal chloroplasts differ in two respects from their higher-plant counterparts. Firstly, they are often associated with finely granular proteinaceous structures, known as *pyrenoids,* which are thought to be associated with carbohydrate metabolism. Secondly, algal chloroplasts show a different type of thylakoid arrangement, which may take the form of single thylakoids (as in the red algae), paired thylakoids (e.g. in the Cryptophyta), or the commoner triple and multiple thylakoids which are found in the Chlorophyta, Chrysophyta and in the Euglenophyta.

Chloroplasts have been isolated from a number of different algae and subjected to detailed biochemical and fine-structure examination. As with the photosynthetic apparatus from bacteria, they are composed principally of protein, lipids and pigments. Chlorophyll *a* (*Figure 2.30*) is found in all algal chloroplasts, and in green algae it occurs together with chlorophyll *b* which is an *accessory photosynthetic pigment*. The accessory pigment in red algae is not chlorophyll *b* but phycoerythrobilin (*Figure 2.30*), and in the brown algae fucoxanthin. Although there has been extensive research on the biochemistry of chloroplasts, physiologists are far from being able to describe the molecular architecture of the basic structure, namely the thylakoid membrane. A model which accommodates most of the data which have come from X-ray and electron-microscope (particularly of freeze-etched material) examination of chloroplasts is shown in *Figure 2.32*. Two types of protein are believed to be located in the thylakoid membrane, and to be associated with molecules of chlorophyll and other pigments. These proteins are envisaged as being embedded in a lipid membrane made up mainly of phospholipids but also including some glycolipids.

2.12 OTHER CYTOPLASMIC INCLUSIONS AND GRANULES

Despite the fact that a vast array of vesicles and similar structures have been detected microscopically, particularly in eukaryotic micro-organisms, very little is known about their molecular architecture.

Lysosomes are often observed in electron micrographs of thin sections through eukaryotic microbes, as spherical structures surrounded by a unit membrane. In *Sacch. cerevisiae* the vacuole, which measures about 0.5 μm in diameter, is thought to have a lysosomal role. Isolated yeast vacuoles contain a variety of hydrolytic enzymes, including lipases, proteinases and ribonucleases, and it would be very illuminating to know how these lytic activities are contained within a membrane-bound organelle. In addition, the vacuole in *Sacch. cerevisiae* contains the amino-acid pool of the organism. Amoebae frequently contain *contractile vacuoles* (as in *Chaos chaos*) as well as food vacuoles when they have been grown on a particulate source of nutrient, but hardly anything is known of the chemical anatomy of these vacuoles. Blue-green algae produce *gas vacuoles*. These are cylindrical structures, up to 0.5 μm long and 80–100 nm wide. Curiously, the vacuole wall is not a unit membrane and is only 2–3 nm thick.

Many cytoplasmic granules in both prokaryotic and eukaryotic microbes contain compounds which serve as *stores of energy* and/or of low molecular-weight organic

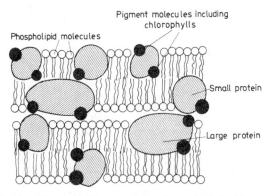

Figure 2.32 Diagram of a model of a thylakoid membrane in a chloroplast in which an attempt has been made to reconcile data obtained from various types of examination. The diagram shows part of a single thylakoid at its periphery and seen in cross-section. No attempt has been made to keep the diagram to scale. See text for details

compounds. These granules can sometimes be seen in unstained preparations under the light microscope, but their presence is not usually a constant feature of a micro-organism and is frequently conditioned to a considerable extent by the environmental conditions under which an organism is grown.

Granules composed of *polyglucan* are produced by many prokaryotes and eukaryotes when grown under conditions where growth is limited by the availability of a nitrogen source. Polyglucan granules can easily be detected in micro-organisms since they stain blue, reddish blue, or brown with iodine. All of the polyglucans so far examined are highly branched molecules with α-1,4 and α-1,6 linkages, and they resemble glycogen obtained from mammalian livers.

Another energy-reserve compound which is accumulated in granular form by several bacteria and by a few eukaryotic micro-organisms is *poly-β-hydroxybutyrate (see Figure 2.22).* A number of bacteria were long ago on record as being able to produce 'lipid' granules which stain with Sudan Black. The granules synthesised by these bacteria are not, in fact, composed of lipid but of poly-β-hydroxy-butyrate, which is a polylactide, and a straight-chain polymer of $D(-)$-3-hydroxybutyrate the formula of which is:

$$HO - \underset{\underset{CH_3}{|}}{CH} - CH_2 - \underset{\underset{O}{||}}{C} \left[O - \underset{\underset{CH_3}{|}}{CH} - CH_2 - \underset{\underset{O}{||}}{C} \right]_n O - \underset{\underset{CH_3}{|}}{CH} - CH_2 - COOH$$

The molecular weight of the polymer ranges from 60 000 to 250 000 daltons, depending on the source. Granules of poly-β-hydroxybutyrate isolated from *Bacillus megaterium* are spherical with a diameter of 0.2–0.7 μm. Each granule must therefore contain several thousand molecules of poly-β-hydroxybutyrate. The granules in many bacteria are surrounded by a membrane which, because it is often only 4–6 nm wide, may not be a unit membrane. Poly-β-hydroxybutyrate is an almost ideal energy-storage compound since it is highly reduced and virtually insoluble in water.

Volutin is the main component of another granular inclusion found in some micro-organisms. When these organisms are treated with toluidine blue and methylene blue, the volutin granules change to reddish violet in colour and are therefore also known as *metachromatic granules*. Volutin granules are made up of polyphosphate. Although there is no question that in some micro-organisms polyphosphate can serve as an energy source, in other organisms these granules probably act primarily as a source of phosphate. Polyphosphate granules have been reported to contain, in addition, RNA and lipid in small amounts. Some micro-organisms synthesise more than one type of energy-reserve granule. In *Myxococcus xanthus*, for example, polyphosphate granules are found together with glycogen-containing inclusions.

There are numerous claims in the literature for the presence of *lipid granules* or *droplets*, particularly in fungi. With only one micro-organism, *Sacch. cerevisiae*, has a study been made of these cytoplasmic inclusions. In this yeast, they are composed not only of triacyl-glycerols, which are true lipids, but also of sterol esters, and they occur as droplets or vesicles with a diameter of about 0.1 μm. It has yet to be demonstrated that lipid droplets in yeast act as an energy source; in fact, it seems more likely that they are involved in growth of the plasma membrane, and possibly also in secretion of polymers including enzymes.

Some sulphur bacteria (the large purple sulphur bacteria such as species of *Beggiatoa* and *Thiothryx*), when growing in the presence of hydrogen sulphide, contain highly refractile droplets of sulphur. This sulphur acts as an energy reserve and is oxidised when the supply of hydrogen sulphide is depleted. The sulphur in these droplets is probably in an unstable form since the stable ortho-rhombic allotopic form cannot exist as moist spherical droplets. In *Chromatium vinosum*, the sulphur droplet is surrounded by a 'membrane' apparently made up entirely of protein.

There exist numerous other cytoplasmic granules in micro-organisms which are not energy reserves. Included among these are crystals of carbonyl diurea which occur in certain amoebae, and which probably represent an end-product of purine catabolism.

2.13   SOLUBLE CYTOPLASMIC CONSTITUENTS

The liquid which remains after all of the particulate material in a suspension of disrupted micro-organisms has been removed by centrifugation constitutes the *soluble fraction* or *cytosol*. The bulk of the large molecular-weight material in this fraction is protein, which includes enzymes that occur in the cytosol in the intact organisms, as well as others which may have become detached from membranes and organelles during cell disruption.

2.13.1   TRANSFER RIBONUCLEIC ACID

Important large molecular-weight components of the cytosol are the *transfer RNA* or tRNA molecules which are involved in protein synthesis (*see* page 286). Well over 20 different tRNA species have been isolated and purified from micro-organisms, and their nucleotide sequences determined. All are relatively small RNA molecules just 70-80 nucleotide residues long, and all those so far examined have many strikingly similar properties. The tRNA molecules consist of a number of loops of different size, once thought to be in the shape of a clover leaf (*Figure 2.33*). The two ends of the nucleotide chain form the end of one of the loops. The 3'-hydroxy terminal end of the molecule always has the sequence -C-C-A, and this is the end of the molecule that accepts the amino acid in ester linkage. The other end is the 5' terminal which has a guanosine residue bearing a 5'

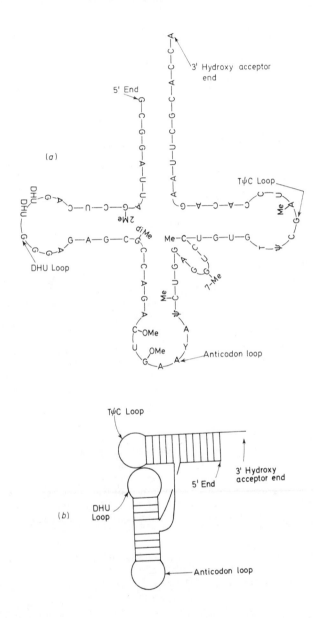

Figure 2.33 *See facing page for details*

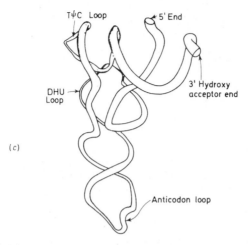

(c)

Figure 2.33 Structure of the phenylalanine tRNA from
Saccharomyces cerevisiae. *(a) The secondary cloverleaf
structure of the molecule showing the sequence of bases
in the nucleotide chain. DHU indicates a residue of
dihydro-uridine, and ψ one of pseudo-uridine; Y indicates
an unidentified base; other abbreviations are explained on
page (ix). (b) Diagrammatic representation of the molecule
showing the way in which the cloverleaf representation
needs to be transformed to show physical connections
between various parts of the molecule. (c) Perspective
diagram in which the tRNA chain is represented as a
continuous coiled tube*

phosphate group. Another fairly large loop frequently
contains 5,6-dihydrouridylic acid (DHU) residues.
A third is the anticodon loop in the middle of which are
three nucleotide residues which form the anticodon which
recognises the amino-acid codon on mRNA. An interesting
feature of tRNA molecules is the presence in the molecules
of rare nucleotide residues. In addition to residues of
dihydrouridylic acid, these include residues of various
methylated nucleotides, pseudouridylic acid (Ψ) and
2-thiocytidine.

Phenylalanine tRNA (tRNA^{phe}) from *Sacch. cerevisiae*
was one of the first tRNA molecules to be sequenced in
full (*Figure 2.33*), the result of work by Robert Holley
and his colleagues at Cornell University, U.S.A.

Moreover, the crystalline form of this tRNA has been submitted to X-ray diffraction to discover the three-dimensional structure of the molecule. This proved to be a more difficult task than expected, but recently these much sought-after data were published. The molecule is L-shaped with the anticodon at one extremity of the L and the acceptor end at the other (*Figure 2.33*). These two major segments of the molecule, each of which contains stretches in which there is a base-paired double helix, are joined by unpaired segments. *Figure 2.33* also shows a perspective diagram of the three-dimensional structure of tRNA^phe from *Sacch. cerevisiae* in which the polynucleotide chain is represented as a continuous coiled tube. Yeast tRNA^phe is therefore the first nucleic acid to yield up its secrets since X-ray crystallography revealed the double helix of DNA.

2.13.2 PIGMENTS

Pigments are found in the soluble fraction of some micro-organisms. These pigments include prodigiosin, which is synthesised by *Serratia marcescens* and some streptomycetes, and the beautiful violet pigment, violacein, which is formed by *Chromobacterium violaceum*. Several pseudomonads synthesise phenazine pigments.

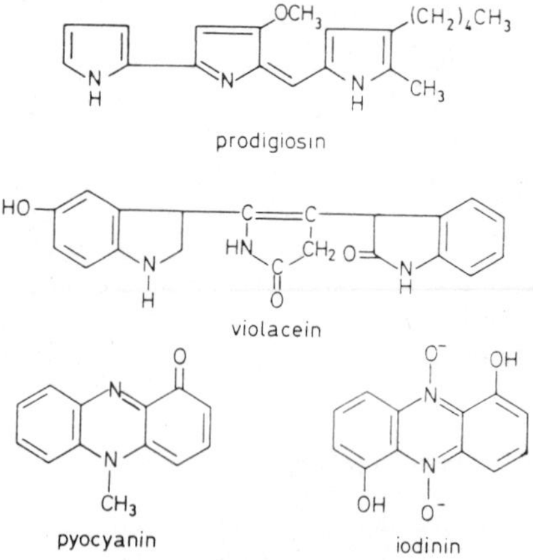

prodigiosin

violacein

pyocyanin

iodinin

Figure 2.34   Structural formulae of some pigments produced by micro-organisms

Pyocyanin is formed by *Pseudomonas aeruginosa,* whereas
Ps. iodinium synthesises the purple phenazine pigment,
iodinin. The structural formulae of some of these
pigments are shown in *Figure 2.34.* The main pigment in
Micrococcus roseus is the carotenoid canthaxanthin
(4,4'-diketo-β-carotene), which is also synthesised by
Corynebacterium michagenense.

2.13.3 LOW MOLECULAR-WEIGHT COMPOUNDS

Many different low molecular-weight compounds are found
in the soluble fraction of micro-organisms. These
compounds can be extracted from organisms using cold
trichloroacetic acid or more conveniently by placing a
suspension of organisms in a bath of boiling water for
10-15 min. The compounds include carbohydrate storage
materials, the chemical nature of which varies in
different organisms. Yeasts and fungi often contain the
disaccharide trehalose. This sugar is also found as a
storage compound in certain algae as are sucrose,
floridoside (2-α-glycerol-α-D-galactoside), mannoglyceric
acid and polyhydroxy alcohols. Mannitol functions as a
storage compound in *Aspergillus clavatus,* and is possibly
concerned with conidiation. Ribitol is also found in
this fungus but it is likely that it acts primarily in
hydrogen-acceptor mechanisms.
   The cytoplasm also contains 'pools' of low molecular-
weight compounds such as amino acids and nucleotides.
One should not, however, envisage these compounds
existing in a single large reservoir. The division of
the cytoplasm into compartments leads inevitably to the
formation of a number of quite separate pools, and
evidence for a heterogeneity in the amino-acid and
nucleotide pools in a number of micro-organisms has come
from kinetic studies on the incorporation of radioactively
labelled precursors into macromolecules. It has already
been noted, however, that in *Sacch. cerevisiae* the amino-
acid pool is concentrated in the vacuole. It has yet to
be shown that this is also the situation in other
eukaryotic micro-organisms.
   The low molecular-weight compounds in the cytoplasm
create an appreciable osmotic pressure difference across

D

the cytoplasmic membrane of the organism, the magnitude
of which varies in different organisms. Cryoscopic
measurements on extracts of micro-organisms have shown
that this osmotic pressure is greatest in Gram-positive
bacteria, in which it may be as much as 30×10^5 Pa
(about 30 atm.). In Gram-negative bacteria it is
usually much lower ($4-8 \times 10^5$ Pa) while in yeast the
value is usually around 12×10^5 Pa.

FURTHER READING

PROKARYOTES AND EUKARYOTES

STANIER, R.Y. (1970). Some aspects of the biology of
cells and their possible evolutionary significance.
Symposium of the Society for General Microbiology,
20, 1-38
STANIER, R.Y. and VAN NIEL, C.B. (1962). The concept of
a bacterium. *Archiv für Mikrobiologie,* 42, 17-35

METHODS USED IN STUDYING THE MOLECULAR ARCHITECTURE OF MICRO-ORGANISMS

HUGHES, D.E., WIMPENNY, J.W.T. and LLOYD, D. (1971).
The disintegration of micro-organisms. In: *Methods
in Microbiology,* Eds. J.R. Norris and D.W. Ribons,
Vol. 5B, pp.1-54. London; Academic Press
SYKES, J. (1971). Centrifugal techniques for the
isolation and characterization of sub-cellular
components from bacteria. In: *Methods in Microbiology,*
Eds. J.R. Norris and D.W. Ribbons, Vol. 5B, pp.55-207.
London; Academic Press

SURFACE APPENDAGES

BRINTON, C.C. (1971). The purification of sex pili, the
viral nature of 'conjugal' genetic transfer systems,
and some possible approaches to the control of bacterial
drug resistance. *Critical Reviews in Microbiology,* 1,
105-160
SMITH, R.W. and KOFFLER, H. (1971). Bacterial flagella.
Advances in Microbial Physiology, 6, 219-339
VALENTINE, R.C., SILVERMAN, P.M., IPPEN, K.A. and
MOHBACH, H. (1969). The F-pilus of *Escherichia coli.*
Advances in Microbial Physiology, 3, 1-52

CAPSULES AND SLIME LAYERS

SUTHERLAND, I.W. (1972). Bacterial exopolysaccharides. *Advances in Microbial Physiology*, 8, 143-213

CELL WALLS

ARCHIBALD, A.R. (1974). The structure, biosynthesis and function of teichoic acids. *Advances in Microbial Physiology*, 11, 53-95

BADDILEY, J. (1972). Teichoic acids in cell walls and membranes of bacteria. *Essays in Biochemistry*, 8, 35-77

BALLOU, C.E. (1976). The structure and biosynthesis of the mannan component of the yeast cell envelope. *Advances in Microbial Physiology*, 14, in the press

BARTNiCKI-GARCIA, S. (1968). Cell wall chemistry, morphogenesis, and taxonomy of fungi. *Annual Review of Microbiology*, 22, 87-108

GANDER, J.E. (1974). Fungal cell wall glycoproteins and peptido-polysaccharides. *Annual Review of Microbiology*, 28, 103-109

GOREN, M.B. (1973). Mycobacterial lipids; selected topics. *Bacteriological Reviews*, 36, 33-64

LEIVE, L. (1973). Ed. *Bacterial Membranes and Walls*. 495 pp. New York; Marcel Dekker Inc.

LUDERITZ, O., GALANOS, C., LEHMANN, V., NURMIMEN, M., RIETSCHEL, E.T., ROSENFELDER, G., SIMON, M. and WESTPHAL, O. (1973). Lipid A: chemical structure and biological activity. *Journal of Infectious Diseases*, 128, S17-S29

PHAFF, H.J. (1971). Structure and biosynthesis of the yeast cell envelope. In: *The Yeasts,* Eds. A.H. Rose and J.S. Harrison, Vol. 2, pp.135-210. London; Academic Press

RATLEDGE, C. (1975). The physiology of the mycobacteria. *Advances in Microbial Physiology*, 13, 115

REAVELEY, D.A. and BURGE, R.E. (1972). Walls and membranes in bacteria. *Advances in Microbial Physiology*, 7, 1-81

ROGERS, H.J. and PERKINS, H.R. (1968). *Cell Walls and Membranes,* 435 pp. London; E. and F.N. Spon, Ltd.

SCHLEIFER, K.H. and KANDLER, O. (1972). Peptidoglycan types of bacterial cell walls and their taxonomic implications. *Bacteriological Reviews,* 36, 407-477

SIEGEL, B.Z. and SIEGEL, S.M. (1973). The chemical composition of algal cell walls. *Critical Reviews in Microbiology*, 3, 1-26

PLASMA MEMBRANES

FINEAN, J.B., COLEMAN, R. and MICHEL, R.H. (1974).
Membranes and their Cellular Functions. 123 pp.
London; John Wiley & Sons
GOLDFINE, H. (1972). Comparative aspects of bacterial
lipids. *Advances in Microbial Physiology*, 8, 1-58

HARRISON, R. and LUNT, G.G. (1975). *Biological membranes:
Structure and Function*. 254 pp. Glasgow; Blackie
and Sons
HUNTER, K. and ROSE, A.H. (1971). Yeast lipids and
membranes. In: *The Yeasts*, Eds. A.H. Rose and
J.S. Harrison. Vol.2, pp.211-270. London; Academic
Press
MACHTIGER, N.A. and FOX, C.F. (1974). Biochemistry of
bacterial membranes. *Annual Review of Biochemistry*,
42, 575-600
OEIZE, J. and DREWS, G. (1972). Membranes of
photosynthetic bacteria. *Biochimica et Biophysica
Acta*, 265, 209-239
RAZIN, S. (1973). Physiology of mycoplasmas. *Advances
in Microbial Physiology*, 10, 1-80
REUSCH, V.M. and BURGER, N.M. (1973). The bacterial
mesosome. *Biochimica et Biophysica Acta*, 300, 79-104
SALTON, M.R.J. (1974). Membrane-associated enzymes in
bacteria. *Advances in Microbial Physiology*, 11, 213-283
SHAW, N. (1970). Bacterial glycolipids. *Bacteriological
Reviews*, 34, 365-377
SHAW, N. (1975). Bacterial glycolipids and glycophospho-
lipids. *Advances in Microbial Physiology*, 12, 141-167
SINGER, S.J. and NICOLSON, G.L. (1972). The fluid mosaic
model of the structure of cell membranes. *Science
(New York)*, 175, 720-730

GENOMES

HOLLIDAY, R. (1970). The organization of DNA in
eukaryotic chromosomes. *Symposium of the Society for
General Microbiology*, 20, 359-380
MESELSON, M., YUAN, R. and HEYWOOD, J. (1972). Restriction
and modification of DNA. *Annual Review of Biochemistry*,
41, 447-466
RICHMOND, M.H. (1970). Plasmids and chromosomes in
prokaryotic cells. *Symposium of the Society for
General Microbiology*, 20, 249-277

RIBOSOMES

KURLAND, C.G. (1972). Structure and function of the
bacterial ribosome. *Annual Review of Biochemistry,*
<u>41</u>, 377-408
NOMURA, M. (1973). Assembly of bacterial ribosomes.
Science (New York), <u>179</u>, 864-873
WITTMANN, H.G. (1970). A comparison of ribosomes from
prokaryotes and eukaryotes. *Symposium of the Society
for General Microbiology,* <u>20</u>, 55-76

MITOCHONDRIA

HUGHES, D.E., LLOYD, D. and BRIGHTWELL, R. (1970).
Structure, function and distribution of organelles in
prokaryotic and eukaryotic microbes. *Symposium of the
Society for General Microbiology,* <u>20</u>, 295-322
LINNANE, A.W., HASLAM, J.M., LUKINS, H.B. and NAGLEY, P.
(1972). The biogenesis of mitochondria in micro-
organisms. *Annual Review of Microbiology,* <u>26</u>, 163-198

LLOYD, D. (1974). *The Mitochondria of Micro-organisms.*
554 pp. London; Academic Press

THE PHOTOSYNTHETIC APPARATUS

ECHLIN, P. (1970). The photosynthetic apparatus in
prokaryotes and eukaryotes. *Symposium of the Society
for General Microbiology,* <u>20</u>, 221-248
EVANS, M.C.W. and WHATLEY, F.R. (1970). Photosynthetic
mechanisms in prokaryotes and eukaryotes. *Symposium of
the Society for General Microbiology,* <u>20</u>, 203-220
KIRK, J.T.O. (1971). Chloroplast structure and
biogenesis. *Annual Review of Biochemistry,* <u>40</u>, 161-196
LASCELLES, J. (1968). The bacterial photosynthetic
apparatus. *Advances in Microbial Physiology,* <u>2</u>, 1-42

OTHER CYTOPLASMIC INCLUSIONS AND GRANULES

DAWES, E.A. and SENIOR, P.J. (1973). The role and
regulation of energy reserve polymers in micro-organisms.
Advances in Microbial Physiology, <u>10</u>, 135-266
SHIVELY, J.M. (1974). Inclusion bodies in prokaryotes.
Annual Review of Microbiology, <u>28</u>, 167-187

SOLUBLE CYTOPLASMIC CONSTITUENTS

GAUSS, D.H., VON DER HAAR, F., MAELICKE, A. and CRAMER, F.
(1971). Recent results of tRNA research. *Annual
Review of Biochemistry*, <u>40</u>, 1045-1078

GERBER, N.N. (1974). Prodigiosin-like pigments. *Critical
Reviews in Microbiology*, <u>3</u>, 469-485

KIM, S.H., QUIGLEY, G.J., SUDDATH, F.L., MCPHERSON, A.,
SNEDEN, D., KIM, J.J., WEINZIERL, J. and RICH, A.
(1973). Three-dimensional structure of yeast
phenylalanine transfer RNA; folding of the polynucleotide
chain. *Science (New York)*, <u>179</u>, 285-288

3

THE ENVIRONMENT

In the previous chapter we considered the types of
molecules that go to make up the various structures and
organelles of micro-organisms. In so far as the
activities of micro-organisms are directed towards the
production of new cell material, these compounds
represent the major end-products of microbial activity.
If growth is to take place, biosynthetic raw materials
and utilisable sources of energy must be present in the
environment; this holds true for all living organisms and
not just micro-organisms. But, as we shall see later,
some micro-organisms are extraordinarily adaptable in
that they can use an enormous variety of substrates for
growth. Biochemists, accustomed to the rather
conservative nutritional requirements of plants and
animals, are often surprised that some micro-organisms
can thrive in environments containing such apparently
noisome compounds as phenol, carbon monoxide or
trichloroacetic acid. However, while a micro-organism
is dependent upon its environment for supplies of
biosynthetic raw materials and energy, it is at the same
time subject to the action of other chemical and physical
factors in the environment, not all of them necessarily
congenial.

The various environmental factors which influence
microbial activity are discussed in this chapter which
also describes ways in which micro-organisms can respond
to and modify their environment.

93

3.1 ENVIRONMENTAL FACTORS

The environmental factors which affect the activities of
micro-organisms can be grouped into two categories,
namely **chemical** and physical, and the more important
factors in each of these categories are discussed in this
section of the chapter. While reading this account, it
should be remembered that there exist very close
interrelationships between many environmental factors,
and varying one factor frequently affects the response of
an organism to other factors.

3.1.1 CHEMICAL FACTORS

A chemical compound can, in general, have one of two
effects on a micro-organism. It may be beneficial which
usually means that it acts as a *nutrient* in promoting
some activity of the organism, usually growth;
alternatively it can have an adverse effect on the
organism and so act as an *antimicrobial compound*.
A third possibility is that the micro-organism may be
completely indifferent to the presence of the compound.
Included in this category are those substances, such as
agar and silica gel, which are used to solidify nutrient
media. There is, however, no hard and fast classification
of chemical compounds into nutrients and antimicrobial
compounds. Micro-organisms differ widely in their
reaction to a particular compound, and a compound that
serves as a nutrient for one organism may exert an
antimicrobial action on another. Phenol, for example,
is a well known antimicrobial compound, and has been used
for many years as a standard with which to compare the
antimicrobial activity of other compounds. Nevertheless,
some micro-organisms are able to use phenol as a nutrient
(*see* page 225). In addition, a compound can often be
either a nutrient or an antimicrobial compound for one
particular micro-organism depending on the concentration
of the compound in the environment. For example sugars,
which at moderately low concentrations (0.5-2.0%) act as
nutrients, usually inhibit growth when present in media
at concentrations of 20-40%.

Nutrients

These can be defined as compounds which must be taken
into a micro-organism from the environment in order to
satisfy the requirements of the organism for biosynthetic
raw materials and for energy.

Water. Water accounts for between 80 and 90% of the
weight of a micro-organism. All chemical reactions that
take place in living organisms require an aqueous
environment, and water must therefore be in the
environment if the organism is to grow and reproduce.
It must be, moreover, in the liquid phase, and this confines
biological activity to temperatures ranging from around
-2°C (or lower in solutions of high osmotic pressure)
to approximately 100°C; this is known as the *biokinetic
zone*.

The need for water to obtain microbial activity has
been recognised for centuries, and forms the basis of
one of the oldest methods for preventing decay of
perishable materials, namely desiccation. The water
requirements of micro-organisms can be expressed
quantitatively in the form of the *water activity* (a_W) of
the environment or substrate; this is equal to p/p_0, p
being the vapour pressure of the solution and p_0 the
vapour pressure of water.

Values for a_W can be calculated using the equation:

$$\ln a_W \;=\; \frac{-vm\phi}{55.5}$$

v = the number of ions formed by each solute molecule,
m = the molar concentration of solute, and ϕ = the molar
osmotic coefficient, values for which are listed for
various solutes in a number of textbooks. Water has an
a_W value of 1.000; this value decreases when solutes are
dissolved in water.

Micro-organisms can grow in media with a_W values between
0.99 and about 0.63. For any one organism, the important
values within this range are the optimum and minimum a_W
values. These have been determined for a number of micro-
organisms, and they seem to be remarkably constant for a
particular species and to be independent of the nature of
the dissolved solutes. The general effect of lowering the
a_W value of a medium below the optimum is to increase the
length of the lag phase of growth and to decrease the
growth rate and the size of the final crop of organisms.

On the whole, bacteria require media of higher a_W value (0.99-0.93) than either yeasts or moulds. Staphylococci and micrococci characteristically have lower optimum a_W values in this range. With *Salmonella oranienburg*, the a_W value of the medium has important effects on the physiology of the bacterium; only at a_W values below 0.97 is proline required for growth of the bacterium. Moreover, in media lacking amino acids, accumulation of potassium ions by this bacterium increases to a maximum as the a_W value of the medium is lowered to 0.975, and then decreases as the water value is lowered further to 0.96. Yeasts also vary in the optimum a_W values required for growth, but the minimum values for these organisms (0.91-0.88) are lower than those for the majority of bacteria. A few yeasts, e.g. *Saccharomyces rouxii*, can grow in media of a_W value as low as 0.73 and these are known as *osmophilic yeasts*. *Saccharomyces rouxii*, unlike non-osmophilic yeasts, contains glycerol and arabitol, and it is believed that these intracellular polyols in some way enable the yeast to grow in media with low water activities. Moulds are in general better able to withstand dry conditions than other micro-organisms, and for some strains, e.g. members of the *Aspergillus glaucus* group, the lower limit of a_W value may be near 0.60. A few moulds, e.g. *Xeromyces bisporus*, have an upper a_W limit of approximately 0.97, and are described as *xerophilic*.

Energy sources. A utilisable supply of energy is required for the growth and activity of all living organisms. Although the ultimate source of energy on this planet is the Sun, only a minority of micro-organisms, including algae and photosynthetic bacteria and protozoa, are able to utilise directly the energy of solar radiation. These organisms are described as *phototrophs* to distinguish them from *chemotrophs* which obtain their energy from chemical compounds.

These groups of micro-organisms are further subdivided on the basis of the nature of the oxidisable substrate used in their metabolism. Organisms which use organic compounds are known as *organotrophs* while those that use inorganic compounds are called *lithotrophs*.

Thus we arrive at the following four nutritional categories of micro-organisms based on their energy-yielding metabolism: *photolithotrophs*, *photo-organotrophs*, *chemolithotrophs* and *chemo-organotrophs*. This classification is summarised in *Table 3.1*.

Table 3.1 Classification of micro-organisms based on their energy-yielding metabolism

---

1. *Phototrophs:* energy provided directly by solar radiation
    (a) *Photolithotrophs:* growth dependent upon exogenous inorganic electron donors
    (b) *Photo-organotrophs:* growth dependent upon exogenous organic electron donors

2. *Chemotrophs:* energy provided by chemical compounds
    (a) *Chemolithotrophs:* growth dependent upon oxidation of exogenous inorganic compounds
    (b) *Chemo-organotrophs:* growth dependent upon oxidation or fermentation of exogenous organic compounds

---

Chemolithotrophs are distinguished from chemo-organotrophs not only by the type of chemical compound used as the oxidisable substrate, but also by the specificity shown towards these compounds. The inorganic compounds oxidised by chemolithotrophic bacteria are relatively small in number and are generally specific for a particular organism (*Table 6.2*, page 231). The energy sources used by chemo-organotrophs, on the other hand, are far more numerous and many of these organisms can obtain their energy from any one of a wide range of organic compounds. The reason for this lack of selectivity among chemo-organotrophs will become apparent when we come to consider the mechanisms by which these organisms obtain energy from organic compounds (*see* Chapter 6, page 181).

Biosynthetic raw materials. Ever since techniques were devised for culturing micro-organisms under laboratory conditions, microbiologists have been interested to learn the nature of those compounds (i.e. nutrients) that are used by micro-organisms as sources of the elements carbon, hydrogen, oxygen, nitrogen, phosphorus and sulphur – elements that account for most of the dry weight of micro-organisms. It is possible to distinguish three stages between the isolation of a micro-organism from its natural habitat and a definition of its nutritional requirements; these stages are: (a) growth in an enrichment culture;

(b) growth in a chemically complex, non-defined medium; and (c) growth in a medium containing only known chemical compounds, i.e. a chemically defined medium. Large numbers of algae, bacteria, fungi and protozoa have been examined in this way, and with many organisms it is possible to list the chemical compounds that can be used as sources of various elements.

There are, however, some micro-organisms which have so far defied all attempts at being grown in chemically defined media. These include the leprosy bacillus, *Mycobacterium leprae,* and many protozoa. Certain protozoa, particularly the parasitic strains, are exceptionally exacting nutritionally since they cannot be grown in the absence of other living organisms (i.e. in *axenic* culture) and will reproduce only in two-membered cultures, that is in media containing live bacteria (bacteria feeders) or other protozoa (carnivorous feeders). Among the *bacteria feeders,* there is not usually a marked selectivity towards the species of bacterium. Strict selectivity is much more common among the *carnivorous feeders;* for example, *Perispira ovum,* the holotrichous ciliate, can use only euglenoid ciliates for food. Strenuous attempts have been made to grow many of these protozoa in axenic culture and to discover the bases of their very exacting nutritional requirements, but so far with only limited success.

The carnivorous habit is also found among a rather esoteric group of fungi that prey on amoebae, rotifers and especially nematodes (eelworms); these are the *predacious* or *nematode-trapping fungi.* Some of these fungi are phycomycetes and there is at least one nematode-trapping basidiomycete, but the majority are members of the Deuteromycetes and include species of *Arthrobotrys, Dactylaria* and *Dactyella.* Many predacious fungi can be grown in non-defined media but a few (e.g. *Stylopage gracilis*) cannot and appear to be obligate predators. Predacious fungi differ with regard to the specificity shown towards their prey. On the whole, the smaller arthropods are the most preyed upon; one fungus, *Arthrobotrys entomopaga,* even catches springtails. Two main types of trap are formed by predacious fungi. The first type are essentially adhesive traps which catch the prey in the same way as flies are caught on a fly-paper. The other type of trap works by entanglement and does not use sticky secretions. In *Dactylaria gracilis* the trap is made of a constricting ring of mycelium, the

cells in which expand to trap the eelworm. After the
prey has been trapped, the fungal mycelium grows inside
the body of the trapped eelworm. The physiology of the
trapping process has so far received only a cursory
examination, but it has been shown that, in *Arthrobotrys
eonoides,* the formation of a mechanical trap of entangled
mycelium is stimulated by a substance known as *nemin*
which is released by the nematode. Nemin is thought to
be a low molecular-weight peptide. Because of the need
to find a substitute for chemical pesticides, it is
likely that there will be an expanded interest in the
future in nematode-trapping fungi, and as a result a
clearer understanding of the physiology of these
organisms may emerge.

For most micro-organisms, however, the nutritional
requirements are now known in some detail. Two main
factors determine whether or not a chemical compound
can be used as a nutrient by a micro-organism. The
first of these is the ability of the compound to enter
the micro-organism which usually means its ability to
penetrate the plasma membrane. The mechanisms by which
solutes are transported across the membrane are rather
selective (page 168). Many micro-organisms, for example,
are impermeable to organic acids of the tricarboxylic
acid cycle, while sugars are usually able to penetrate
far more readily. Molecular size is an important factor
in the ability of a molecule to enter the cell. Many
high molecular-weight compounds cannot penetrate the
plasma membrane, and some are even unable to pass across
the cell wall. However, this does not mean that all
high molecular-weight compounds cannot be used as
nutrients by micro-organisms. Organisms often produce
extracellular or cell-bound hydrolytic enzymes, which
catalyse hydrolysis of certain high molecular-weight
substrates, such as polysaccharides, to compounds of
lower molecular-weight (sugars) that can readily be
transported across the plasma membrane. Production of
extracellular enzymes by micro-organisms is considered
more fully on page 144.

Although a compound may be capable of entering a
micro-organism, it can act as a nutrient only if the
organism is capable of synthesising the enzymes required
for its metabolism. If a compound is unable to enter the
cell, a micro-organism does not usually synthesise those
enzymes specifically concerned with metabolism of the
compound. Exceptions to this generalisation have been
reported with certain micro-organisms, and these organisms

are said to be *cryptic* with regard to utilisation of the
particular compound. For example, strains of *Escherichia
coli* have been described which, although they cannot
transport lactose across the plasma membrane, nevertheless
produce the enzyme β-galactosidase which catalyses
hydrolysis of lactose to glucose and galactose.

CARBON, HYDROGEN AND OXYGEN. These elements are frequently
made available to a micro-organism in the form of one
compound and they are therefore considered together. They
account for the bulk of the dry weight of micro-organisms
so that, if growth is to take place, utilisable compounds
containing these elements must be available in the
environment in relatively high concentrations (usually
between 0.2 and 1.0%).
    Probably all micro-organisms can use or 'fix' the
simplest of all carbon compounds, carbon dioxide. A few
can use carbon dioxide as the sole carbon source, and
this property is used to subdivide micro-organisms into
autotrophs (literally 'self-sustaining' or 'self-nourishing')
which have this ability to synthesise all of their carbon
compounds from carbon dioxide, and *heterotrophs* which
cannot synthesise all of their carbon compounds from
carbon dioxide and require in addition some organic source
of carbon in the environment.
    Not unexpectedly, this subdivision into autotrophs and
heterotrophs is not clear-cut, and organisms are known
which could quite justifiably be placed in both categories.
For example, some rumen bacteria, including *Bacteroides
ruminicola,* have an absolute requirement for carbon
dioxide, although they are heterotrophs. Also, strains of
Hydrogenomonas are able to synthesise all of their carbon
compounds from carbon dioxide with the exception of very
small amounts of certain vitamins which **must be**
supplied by the environment. It is difficult to decide,
therefore, whether these bacteria should be considered as
autotrophs, which basically they are, or as heterotrophs
which is correct if one adheres strictly to the definition
given above. The term 'mixotrophy' has been used to refer
to a commingling of autotrophic and heterotrophic modes of
carbon assimilation.
    The number of different organic compounds that are known
to be used as carbon sources by heterotrophic micro-
organisms is extremely large. Indeed, it would seem that
for every carbon compound formed in Nature, there exists
some microbial agency for its decomposition. This explains

the important role played by micro-organisms in the
natural cycling of elements, not only carbon but also
nitrogen, sulphur and phosphorus, so that these elements
can be used over and over again to sustain life of other
organisms. However, certain commercially important
synthetic carbon compounds, including some herbicides and
detergents that have been manufactured in recent years,
are relatively resistant to microbial attack, and this
has caused problems with regard to their breakdown in
Nature. Heterotrophic micro-organisms vary widely with
respect to the types of organic compound that are preferred
as carbon sources. Quite often, these selectivities are
taxonomically useful, as for example in the time-honoured
sugar utilisation tests which are often used for subdividing micro-organisms into genera and species. Some
groups of micro-organisms, such as the pseudomonads, are
able to use any one of a very large number of different
organic compounds, whereas with other groups the choice
is far more restricted. Whatever carbon compound is
used, it is usually broken down to low molecular-weight
compounds that enter one or more of a relatively small
number of biosynthetic pathways which are common to many
and sometimes all living organisms. During this breakdown, or catabolism, energy is often extracted from the
compounds and made available to the micro-organisms in
the form of ATP (*see* Chapter 6, page 181).

Carbohydrates are among the most commonly available
sources of carbon for micro-organisms. Monosaccharides,
particularly hexoses, are widely used by micro-organisms,
although there are exceptions as, for example, the
fungus *Leptomitus lacteus* and the 'acetate' flagellates
such as *Polytomella caeca*. Heptose utilisation by micro-
organisms has not been extensively studied; of the
pentoses, xylose appears to be the one most frequently
used. Polyhydric alcohols are often good carbon sources;
mannitol, for example, is readily utilised by many fungi
and actinomycetes, although on the whole it is poorly
used by phycomycetes and yeasts. Glycerol is a
particularly good carbon source for actinomycetes.

Micro-organisms are often impermeable to *organic acids,*
especially keto acids, so that these compounds cannot be
used as carbon sources. When examining utilisation of
organic acids by micro-organisms, two factors **must be**
borne in mind. First, uptake of a neutralised organic
acid can cause a rise in the pH value of the culture which
may adversely affect growth of the organism. Secondly,
some organic acids (e.g. citrate, tartrate) have powerful

cation-chelating effects so that growth of the organism may be affected by a deficiency of metal ions.

Some of the traditional media of the microbiological laboratory, such as nutrient broth, provide *amino acids* as the major carbon source for micro-organisms. The fact that many organisms grow well in these media shows that these organisms can reaily use amino acids as carbon sources.

Relatively few studies have been made on the availability of lipids as carbon sources for micro-organisms, mainly because of the insolubility of these compounds in water. Some organisms are, however, known to use these compounds. Hydrolysis of triglycerides by extracellular or intra-cellular lipases gives rise to glycerol and fatty acids, either or both of which can then be used as carbon sources by the micro-organism.

Other groups of organic compounds are used as sources of carbon by a more restricted range of micro-organisms. Utilisation of *hydrocarbons,* for example, is most frequently found among members of the genera *Corynebacterium, Mycobacterium* and *Pseudomonas.* The ability to utilise *aromatic compounds* is distributed rather more widely, but is probably most extensive among the pseudomonads. Breakdown of aromatic compounds usually occurs mainly under aerobic conditions, and where these compounds have accumulated under anaerobic conditions (e.g. as lignin in peat and coal) they usually survive decomposition.

NITROGEN. A utilisable source of the element nitrogen must be present in the environment in order that a micro-organism can synthesise amino acids (and proteins), nucleotides and certain vitamins. The nitrogen atom exists in natural compounds in a variety of oxidation states, ranging from +6 (as in NO_3) to -3 (as in NH_4^+). With the exception of the +6 oxidation state, inorganic compounds containing the nitrogen atom in each of these states have been reported to be utilised by different micro-organisms. Not surprisingly, the preferred form is usually the ammonium ion, since the nitrogen atom is incorporated into organic compounds in the form of this ion. Nitrate (NO_3^-; oxidation state +5) can be used by many algae and fungi and, although not so extensively, by bacteria and yeasts. The ability to fix nitrogen gas or dinitrogen (oxidation state 0) is confined to prokaryotic micro-organisms. Earlier reports of fixation

by eukaryotes have been discounted following the
introduction of more rigorous techniques, such as the
acetylene reduction method (*see* page 264), for
establishing nitrogen-fixing ability. Nitrogen fixation
is carried out by both free-living bacteria and by
bacteria in symbiotic associations. Obligately aerobic
free-living nitrogen-fixing bacteria are restricted to just
a few genera including *Azotobacter, Beijerinckia* and
Derxia. On the other hand, facultatively aerobic
nitrogen-fixing bacteria are more widespread, and include
Klebsiella species. Other groups of bacteria that
encompass nitrogen fixers include obligate anaerobes
(*Clostridium* and *Chloropseudomonas* species). Blue-green
algae are represented by species of *Anabaena, Nostoc* and
Tolypothrix. Species of *Rhizobium* and *Frankia* are the
more important representatives that fix dinitrogen in
symbiotic associations.

Organic nitrogenous compounds may be used as a source
of nitrogen by microbes that can break down these
compounds to ammonia. Many amino acids can be deaminated
by micro-organisms and in general are excellent sources
of nitrogen for microbial growth. Aliphatic amides, too,
can be used as nitrogen sources by a wide variety of
microbes, and the ability to use a range of amides has
been employed to subdivide groups of micro-organisms.
Purines and pyrimidines on the whole are less widely
used. Thiocyanate can be used as the sole source of
both nitrogen and sulphur by a few species including a
thiobacillus and *Pseudomonas stutzeri*.

PHOSPHORUS. The element phosphorus occurs in living
organisms as phosphate, principally in the form of sugar
phosphates in nucleotides and nucleic acids. Since these
compounds include such important cellular constituents
as DNA, RNA and ATP, it is clear that phosphate plays a
very important role in cell metabolism. Phosphates,
usually inorganic, need to be present in the environment
if a micro-organism is to grow. Chemically defined media
often include phosphates as buffering substances;
potassium phosphates (KH_2PO_4 and K_2HPO_4), in suitable
concentrations, can give any pH value in the range 4.5-
8.0. The phosphate in non-defined media, e.g. nutrient
broth, arises mainly from nucleic acids present in the
constituents used for preparing the media.

SULPHUR. The most important sulphur-containing cell
constituent is the amino acid cysteine which occurs
mainly as an amino-acid residue in proteins. The sulphur
atom in cysteine is in the form of a thiol (-SH) group,
and the sulphur atoms in most other sulphur-containing
cell constituents (e.g. methionine, biotin, thiamine,
glutathione) originate from the -SH group of cysteine.
Most micro-organisms take up sulphur from the environment
in the form of the sulphate ion. In this ion, the
sulphur atom is in an oxidation state of +6, and it must
be reduced by the organism to an oxidation state of -2
(as in -SH). This process is known as *assimilatory
sulphate reduction* to distinguish it from the much less
widely used process of dissimilatory sulphate reduction
in which sulphate is used as a terminal electron
acceptor (*see* page 207). Thiosulphate ($S_2O_3^{2-}$) can also
serve as the sole source of sulphur for many micro-
organisms, and it is thought that this ion is reductively
cleaved to sulphite and thence to sulphide. Elemental
sulphur can be used as a sulphur source by a small number
of microbes. A few micro-organisms have lost the capacity
to reduce the sulphur atom and require, instead, compounds
containing the sulphur atom in a reduced form (e.g.
hydrogen sulphide, cysteine).

Growth factors. Some micro-organisms have lost (or have
never acquired) the ability to synthesise sufficient
quantities of all the organic compounds (amino acids,
purine and pyrimidine nucleotides, vitamins) that are
required for synthesis of new cell material. These
compounds therefore must be supplied in the
environment in order for growth to occur; consequently
they are known as *growth factors*. A micro-organism which
requires a particular growth factor is said to be
auxotrophic for that compound, to distinguish it from
prototrophic organisms that do not require the compound.
The requirement of a micro-organism for a growth factor
is usually detected by its ability to grow in complex
non-defined media but not in simple chemically defined
media. By fractionating the constituents of the non-
defined medium it is possible to obtain fractions rich
in the growth factor and, using these fractions, to
isolate the growth-promoting compound in a purified form.
Confirmation of the growth-promoting properties of the
compound can then be sought by testing the ability of the
natural and synthetic compound to permit growth of the
micro-organism in chemically defined media.

The growth factor requirements of a large number of micro-organisms have now been established. Some micro-organisms are far more demanding than others for growth factors. Lactic-acid bacteria (which often require several amino acids and vitamins) and protozoa are among the most fastidious; other micro-organisms such as *E. coli* rarely have a requirement for growth factors. Auxotrophic micro-organisms vary with regard to their degree of dependence upon the environment for growth factors. Some auxotrophs are completely unable to synthesise a compound which therefore becomes an *essential growth factor*. Others are able to synthesise some of the compound but not sufficient to satisfy all of the metabolic needs; under these conditions, the compound acts as a *stimulatory growth factor*.

The requirements of a micro-organism for growth factors are not fixed but may vary with the conditions under which the organism is grown. *Mucor rouxii,* for example, requires the vitamins biotin and thiamine only when grown under anaerobic conditions; when grown aerobically, the mould is able to synthesise adequate amounts of both compounds. Anaerobiosis also induces additional nutritional requirements in *Sacch. cerevisiae.* When grown in the absence of molecular oxygen, strains of this yeast require a sterol and an unsaturated fatty acid, compounds which are not necessary for growth of the strains under aerobic conditions. Variations in growth-factor requirements have been observed when organisms are grown in media of different pH value. Temperature of incubation, too, can affect the growth-factor requirements of organisms and there is a tendency for organisms to be more exacting nutritionally when incubated at temperatures above the optima for growth. The chemical composition of the environment is also important in relation to the growth-factor requirements of micro-organisms. Most vitamins are incorporated into coenzymes and it is often possible for an organism to dispense with the need for a particular vitamin by including in the medium the products of the reactions catalysed by the appropriate enzyme. In this way, a mixture of L-amino acids has been shown to spare the pyridoxine requirement of certain lactic-acid bacteria. By studying the specificity of a particular growth factor requirement, it is often possible to pinpoint the step or steps in the biosynthesis of the growth factor which cannot be carried out by an auxotrophic micro-organism. Pantothenic acid, for example, is known to be synthesised from pantoic acid and β-alanine (*see*

page 317). The pantothenate requirement of some
organisms (e.g. *Corynebacterium diphtheriae*) is satisfied
by β-alanine alone, indicating that these organisms can
synthesise pantoic acid and, when supplied with β-alanine,
can use the pantoic acid in the synthesis of pantothenate.
Similar studies on the specificity of the requirements of
auxotrophic micro-organisms for amino acids have been
useful in charting pathways for the biosyntheses of these
compounds (*see* page 268).

The various growth factors required by micro-organisms
are briefly discussed in the following paragraphs.

AMINO ACIDS. Micro-organisms can be auxotrophic for one
or more of the 20 amino acids that are incorporated into
proteins. Probably the most exacting amino acid-requiring
micro-organism is *Leuconostoc mesenteroides* P.60 which
requires no fewer than 17 amino acids for growth. Most
strains of lactic-acid bacteria are, in fact, auxotrophic
for amino acids. Usually the requirement of a micro-
organism is for L-amino acids, but a few bacteria have
been shown to require D-alanine presumably for the synthesis
of cell-wall peptidoglycan or teichoic acid (*see* page 298).
The concentrations of amino acids required for maximum
growth of auxotrophic micro-organisms are usually between
20 and 50 μg amino acid ml^{-1}.   Quite often, amino-acid
requirements can be met by low molecular-weight peptides
containing the essential amino acid. The growth response
of an auxotrophic microbe to a peptide may be less or
equal to that brought about by the essential amino acid.
With a few micro-organisms, the response to peptides has
been shown to be greater and this is thought to be because
the peptide acts as a more effective vehicle for transport
of the essential amino acid into the microbe.

PURINES, PYRIMIDINES, NUCLEOSIDES AND NUCLEOTIDES. Some
micro-organisms require one or more of these breakdown
products of nucleic acids. Usually the requirement is
for a purine or a pyrimidine which the organism
incorporates into nucleotides. Purine and pyrimidine
requirements are most commonly encountered among
protozoa and in certain bacteria, especially lactobacilli.
The concentrations required for maximum growth are usually
of the order of 10-20 μg ml^{-1}. Some organisms are not only
unable to synthesise a purine or pyrimidine nucleotide but
are also incapable of incorporating exogenous purines and

pyrimidines into these nucleotides. These micro-organisms
therefore need to be provided with preformed nucleosides
or occasionally nucleotides. Included among the nucleoside-
requiring micro-organisms are species of *Thermobacterium*
and *Lactobacillus*. Requirements for nucleotides have been
reported among species of *Gaffkya*, *Mycoplasma* and
Tetrahymena. For maximum growth these organisms require
rather high concentrations of nucleosides or nucleotides,
usually around 200-2000 μg ml^{-1}.

VITAMINS. Members of the B group of vitamins were the
first growth factors reported for micro-organisms. Almost
all of these vitamins function as coenzymes, or are
incorporated into coenzymes, and some examples of the
metabolic functions of vitamins are given in Chapters 6
and 7. The number of vitamins for which micro-organisms
have been shown to be auxotrophic is about 20. The
requirements of individual auxotrophs differ considerably,
as shown in *Figure 3.1*.
   Among the more exacting micro-organisms are protozoa
and lactic-acid bacteria which may require as many as
five or six vitamins for growth; biotin and thiamine are
the most commonly required vitamins. Among algae and
phytoflagellates, a requirement for cobalamin is quite
common. The concentration of a vitamin required for
maximum growth varies, but in general is quite low and
of the order of 1-50 ng vitamin per ml. The riboflavin
requirement is often comparatively large whereas, at the
other end of the scale, the amounts of biotin which need
to be provided are usually exceedingly small (about 0.2
ng ml^{-1}).
   Figure 3.1 lists some of the more important vitamins
commonly required by auxotrophic micro-organisms,
together with their structural formulae and examples of
dependent organisms. Several of these vitamins act as
coenzymes; others, such as *p*-aminobenzoic acid, are
incorporated into coenzymes (tetrahydrofolic acid).
Cobamide (vitamin B$_{12}$) has no coenzyme activity itself;
the active form is the 5-deoxyadenosyl derivative (the R
substituent in the formula in *Figure 3.1*).

Vitamin	Structural formula	Examples of dependent organisms
Para-Aminobenzoic acid	H_2N—⟨ ⟩—COOH	*Blastocladia ramosa* *Chlamydomonas moewusii* *Clostridium aceto-butylicum*
Biotin		*Lactobacillus plantarum* *Oxyrrhis marina* *Saccharomyces cerevisiae*
Cobalamin (Vitamin B_{12})		*Euglena gracilis*
Haem		*Haemophilus parainfluenzae* *Pilobolus* sp.

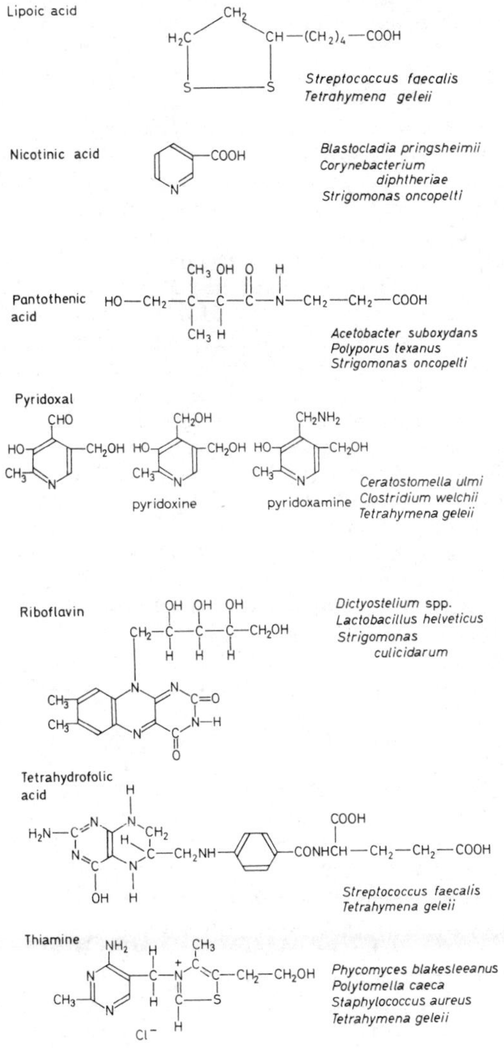

Lipoic acid

Streptococcus faecalis
Tetrahymena geleii

Nicotinic acid

Blastocladia pringsheimii
Corynebacterium
diphtheriae
Strigomonas oncopelti

Pantothenic acid

Acetobacter suboxydans
Polyporus texanus
Strigomonas oncopelti

Pyridoxal

pyridoxine   pyridoxamine

Ceratostomella ulmi
Clostridium welchii
Tetrahymena geleii

Riboflavin

Dictyostelium spp.
Lactobacillus helveticus
Strigomonas
culicidarum

Tetrahydrofolic acid

Streptococcus faecalis
Tetrahymena geleii

Thiamine

Phycomyces blakesleeanus
Polytomella caeca
Staphylococcus aureus
Tetrahymena geleii

Figure 3.1   Some vitamins commonly required by auxotrophic micro-organisms

LIPIDS. Micro-organisms are known which are auxotrophic
for lipids and related compounds that are specifically
required for biosynthesis of membranes. This type of
growth-factor requirement is especially common among
mycoplasmas and protozoa and, as already indicated, is
encountered in *Saccharomyces cerevisiae* grown under
strictly anaerobic conditions.

A requirement for specific phospholipids has not yet
been reported for micro-organisms, but compounds that
are involved in phospholipid biosynthesis are required as
growth factors by some organisms. A requirement for
glycerol is not common but it has been encountered in
Mycoplasma mycoides var. *mycoides*. Fatty-acid requirements
are more commonly encountered, the need usually being for
a long-chain (C_{16} or C_{18}) unsaturated acid. With
mycoplasmas, protozoa and anaerobically grown *Sacch.
cerevisiae*, the fatty-acid requirement is not usually too
specific. On the other hand, some bacteria, such as
Actinomyces israeli, have a more or less specific
requirement for oleate, while the yeast *Pityrosporum
ovale* requires C_{14} or C_{16} acids for growth. At
concentrations greater than those required for optimum
growth, many of these fatty acids have an antimicrobial
action, which can often be overcome by addition of serum
albumin or phosphatidylcholine. **Shorter-chain fatty acids
are required by some rumen bacteria.**

Meso-Inositol is a growth factor for several yeasts and
fungi in which it is used in the synthesis of
phosphatidylinositol. The concentrations of inositol
required for optimum growth (10-20 μg ml^{-1}) are rather
higher than with B-group vitamins. Choline is required
by some strains of pneumococci, but is thought not to be
used in the synthesis of phosphatidylcholine since this
phospholipid is not found in pneumococci. Rather, it is
incorporated into pneumococcal C substance, a cell-wall
ribitol teichoic acid that contains choline phosphate.

Sterols are growth factors for many strains of mycoplasma,
some protozoa and anaerobically grown *Sacch. cerevisiae*.
The specificity of this requirement is fairly broad,
provided there is a hydroxyl group at C-3 of the steroid
nucleus and a long alkyl side chain at C-17 (*see* page **49**).
The sterol (and unsaturated fatty-acid) requirement of
mycoplasmas and anaerobically grown yeast has been
successfully exploited in studies on relationships between
composition and function in cellular membranes.
Mevalonate, which is a precursor in the biosynthesis of
sterols (*see* page **314**), is a growth factor for certain
lactobacilli.

A requirement for diamines and polyamines, including spermine and spermidine, is found with certain bacteria including *Haemophilus parainfluenzae* and *Neisseria perflava*. These polycations are associated with ribosomes and may help to stabilise these organelles. It has also

$$NH_2-(CH_2)_3-NH-(CH_2)_4-NH-(CH_2)_3-NH_2 \quad \text{spermine}$$

$$NH_2-(CH_2)_3-NH-(CH_2)_4-NH_2 \qquad\qquad \text{spermidine}$$

been suggested that they protect the plasma membrane in media of low tonicity.

Recent research has focussed attention on an extremely interesting group of microbial growth factors, namely the *mycobactins*. Although strictly speaking these compounds are not lipids, they are very insoluble in water and frequently have a fatty-acyl chain in the molecule. In the general formula for ferric mycobactins, R_1 is usually a fatty-acyl chain with 11-20 carbon atoms, R_2, R_3 and R_5 are either hydrogen atoms or methyl groups, and R_4 either a methyl or ethyl group or a saturated alkyl group with 15-17 carbon atoms. Mycobactins are growth factors for mycobacteria and are thought to sequester iron from the growth medium (*see* page 112).

Inorganic nutrients. Small amounts of many inorganic cations and anions are required for growth of all micro-organisms. These inorganic nutrients fall into two classes. Some are required in relatively high concentrations (around mM or 0.1 mM) and are termed *macronutrient elements*. They are customarily included in chemically defined media. Macronutrient elements include Mg^{2+}, K^+, Fe^{3+}, Mn^{2+}, Zn^{2+}, Na^+, Ca^{2+} and Cl^-, and the metabolic functions of most of these ions are reasonably well understood. Other inorganic nutrients are required in very much lower concentrations ($10^{-3}-10^{-5}$ mM), and are often referred to as *micronutrient elements*. It is frequently very difficult to establish a requirement for a micronutrient element because of contamination in even the purest of medium constituents. The metabolic functions of some of these micronutrients (such as cobalt) are known, but the need for others, such as vanadium which is required by green algae and by *Aspergillus niger*, and nickel which is essential for maximum growth of *Hydrogenomonas*, has yet to be explained.

Mineral elements function in microbial metabolism mainly

Ferric mycobactins; a general formula

as activators of various enzymes. Iron, for example, is required for iron porphyrin enzymes (e.g. catalase), while zinc is essential for the action of alcohol dehydrogenase. Sodium ions are required for the activity of oxaloacetate decarboxylase in *Aerobacter aerogenes*. Magnesium is of more than passing interest for, in addition to activating certain enzymes such as hexokinase, it has the important role of regulating the degree of association of ribosomal particles. The activation of enzymes by metal ions is not always absolutely specific. The isocitrate lyase of *Pseudomonas aeruginosa*, for example, is activated by Mg^{2+}, Mn^{2+}, Fe^{2+} or Co^{2+}. Ion antagonism, which is a counteraction of the stimulatory effect of one ion by another, has also been observed. Thus, sodium ions will inhibit growth of *Lactobacillus casei*, but this inhibition can be overcome by K^+. Presumably this effect is caused by a competition between these two ions for a single site on an enzyme or coenzyme.

One group of micro-organisms, *halophiles*, are especially interesting in that, unlike other micro-organisms, they have a specific requirement for Na^+ and often for Cl^-. Halophilic (i.e. salt-loving) micro-organisms can be grouped into three categories: (a) slightly halophilic organisms which grow best in media containing 2-5% (w/v) sodium chloride; this group includes many marine micro-organisms and representatives of the group often have a specific requirement for Na^+; (b) moderately halophilic micro-organisms which prefer media containing 5-10% (w/v) sodium chloride; included in this group are certain species of *Achromobacter*, *Pseudomonas*, lactobacilli and some protozoa; moderate halophiles usually have a specific requirement for sodium chloride; (c) extremely halophilic organisms which require media containing 20-30% (w/v) sodium chloride; this group includes species *Halobacterium*, *Micrococcus* and *Sarcina*. The alga *Dunaliella viridis* has also been reported to be extremely halophilic. Extremely halophilic bacteria are obligate aerobes, and are red-orange in colour. All extreme halophiles have a specific requirement for sodium chloride.

The biochemical basis of the halophilic habit has been studied by several groups of workers. Measurements have been made on the content of intracellular sodium chloride in various halophilic micro-organisms, and it has been shown that, in moderate and extreme halophiles, this is often high and may approach that of the medium. Enzymes in extremely halophilic bacteria have apparently

become adapted to function in the presence of high
concentrations of sodium chloride; some show maximum
activity in the presence of as much as 4 M sodium chloride.
In contrast, the enzymes from slightly halophilic bacteria
do not usually show optimum activity in the presence of
sodium chloride. Ribosomes in extremely halophilic
bacteria, although similar in size to those in other
bacteria, differ from other bacterial ribosomes in that
they require high concentrations of K^+ for stability.
It has been suggested that potassium ions are necessary
to neutralise the charge on ribosomal proteins which,
unlike those in other bacterial ribosomes, contain a high
proportion of acidic amino-acid residues.

There is evidence that Na^+ may be required for uptake
of solutes in all classes of halophiles. Uptake of
glutamate by *Halobacterium salinarium* has a specific
requirement for Na^+, as does uptake of sugars and certain
amino acids by a marine pseudomonad. Some representatives
of each of the three groups of halophiles lyse when
suspended in hypotonic solutions of sodium chloride.
Species of *Halobacterium,* when suspended in these solutions,
have been shown to change from rods to spheres which
subsequently lyse. It has been suggested that, in these
extreme halophiles, high concentrations of sodium chloride
are required to keep the cell-wall proteins together.
By exposing the organisms to hypotonic solutions of sodium
chloride, the protective effect of the salt is diminished
and the cell wall disintegrates leading to lysis of the
bacteria. Extremely halophilic bacteria synthesise several
unusual lipids (*see* page 50), but nothing is known of the
role, if any, of these lipids in solute uptake or protein
synthesis by these bacteria.

Microbiological assays. The requirement of auxotrophic
micro-organisms for small quantities of growth factors
and mineral elements has been put to practical use by
the analytical chemist in the form of the microbiological
assay. These assays are based upon the fact that, if a
micro-organism which is auxotrophic for a particular
growth factor (or mineral element) is grown in a medium
containing superoptimum concentrations of all essential
nutrients but a suboptimum concentration of the growth
factor, then the rate of growth or the total amount of
growth of the micro-organism can frequently be related to
the concentration of the growth factor by a standard
dose-response curve (*Figure 3.2*). Microbiological assays

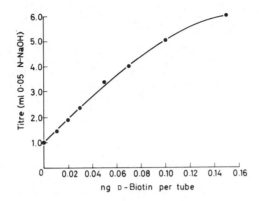

Figure 3.2 A standard dose-response curve for the microbiological assay of biotin using Lactobacillus plantarum

for B-group vitamins and amino acids are widely used and are frequently recommended as standard methods by authoritative groups of analysts.

Oxygen. Although almost all higher plants and animals are dependent upon a supply of molecular oxygen, this does not hold true for all micro-organisms. Some micro-organisms, including mycobacteria and several micrococci, resemble higher organisms in requiring molecular oxygen; these are obligately *aerobic* micro-organisms. There are also obligately *anaerobic* micro-organisms, such as the clostridia, which grow only in the absence of oxygen. In between these extremes of oxygen requirement are large numbers of micro-organisms which are capable of growing well in either the absence or presence of oxygen. The lactic-acid bacteria form a rather special group, for they develop best in the presence of low concentrations of molecular oxygen and so are termed *microaerophiles*.
  Molecular oxygen, unlike other major nutrients, is relatively insoluble in water and so must be continuously made available to aerobic micro-organisms if they are to grow. Micro-organisms growing aerobically in static batch culture are dependent on oxygen dissolved in the culture fluid, supplemented by the small amount which is absorbed from the atmosphere during incubation of the culture. This amount of oxygen is usually insufficient to meet

fully the demands of the micro-organism which tends,
therefore, to be subjected to increasingly anaerobic
conditions. The extent to which this shortage of
oxygen limits the amount of growth of an aerobic micro-
organism in a static batch culture depends to some extent
on the temperature of incubation. Since the solubility
of oxygen in water increases as the temperature is
decreased, organisms grown at lower temperatures will not
be limited by the availability of oxygen to the same
extent as when incubated at higher temperatures. This
explains why the total crop of an organism grown at a
lower temperature can be greater than when grown at a
higher temperature although the rate of growth may be
faster at the higher temperature. Some aerobic micro-
organisms, including bacilli, moulds and actinomycetes,
are well adapted to growing in liquid batch culture and
form a felt or pellicle on the surface of the culture.

Several methods, including shaking and bubbling in
sterile air or oxygen, are used for increasing the
availability of oxygen to micro-organisms growing in
liquid batch culture. When moulds are grown under these
conditions, the characteristic dense mat of interwoven
hyphae which forms on the surface in static culture is
replaced by small pellets of mycelium which grow in the
culture fluid. A similar type of submerged growth is
obtained with actinomycetes. For a long time it was
technically very difficult to control and regulate the
oxygen concentration or oxygen tension in a culture.
However, with the advent of reliable oxygen electrodes,
quantitative studies on this problem have been made
possible. It is now well established that aerobic
micro-organisms differ considerably with regard to the
amount of oxygen required for maximum growth. Also the
amount required for maximum growth can differ from that
required for other metabolic processes. For example,
during citric acid production by industrial strains of
Aspergillus niger, the amount of oxygen required for
production of maximum amounts of citric acid is greater
than that required for maximum growth of the mould.

The presence of molecular oxygen prevents growth of
obligately anaerobic micro-organisms. The gas has a
toxic, and often lethal, effect but the physiological
basis for this action is not fully understood. At least
four hypotheses have been proposed. Firstly, molecular
oxygen may have a directly toxic and lethal action by
some as yet unknown mechanism. Secondly, since anaerobes
flourish in media with low redox potential (E_h values of

-150 mV to -400 mV at pH 7.0), it is possible that molecular oxygen prevents the attainment of this E_h value. Alternatively, molecular oxygen may be a more avid electron acceptor than the terminal acceptors that operate on fermentation pathways so that, in the presence of oxygen, a microbe cannot attain the necessary intracellular concentrations of reduced nicotinamide adenine nucleotides. Finally, organisms which cannot synthesise catalase, and these include clostridia, may be poisoned by hydrogen peroxide which is formed by reduction of molecular oxygen. Further research is needed to decide which is the most likely explanation.

Hydrogen ions. The optimum concentration of hydrogen ions required for growth of a micro-organism is usually quite low; in higher concentrations the ion can have a toxic or lethal effect. The limits of hydrogen ion concentration for growth of micro-organisms are from around pH 4.0 to 9.0. For most organisms there is a fairly narrow range within these limits that is most favourable for growth. Bacteria, in general, prefer media of pH value near neutrality, and cannot usually tolerate pH values much below 4-5. There are some exceptions to this generalisation including acetic-acid bacteria, and some sulphur bacteria which oxidise sulphur to sulphuric acid. Animal pathogens are usually favoured by an environment at pH 7.2-7.4. At the opposite extreme, bacteria that infect the human urinary tract and hydrolyse urea to give ammonia can grow at pH 11. Slightly acid conditions (pH 4-6) usually favour growth of yeasts. Still higher concentrations of hydrogen ions can be tolerated by a few fungi, some of which are found growing in laboratory reagent bottles containing dilute mineral acid. Actinomycetes in general prefer slightly alkaline conditions.

The pH value of the environment can materially affect metabolic processes other than growth. It can, for example, affect the morphology of a micro-organism. As the pH value of the growth medium is increased above 6.0, the length of the hyphae of *Penicillium chrysogenum*, grown in continuous culture, decreases and, at about pH 6.7, pellets of mycelium rather than free hyphae are formed. On the whole many more studies are urgently required on the effect of pH value on the physiological activities of chemostat-grown organisms.

The plasma membrane in micro-organisms is relatively impermeable to hydrogen ions and hydroxyl ions, so that the concentrations of these ions in the cytoplasm probably remain reasonably constant despite wide variations in the pH value of the media surrounding the micro-organism. The pH value of the environment affects microbial activity as a result of the interaction between hydrogen ions and enzymes (and presumably transport proteins) in the plasma membrane. The pH value of the environment also affects the activities of enzymes in the cell wall.

Antimicrobial compounds

In this section, I shall deal with chemical compounds that are characteristically antimicrobial in their action and which rarely, if ever, have a beneficial effect on micro-organisms. The existence of antimicrobial compounds has been recognised for many years, and several of these compounds have been used as disinfectants and antiseptics while others have been employed chemothera-peutically in combating diseases caused by micro-organisms. The term *disinfectant* is usually reserved for antimicrobial compounds that are used in the presence of fairly dense microbial populations, such as when cleaning drains or animal quarters, whereas the word *antiseptic* is customarily taken to denote a compound that can be applied topically to the human body in order to minimise infection with surface organisms. A related term is *biocide,* the name given to a compound that is used to prevent microbial attack on a perishable material such as wood or gasolene. Finally a *sterilant* is an antimicrobial compound, often a gaseous one, used to sterilise a restricted or confined area or space. However, these terms are often used quite interchangeably, and indeed are to some extent overlapping. Although Louis Pasteur and Robert Koch, working in the late nineteenth century, established that microbes are the cause of infectious diseases, it was Paul Ehrlich in the early part of the present century who suggested that chemical compounds could be used to combat infectious disease. In fact, he coined the term *chemotherapy*. Many synthetic chemical compounds were used very successfully in combating infectious disease in the first half of this century, but a major breakthrough in the chemotherapeutic application of antimicrobial compounds came with the discovery of *antibiotics,* which are chemical compounds produced by micro-organisms and which in low concentration inhibit growth and activity of other micro-organisms. The mechanisms of action of antibiotics have been intensively studied mainly in the hope that this information will enable biochemists to tailor-make drugs that are even more effective than the natural antibiotic.

Antimicrobial compounds are a truly motley collection of compounds. Furthermore, there are often conflicting suggestions as to the mechanism of action of any one of these compounds, although this often comes from failure on the part of the microbial physiologist to distinguish between primary and secondary effects on microbial metabolism. In the account that follows, the effects of antimicrobial compounds on microbes are discussed from a basically physiological standpoint, with the aim of showing how these compounds act on intact organisms as well as the manner in which they perturb individual reactions in the microbial metabolic machinery.

Compounds which act on microbial plasma membranes. Many antimicrobial compounds do not need to penetrate a micro-organism in order to exert their antimicrobial action. These compounds act not on the cell wall or extramural layers, which are comparatively inert metabolically, but on the plasma membrane which is without doubt the Achilles Heel of a micro-organism and the primary target of many antimicrobial compounds. Those compounds which act on the plasma membrane do so by combining with proteins, lipids or other compounds in the membrane, with the result that one or more vital physiological roles, or indeed the integrity of the membrane, are destroyed, and the microbe loses the capacity to multiply. There is ample evidence that many antimicrobial compounds which react with plasma-membrane components also have an effect on intracellular proteins and lipids. Nevertheless, there is little doubt that the plasma membrane represents the primary site of action of many if not all of these compounds.

A great variety of compounds react with *membrane-bound proteins,* and with many of these the primary action is on thiol groups which are necessary for the enzymic or transporting activity of the protein (*Table 3.2*). Many of the old-established but still widely used disinfectants, such as phenol and hydrogen peroxide, are strong oxidising agents and probably have their main action on thiol groups in membrane proteins. Nowadays, for more refined

E

Table 3.2 Types of reaction that occur between thiol groups and inhibitors

Reaction	Examples of inhibitors
Oxidation:	
$2R\text{-}SH + X \rightarrow R\text{-}S\text{-}S\text{-}R + XH$	Hydrogen peroxide; iodine; permanganate; sulphite
Mercaptide formation:	
$R\text{-}SH + X^+ \rightarrow R\text{-}S\text{-}X + H^+$	Arsenite; organic arsenicals; heavy metal ions (e.g. Cu^{2+}, Pb^{2+}); organic mercurials
Alkylation:	
$R\text{-}SH + X\text{-}R' \rightarrow R\text{-}S\text{-}R' + XH$	Iodoacetate; sulphur and nitrogen mustards
Addition of thiol groups to double bonds:	
$R\text{-}SH + \begin{matrix} CH\text{-}R' \\ \| \\ CH\text{-}R'' \end{matrix} \rightarrow \begin{matrix} R\text{-}S\text{-}CH\text{-}R' \\ \| \\ CH_2\text{-}R'' \end{matrix}$	Acrolein; maleate; quinones

applications as disinfectants, chlorinated cresols or xylenols are commonly used, as in the commercial product Dettol. Another widely used antimicrobial compound, particularly for combating Gram-positive bacteria, is hexachlorophene (*Figure 3.3*). Others of the thiol-group inhibitors listed in *Table 3.2* also act preferentially on certain groups of micro-organisms. Quinones, for example, are useful as fungicides, especially in the form of halogenated derivatives such as 2,3-dichloro-1,4-naphthaquinone (Dichlone, Phygon). Arsenicals are often used to combat protozoa. Growth of *Entamoeba histolytica*, for example, is inhibited by 0.025 mM arsenite; yeasts, in contrast, are in general relatively resistant to arsenite. Alkylating agents are widely used as mutagenic agents, and more information on this effect is given on page 333.

Many of the antimicrobial compounds which react with *membrane lipids* do so because of their surface-active properties, as a result of which the integrity of the plasma membrane is destroyed. These compounds include *detergents* which are compounds that have in common basic

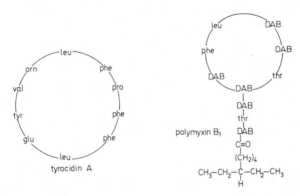

Figure 3.3 Structural formulae of some antimicrobial compounds that act on the plasma membrane. DAB indicates a 2,4-diaminobutyrate residue; other abbreviations are explained on page (x)

or acidic hydrophilic groups attached to fairly large non-polar groups. Commercially useful cationic detergents include cetyltrimethylammonium bromide or cetrimide, and chlorhexidine (*Figure 3.3*), while sodium dodecyl sulphate is a representative anionic detergent and one which is often used in experiments in membrane physiology. Some antibiotics act on the plasma membrane by perturbing the structure of the membrane. They include the polypeptide antibiotics, such as the polymyxins and tyrocidins (*Figure 3.3*), as well as polyene antibiotics such as nystatin and amphotericin B. Polyene antibiotics are thought to act by combining with sterols in the plasma membrane, which explains why they are inactive against prokaryotic microbes. Inorganic cations have an important role in maintaining the integrity of plasma membranes. The antimicrobial action of metal-chelating agents, such as EDTA, is probably explained in part by the ability of these compounds to combine with divalent cations, such as Mg^{2+}, in plasma membranes. There are a few antimicrobial compounds which act by reacting specifically with phospholipids in the plasma membrane. Uranyl ions, for example, prevent solute transport across the plasma membrane by combining with phosphate groups in phospholipids.

Metabolic inhibitors. Antimicrobial compounds which gain access to the inside of a micro-organism inhibit one or more of the thousands of metabolic reactions that occur in a growing microbe. With many of these *metabolic inhibitors,* physiologists have managed to identify the reaction or reactions that are affected. Many examples of specific metabolic inhibitors are quoted in Chapters 5 and 6; they include compounds that inhibit oxidative phosphorylation (e.g. antimycin, carbon monoxide and cyanide), synthesis of cell walls and membranes (penicillins), and transcription and translation of genetic information (rifamycins, streptomycin).

Once the reaction on which a metabolic inhibitor acts has been identified, the next step is to establish the mechanism of action of the inhibitor. This is frequently a difficult and time-consuming task, which explains why the mechanism of action of only a handful of inhibitors is as yet understood.

One of the most subtle types of antimicrobial action is that caused by *metabolic antagonists.* These are compounds which have such a close structural similarity to metabolic precursors and intermediates that they are able to mimic the action of the precursor or intermediate, and thus become incorporated into the metabolic machinery of the micro-organism.

Metabolic antagonists can act in one of several different ways. Some combine at the active site of an enzyme to the exclusion of the natural substrate although they cannot participate in the enzyme reaction. Others can take part in the enzyme reaction and become incorporated into the product of the reaction. However, as the reaction product then differs albeit only slightly from the natural product, it often cannot be metabolised to the same extent. Another group of metabolic antagonists includes compounds that act by causing *false feedback inhibition,* that is, by mimicking the inhibitory action of the intermediate on the activity of the enzyme that catalyses the first reaction on a metabolic pathway (*see* page 331). These antagonists do not usually act by mimicking the action of the intermediate in repressing synthesis of enzymes. The addition to a growing culture of one of these metabolic antagonists can therefore cause a transient microbiostasis, although the effect will continue only until derepression mechanisms take over and synthesis of the intermediates is restored. An example is provided by the action of the histidine antagonist, 2-thiazole-alanine.

Metabolic antagonists compete with an intermediate, or a precursor, for a position on the active site of an enzyme. This competition has been put on a quantitative basis in the form of the *inhibition index*, which is equal to the ratio of the concentration of an antagonist to that of the intermediate or precursor which allows half maximum growth of a micro-organism. Inhibition by metabolic antagonists may be overcome by increasing the concentration of the intermediate or precursor, by supplying the organism with the product of the inhibited enzyme reaction, or by adding to the environment a compound which reacts with the antagonist and prevents its combining with the enzyme. Compounds which overcome the inhibitory action of metabolic antagonists can do so in one of two ways: competitively (as with the substrate of the enzyme) when the amount of relief of inhibition is proportional to the concentration of the compound, or non-competitively (such as by the product of the reaction) when this proportionality is not observed.

Many different compounds act as metabolic antagonists in micro-organisms. Research into these compounds was sparked off by the observation of the late Donald D. Woods, working in the University of Oxford in 1940, that the inhibitory action of sulphanilamide could be competitively reversed by *p*-aminobenzoic acid, a previously unrecognised bacterial growth factor. Sulphonamides act by preventing incorporation of *p*-aminobenzoate into dihydropteroate, which is a precursor of tetrahydropteroylglutamate. The discovery

H_2N —⟨ ⟩— COOH H_2N —⟨ ⟩— $SO_2 - NH_2$

p-aminobenzoic acid   sulphanilamide

led to the hope that the development of new
chemotherapeutic agents would cease to depend upon an
empirical approach, but could henceforth be based
rationally on the synthesis of structural analogues of
intermediates.   Thousands of potential metabolic
antagonists have been synthesised, but only a few have
been found to be sufficiently selective in their *in vivo*
action on pathogenic micro-organisms to be clinically
useful.

Many amino-acid analogues have been tested as metabolic
antagonists.   Some, such as 7-azatryptophan, can replace
up to half of the amino acid in microbial protein, whilst
others (e.g. 5-methyltryptophan) inhibit growth without
being incorporated into protein.   Similar types of
behaviour are found with purine, pyrimidine and vitamin
analogues.

Some of the most interesting metabolic antagonisms are
caused by antibiotics.   Only a few antibiotics have
definitely been shown to act in this way, and they include
azaserine and 6-diazo-5-keto-L-norleucine, both of which
are metabolic antagonists of glutamine.   Other antibiotics
which act as metabolic antagonists include cycloserine
(which antagonises D-alanine; page 301) and puromycin
(an aminoacyl-tRNA antagonist; page 293).   Other
antibiotics may well be found on more detailed examination
to act in this way.

glutamine   azaserine   6-diazo-5-keto-L-norleucine

3.1.2 PHYSICAL FACTORS

Radiations

Most of the radiation reaching the Earth lies in the near
ultraviolet, visible or infrared, but Man's recent
success in producing radiation artificially, by atomic
fission and nuclear fusion reactions, has prompted a
large number of studies on the effect of shorter wavelength radiations on living cells including micro-organisms.
The names given to the various wavelengths of radiation in
the spectrum of electromagnetic radiation are shown in
Figure 3.4. For radiation to affect matter, including

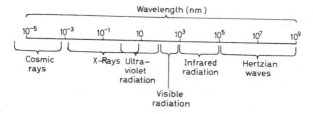

Figure 3.4 The spectrum of electromagnetic radiation

micro-organisms, it must be absorbed. The energy in
radiation is present as discrete packets or quanta. The
energy of each quantum is the same for radiation of a
particular wavelength and varies inversely with the wavelength of the radiation. With cosmic rays and X-rays,
the energy of the quanta is so great that, when these
radiations encounter an absorbing material, electrons
are ejected from all types of atoms in the molecules of
absorbing material. As a result, the molecules containing
these atoms become ionised and this explains why these
shorter wavelength radiations are referred to as *ionising
radiations*. The action of longer wavelength radiations
on the molecules of absorbing material is more selective.
Visible radiation, for example, is absorbed by only a
small proportion of molecules in living cells and, because
of their selective absorption of certain wavelengths of
light, these molecules appear coloured, i.e. they are
pigments. The energy content of infrared radiation is
so low that, on contact with absorbing material, it is
immediately converted into heat. Some of the effects

which various types of radiation have on micro-organisms are discussed in the following pages.

Cosmic rays, X-rays.   These ionising radiations (wavelength below 10 nm) in general have a lethal or mutagenic effect on micro-organisms. The damage caused can result either from a direct effect on susceptible molecules within the organism or from indirect effects caused by chemical reactions involving compounds, particularly free radicals and epoxides, produced by the primary radiation. A number of environmental factors influence the type of damage caused by ionising radiations. In particular, the presence of molecular oxygen makes organisms susceptible to damage by these radiations. Some measure of protection can be afforded by compounds containing thiol groups. Micro-organisms differ widely in their sensitivity to ionising radiations. The red-pigmented bacterium, *Micrococcus radiodurans,* has been studied extensively since it is extremely resistant to both ionising and ultraviolet radiations.

The nature of molecules in micro-organisms that are affected by ionising radiations has not been fully established, although it is known that they include DNA. Some micro-organisms possess a mechanism by which damaged DNA can be repaired. The repair mechanism involves the action of endo- and exonucleotidases, a DNA polymerase and a ligase.

Ultraviolet radiation (10-300 nm). This radiation acts on an absorbing molecule by exciting an electron in an atom in the molecule, thereby raising it to a higher energy level. Ultraviolet radiation resembles ionising radiation in that its action can be either lethal or mutagenic depending upon the organism, the wavelength of the radiation, and the dose administered.

Vegetative microbes are extraordinarily varied in their response to ultraviolet radiation. The dose of ultra-violet radiation needed to inactivate 90% of a population of *E. coli,* for instance, is less than 10^{-7} Jmm^{-2} (i.e. 1 erg mm^{-2}), while a dose of 7×10^{-4} Jmm^{-2} is needed to achieve the same effect with the highly resistant bacterium *M. radiodurans*. The lethal effect of ultraviolet radiation is greatest at a wavelength just below 280 nm which is the wavelength of maximum absorption by purine and pyrimidine bases in DNA and RNA.

Ultraviolet radiation causes a number of different effects on DNA when examined *in vitro,* effects such as chain breakage, intrastrand cross-linking and formation of DNA-protein linkages. However, these effects tend to be brought about by doses of ultraviolet radiation that are much larger than those needed to inactivate micro-organisms. The inactivating effect of the radiation on microbes is attributable to formation of covalent linkages between pairs of pyrimidine residues in DNA; this is known as *pyrimidine dimerisation*. The reaction between a pair of thymine residues is as shown in *Figure 3.5*. The

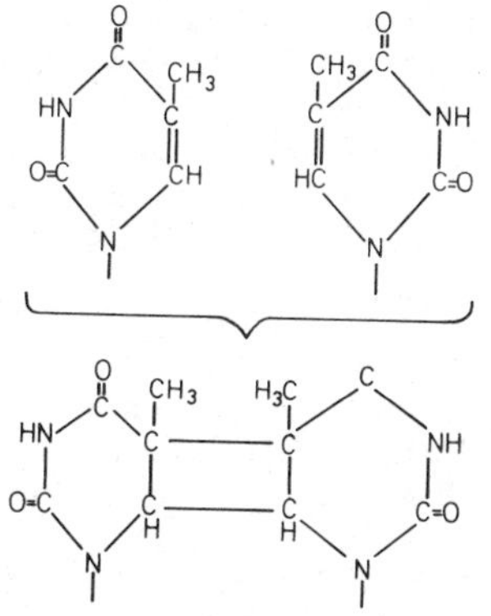

Figure 3.5 The formation of a thymine dimer between two adjacent thymine residues in a segment of DNA

reaction involves the forging of C-C bonds between the respective C-5 and C-6 atoms to give a cyclobutane ring ·between the two thymine residues. There is a wavelength **dependence** in this process, longer wavelengths (around 280 nm) being more effective than shorter wavelengths (240 nm).

The effects of ultraviolet radiation on micro-organisms are not entirely irreversible. With many microbes, the effects can to some extent be reversed by exposing irradiated organisms to visible radiation, a phenomenon known as *photoreactivation*. It involves the action of an enzyme on regions of the DNA strand containing pyrimidine dimers, an action for which the energy of visible radiation is necessary. Photoreactivation never restores the DNA in cells to full activity; there is always some residual damage after photoreactivation has reached a maximum level.

Repair of DNA after ultraviolet irradiation can also take place in the dark, a process known as *dark repair*. This

phenomenon largely explains the differences in the
susceptibility of microbes to ultraviolet irradiation;
those which are more resistant (such as *M. radiodurans*)
have a more efficient dark-repair mechanism.   Dark repair
involves a sequence of enzyme-catalysed processes.   In
the first, an incision is made in the backbone of the DNA
strand of the region that contains pyrimidine dimers.
This step is catalysed by an endonucleotidase, and is
followed by excision or removal of the damaged region,
a process that is catalysed by specific exonucleotidases.
There follows a repair replication, which involves
synthesis of a new strand of DNA using the corresponding
single-strand stretch in the helix as a template; it is
catalysed by DNA polymerase.   Finally, a polynucleotide
ligase closes the gap in the DNA strand.   Dark repair in
microbes can take place following damage to DNA from
other causes, such as ionising radiation and mutagenic
compounds.

Visible radiation (300–1000 nm).   The ability to harness
the energy of visible radiation is restricted to pigmented
organisms.   The best known of these are the photosynthetic
micro-organisms which contain chlorophylls, carotenoids
and, with algae, biliproteins.   A wide range of pigments
are produced by other micro-organisms but the physiological
role, if any, of these pigments remains unknown.   Non-
pigmented micro-organisms can be made sensitive to
visible radiation by staining, and it has been shown that
the energy absorbed by these stained organisms leads to
an increased rate of mutation.

The elegant experiments carried out by Walter Engelmann
at the end of the last century first demonstrated the
wavelengths of visible radiation that are preferentially
absorbed by photosynthetic micro-organisms.   Using motile
photosynthetic bacteria and algae, Engelmann showed that
these organisms tend to migrate to an illuminated region
on a microscope slide and that, when white light is
resolved into a spectrum of wavelengths, the micro-
organisms congregate into certain bands in the spectrum.
The bacterium *Rhodospirillum rubrum* congregated into three
bands at wavelengths around 800–900 nm, 590 nm and 470–530
nm respectively, which are now known to correspond
approximately to the wavelengths of maximum absorption of
the bacteriochlorophylls and carotenoids in the intact
bacterium.   Similar results were obtained in experiments
with motile algae, except that the algal chlorophylls

absorbed maximally at shorter wavelengths (around 650-700 nm) than the bacteriochlorophylls. The way in which energy of visible radiation is converted by micro-organisms into energy of ATP is described on page **234**.

Many reports have described the effect of visible radiation on various other activities of micro-organisms, but very little is known about the way in which radiation acts in these processes. Among the effects reported are light-induced changes in growth rate, induction of pigment formation, and both positive and negative tropic effects. A few responses of non-photosynthetic micro-organisms to visible radiation have been studied in some detail, notably the tropic effect of light on sporangiophores in the fungi *Phycomyces* and *Pilobolus*. The light-sensitive region for the response of these fungi is a zone, some 3 mm long, just below the sporangium. The zone appears to behave as a cylindrical lens which focuses light on to the far wall of the sporangiophore. On the inside of this wall is a 'retina' or region of susceptible molecules, probably of riboflavin-protein, which absorbs the visible radiation and, by some unknown mechanism, causes the sporangiophore to be kept aligned with the general direction of maximum illumination.

Infrared radiation (1000-100 000 nm). Little is known of the effect of infrared radiation on micro-organisms since the energy of this radiation is immediately converted into heat or thermal energy on contact with absorbing materials. Also, the effect of Hertzian and other long wavelength radiations on micro-organisms has been little explored.

Effect of temperature on micro-organisms

To the biochemist and physiologist, growth and reproduction of living organisms are the manifestation of an intricately co-ordinated series of metabolic reactions and, in order that these reactions can proceed at a satisfactory rate, the organism must be supplied with heat. This heat comes mainly from the environment and also partly from heat generated during metabolism.

In general, rates of chemical reactions increase as the temperature is raised, and biochemical reactions are no exceptions to this generalisation. Biologists often describe the effect of heat on a biochemical reaction or process in terms of the *temperature coefficient* or Q_{10} value, which is equal to the rate of a reaction or process at one temperature compared with the rate at a temperature $10°C$ lower, as given by the following equation:

$$Q_{10} = \frac{k_{t+10}}{k_t}$$

In this equation, k is the rate constant and t the temperature. Temperature coefficient values have been determined for many biological processes and, like the values for chemical reactions, they usually lie between 3.0 and 4.0 at room temperature (18-$22°C$) and decrease as the temperature is raised. For example, Q_{10} values for growth of *E. coli* range from 4.2 in the range 15-$25°C$ to 1.04 in the 35-$45°C$ range.

It has already been noted that microbial activity is restricted by the availability of water in the liquid phase to temperatures within the biokinetic zone (page 95). Before discussing further the ways in which micro-organisms are affected by temperatures within the biokinetic zone, let us consider briefly the effects of temperatures beyond the extremes of the zone.

High temperatures. Micro-organisms are killed at temperatures above the upper limit of the biokinetic zone, that is, above about $100°C$. Indeed, with most micro-organisms, death follows when they are subjected to temperatures much above about $50°C$. This is a process of considerable economic importance since it is the basis of one of the main methods of sterilisation. The biochemical basis of *thermal death* is not well understood although, since enzymes are the most heat-labile constituents in living organisms, it is generally assumed that death of the organism is caused by inactivation of certain enzymes. There is evidence that respiratory enzymes, particularly those that catalyse reactions in the TCA cycle, are especially susceptible to heat denaturation, but it has yet to be shown that denaturation of these enzymes leads to death of an organism. It is also likely that the death of micro-organisms at high temperatures results at least in part from thermal inactivation of RNA and from damage to the plasma membrane.

Low temperatures. If a culture of micro-organisms is frozen and thawed, only a certain proportion of the organisms in the population survives. The remainder suffer *death by freezing* which, like thermal death, has a very incompletely understood biochemical basis. From studies on bacteria and yeasts, it appears that, when a suspension of organisms is frozen, the intracellular water remains supercooled despite the formation of ice in the suspending liquid. Ice formation in the suspending liquid causes water to leave the cells and this can lead to death of the organisms. This metastable situation does not continue, and the intracellular water quickly freezes. At -20°C about 90% of the intracellular water is frozen, the remaining 10% being bound water.

Temperatures in the biokinetic zone. Micro-organisms differ considerably with regard to their response to temperatures within the biokinetic zone. Ever since microbiologists first grew organisms in pure culture it has been appreciated that growth of a particular organism is favoured only over a restricted range of temperatures, the most favourable of which is referred to as the *optimum temperature.* The value for the optimum temperature for growth of a micro-organism depends upon whether it is based on measurements of growth rate or of total growth. It is usual to define the optimum temperature for growth as that at which a micro-organism grows most rapidly. Quite often, the optimum measured on the basis of growth rate is a few degrees higher than when based on the size of the crop of organisms formed. The *maximum temperature* for growth is the highest temperature above the optimum at which growth occurs, and the *minimum temperature* is the lowest temperature below the optimum at which growth takes place. While these are considered cardinal temperatures by microbiologists, it should be stressed that, with any one organism, their values may vary with the chemical composition and physical state of the environment.

Microbes have traditionally been divided into three groups based on the values for their optimum and minimum temperatures for growth (*Figure 3.6*). Many micro-organisms, including almost all of those favoured for laboratory studies by the microbial physiologist, have optimum temperatures in the range 25-40°C, and are known as *mesophiles.* Organisms which grow best at temperatures above 40°C are known as *thermophiles,* a group which

includes bacteria and blue-green algae but curiously few
fungi. For many years, the thermophilic microbe which
received most attention from physiologists interested in
temperature responses in microbes was the food-spoilage
bacterium *Bacillus stearothermophilus,* which grows
optimally around 65-70°C. Really, this bacterium is
only a moderate thermophile, and recently microbial
physiologists interested in thermophily have turned to
Thermus aquaticus, a bacterium which was isolated from
hot springs in Yellowstone National Park in Wyoming,
U.S.A., by Thomas Brock and his colleagues. *Thermus
aquaticus* grows best at 80-85°C which, at the elevation
of Yellowstone National Park, is not too far removed from
the boiling point of water. Doing experiments at these
high temperatures brings with it some hazards, not the
least of which is the need to wear asbestos gloves when
removing cultures from the incubator!

Micro-organisms which prefer lower temperatures for
growth as compared with mesophiles are known as
psychrophiles. Some confusion has arisen regarding the
definition of psychrophilic micro-organisms, but the one
distinguishing feature possessed by these organisms is
their ability to grow reasonably rapidly at temperatures
between 0°C and 5°C. Psychrophilic micro-organisms are
therefore best defined by their having a lower minimum
temperature for growth as compared with mesophiles and
thermophiles. The optimum temperatures for growth of
many psychrophilic micro-organisms are in the same range
as that for mesophiles (i.e. 25-40°C).

The optimum temperature for growth of a micro-organism
is presumably determined by the integrated effects of
temperature on the thousands of enzyme reactions that take
place in the organism. There may be certain enzymes which
are rate-limiting and act as 'pace-makers' in this respect,
but little has yet been reported on them.

The sharp drop in the growth rate of micro-organisms at
temperatures above the optimum for growth is the result
of denaturation, reversible or irreversible, of certain
rate-limiting and possibly other enzymes. It has been
suggested, for example, that when a culture of *E. coli*
growing at 37°C is shifted to 45°C, the drop in growth
rate to 20% of that at the optimum temperature is due
mainly to decreased activity of the first enzyme on the
branch of the pathway leading to methionine synthesis,
namely homoserine transuccinylase (*see* page 270). Above
the maximum temperature for growth, these enzymes are
presumably denatured to such an extent that growth is not
possible.

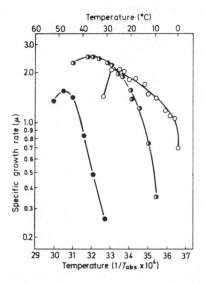

Figure 3.6 Arrhenius plots of the specific growth rates of a psychrophilic pseudomonad (o), a mesophilic strain of Escherichia coli *(◐) and a thermophilic strain of* Bacillus circulans *(●)*

The biochemical basis of thermophily in micro-organisms has been studied by a number of physiologists in recent years, and their work has revealed several possible explanations for the ability of thermophilic microbes to grow at high temperatures in the biokinetic zone. There is abundant evidence that thermophilic microbes synthesise enzymes that are inherently more heat-stable than those produced by mesophiles. This increased stability is basically attributable, with some enzymes, to a particularly favourable secondary and tertiary structure, although heat stability can also be influenced by combination of the enzyme with low molecular-weight effector molecules. Other workers have examined ribosomes from thermophilic microbes, and shown them to be more heat-stable than those from mesophiles, although a convincing explanation of this stability has not so far been forthcoming. Finally, it has been established that the plasma membrane in thermophilic micro-organisms is less easily damaged by heat than that in mesophilic microbes, although again the molecular basis of this stability is not understood.

Very little more is known about the biochemical basis
of the minimum temperature for growth of microbes, and
the ability of psychrophilic organisms to grow at near-
zero temperatures. The problem was stated quite
succinctly many years ago by the American physiologist
Otto Rahn when he observed that, since growth and
respiration are chemical reactions, they should continue
though at a greatly diminished rate as the temperature
of incubation is lowered, and should cease only when the
medium in which the micro-organisms are suspended
freezes solid. This behaviour is found with psychrophilic
micro-organisms which frequently have minimum temperatures
for growth below 0°C. Mesophiles, however, fail to
reproduce when incubated at temperatures below 5–10°C,
although they continue to respire endogenous reserves at
temperatures as low as 0°C. It has been suggested that
the minimum temperature for growth of many mesophilic
micro-organisms is determined by the temperature at which
solute transport across the plasma membrane ceases.
Several workers have shown that the minimum temperatures
for growth of certain mesophiles coincide approximately
with the temperatures at which transport of sugars across
the plasma membrane is arrested. The biochemical basis
of this inactivation of solute transport in mesophiles
at low temperatures is still being explored. Any one,
or combination, of three mechanisms may be operating.
Individual transport proteins in mesophiles may be
inactivated by conformational changes that are induced
at low temperatures; the functioning of transport proteins
within the membrane may be prevented as a result of changes
in membrane architecture at low temperatures; finally,
there is the possibility that production and expenditure
of ATP required for active uptake of solutes is prevented
at low temperatures. At the same time it has been
suggested that, because they possess a specially
favourable secondary and tertiary structure, some enzymes
in psychrophilic microbes are able to act more efficiently
at near-zero temperatures than their mesophilic counterparts.

Temperature can affect other activities of micro-organisms
quite apart from its effect on growth. Production of
pigments by some organisms is affected by temperature.
For example, strains of *Serratia marcescens* produce much
greater amounts of the red pigment prodigiosin (*see* page
86) when grown at 20–25°C, than when grown at 37°C,
although the rate of growth is faster at the higher
temperature. Flagella production in some organisms is

also favoured at low temperatures within the growth
range. Various metabolic activities of micro-organisms
can also be differently affected by temperature. With
many micro-organisms synthesis of polysaccharides, either
capsular or intracellular, is greater at lower than at
higher temperatures within the growth range.

Pressure

Many micro-organisms thrive and multiply in environments
that are at atmospheric pressure. But there are also
large numbers of micro-organisms in the deep ocean and
in subterranean locations which apparently grow just as
well under conditions of much greater *hydrostatic
pressure*. Over half of the World's surface is covered
with 3800 m or more of water, and organisms living in
the depths of the ocean will be subjected to hydrostatic
pressures of $300-400 \times 10^5$ Pa (about 300-400 atm).
    Unfortunately, the number of laboratory studies made
on the effects of hydrostatic pressure on the physiology
of micro-organisms is quite small. Nevertheless, certain
basic responses are well documented. Hydrostatic
pressures greater than about 10^8 Pa inactivate vegetative
cells of most micro-organisms. Bacterial endospores,
however, can survive after being subjected to 12×10^8 Pa
pressure. Moderate pressures $(1-5 \times 10^7$ Pa) generally
inhibit growth and reproduction of micro-organisms,
although strains have been isolated from the ocean that
are not adversely affected by these pressures; the term
barophilic has been used to describe these organisms.
One effect of moderate pressure is to prolong the lag
phase of growth. Also cell division seems to be more
susceptible than growth to inactivation by these pressures,
and many unicellular micro-organisms when grown under
increased hydrostatic pressure form long filaments.
    A few studies have been made on the effects of high
hydrostatic pressures on individual physiological
activities of microbes, in an attempt to explain the
effects of these pressures on intact organisms. Several
enzymes are known to be denatured by moderate hydrostatic
pressures, while energy production in isolated mitochondria
of the fungus *Allomyces macrogynus* is decreased when these
organelles are subjected to hydrostatic pressures. However,
there is a growing belief, advocated mainly by Robert
Marquis and his colleagues at The University of Rochester
in New York State, U.S.A., that the primary effect of

increased hydrostatic pressures on microbes is to
inhibit many of the reactions on biosynthetic
pathways. It has to be admitted that there is little
likelihood of a clearer understanding of the effects of
high hydrostatic pressures on microbes emerging in the
near future, mainly because of the difficulties
encountered in carrying out experiments on these effects.

   High partial pressures of gases, particularly oxygen,
have an inhibitory effect on growth of many micro-
organisms. In general, fungi are more resistant to
hyperbaric oxygen than animal and plant cells, although
they are not as resistant as bacteria. Hyperbaric
oxygen is used to combat infections caused by anaerobic
bacteria, and this use has renewed interest in the
effects of high partial pressures of oxygen on microbial
activity. Many of the observed effects are conceivably
caused by oxidation of thiol groups in proteins.

3.2   RESPONSES TO THE ENVIRONMENT

The responses which a micro-organism makes when it
encounters a particular environment will clearly depend
on its response to the individual factors in that
environment. In general, the overall responses can be
grouped under the headings listed below.

3.2.1   GROWTH AND REPRODUCTION

These are the most commonly observed responses of a
micro-organism to a particular environment, but the rate
and extent of growth depend on the chemical and physical
properties of the environment. Chapter 8 describes ways
in which synthesis and activity of enzymes can be
regulated by the chemical composition of the environment,
and these regulatory phenomena may cause marked changes
in the chemical composition and sometimes the morphology
of micro-organisms. Occasionally, when the chemical
composition of the environment is subject to extreme
change, as for example when the organisms are starved
of a particular nutrient, then the chemical composition
of a microbe can be altered in a quite dramatic fashion.

   Some of the best examples of alterations to the chemical
composition of microbes caused by changes in the
composition of the environment have come from studies
made by David Tempest and his colleagues working until

recently at the Microbiological Research Establishment
at Porton, England. This group conducted an extensive
study on the physiology of microbes subjected to the
stresses of substrate limitation in chemostat cultures
(*see* page 389). A good illustration is provided by
changes in the wall composition of *Bacillus subtilis* var.
niger when the bacterium is grown under conditions of
nutrient limitation. Bacteria grown under conditions of
ammonia-, magnesium- or sodium-limitation have walls
which contain a glucosylated glycerol teichoic acid.
However, when the bacterium is grown in phosphate-limited
cultures, the wall glycerol teichoic acid ceases to be
synthesised. Instead, there appears in the walls a
phosphate-free polymer, a *teichuronic acid,* which is a
polymer containing glucuronic acid residues. This change
in wall composition represents a 'sensible' reaction to
a shortage of phosphate in that the bacterium continues
to produce an anionic polymer in the wall, one which,
like a teichoic acid, has the capacity to bind divalent
ions. Interestingly, the bacterium continues to
synthesise a membrane-bound teichoic acid under
conditions of phosphate limitation.

Phosphate limitation, can however, cause changes in
the composition of lipids in the plasma membrane. When
pseudomonads are grown in chemostat culture under
conditions of phosphate limitation, they synthesise very
small amounts of phospholipid and produce instead
phosphate-free glycolipids. Almost nothing is known,
however, of the way in which glycolipids take over the
physiological functions of phospholipids in the membrane.

3.2.2 GROWTH INHIBITION AND DEATH

When an organism encounters unfavourable environmental
conditions, growth may be inhibited or the organism may
be killed. A chemical compound or an environmental
factor which has a detrimental effect on a micro-organism
either temporarily prevents the organism from reproducing
(that is, it is *microbiostatic*) or it causes death of the
organism (*microbicidal*).

When a microbial population is stressed by the presence
of an antimicrobial compound, the small number of cells
in the population which have the ability to resist the
action of the compound and to grow in its presence
gradually become the predominant cell type as the
population grows. *Resistance* to antimicrobial compounds

was noted as far back as 1907 by Paul Ehrlich, who observed that some trypanosomes were able to grow in the presence of arsenicals used to treat diseases caused by these protozoa. The ability to grow in the presence of an antimicrobial compound is acquired as a result of a change in the genetic information in the cell, the information being located either on the genome or on extrachromosomal plasmids (*see* page 67). Whether or not resistant strains arise in a population depends to some extent on the chemical nature of the compound and the micro-organism. Resistance to the more powerful antimicrobial compounds, such as phenols, is rarely encountered. Even with antibiotics, which as far as resistant strains are concerned are the most intensively studied antimicrobial compounds largely for medical reasons, production of resistant strains depends upon the nature of the organism. About 70% of the staphylococci nowadays isolated in hospital laboratories are resistant to penicillin-G, whereas Group A streptococci that are isolated are about as sensitive as they were 20 years ago.

In biochemical terms, resistance to antimicrobial compounds can be acquired in one of four main ways. Although, in principle, the easiest way in which a micro-organism might become resistant to the action of an antimicrobial compound is to acquire the capacity to prevent entry of the compound into the cell, strangely this type of resistance is not often encountered. However, resistance to the action of tetracycline antibiotics by some bacteria does result from a change in the ability of the plasma membrane to transport the antibiotic. Resistance to tetracyclines, which is determined by acquisition of the appropriate plasmid-borne genetic information, results from the ability of the bacteria to excrete the tetracycline that has been transported into the cell. Bacteria which are not resistant to the antibiotics are able to transport the compounds into the cell, but lack the capacity to excrete the molecules.

A second way in which resistance to an antimicrobial compound can be acquired is for the microbe to modify the properties of the target enzyme. Resistance of pneumococci to the action of sulphonamides is often explained by the production of mutant strains of the bacteria which are altered in the gene coding for the enzyme tetrahydropteroate synthetase (*see* page 123) which directs synthesis of an enzyme that has a much lower affinity for the sulphonamide compared with *p*-aminobenzoate. Similarly, some examples of streptomycin resistance in bacteria are explained by a mutation in the gene which codes for the ribosomal protein (dubbed P-10 by students of ribosome structure) which binds streptomycin to the 30S ribosomal subunit.

A third way in which resistance to the action of an antimicrobial compound can be acquired is for the microbe to bring about changes in its metabolism with the result that the physiological importance of the target enzyme is diminished. Examples of this type of effect are not well documented, although it has been suggested that resistance to penicillin in some bacteria may be explained by the presence of a decreased proportion of peptidoglycan in the cell wall; under these conditions it is conceivable that the structural role of the peptidoglycan in the wall is taken over by other polymers. A different type of example which also comes under this heading are certain mutants which are resistant to the purine nucleoside antibiotic, psicofuranine. These mutants synthesise a defective form of the enzyme IMP dehydrogenase, and as a result are unable to synthesise a normal supply of XMP (page 276). This change causes a derangement in the regulatory mechanisms that operate on these pathways, and as a consequence biosynthesis of XMP aminase, which is the target enzyme for psicofuranine, is increased because of the presence of lower intracellular concentrations of GMP which represses synthesis of the target enzyme.

Resistance to the action of antimicrobial compounds is quite frequently explained by the acquisition on the part of the microbe of a means for chemically inactivating the compound. Two main types of inactivating mechanism have been recognised. Some resistant organisms secrete an enzyme which converts the antimicrobial compound to an inactive form. Probably the best documented examples of these types of enzymes are the β-lactamases, which are essentially peptidases and catalyse the hydrolytic breakdown of penicillins and cephalosporins. Hydrolysis of penicillin-G by a β-lactamase, as illustrated, leads to production of an antibiotically inactive acid.

penicillin G   penicilloic acid G

A second type of inactivation results from a chemical modification of the antimicrobial compound. A number of examples have been recorded in which inactivation results from acetylation of an antibiotic. The best known example is conversion of chloramphenicol into the inactive 3-acetoxychloramphenicol. The reaction is catalysed by the enzyme chloramphenicol transacetylase. Other examples

of metabolic inactivation of antimicrobial compounds include phosphorylation and adenylylation, both involving ATP, which lead to inactivation of streptomycin in various bacteria. The intriguing feature of these inactivation processes is that they call for the production on the part of the microbe of a completely new enzyme, a process that represents a considerable feat in terms of molecular evolution. Synthesis of these inactivating enzymes in bacteria is directed by plasmid-bound genes which can be transferred from one bacterium to another (*see* page 67).

3.2.3 MOVEMENT

Movement of organisms or of individual organelles can be caused by a number of environmental factors, both chemical and physical. These movements are referred to as *taxes*. When the environmental stimulus attracts the organism, the taxis is positive; when the organism is repelled the taxis is negative. Different taxes are distinguished according to the nature of the stimulus.

Chemotaxis involves an attraction or repulsion of organisms in concentration gradients of chemical compounds.

In general, compounds with a high nutritional value
evoke a positive chemotaxis, while antimicrobial compounds
have the opposite effect. Chemotaxis is often important
in the attraction of mating types of micro-organisms.
Phototaxis is often observed among photosynthetic micro-
organisms which, when exposed to a spectrum of light,
accumulate in light of certain wavelengths (page 128).
A few reports have appeared showing that certain micro-
organisms (e.g. *Paramecium* spp.) can move in response to
a magnetic field (*magnetotaxis*). Other microbial taxes
which have been reported include *thermotaxis* (in response
to heat stimuli) and *galvanotaxis* (movement in an
electrical field). The suggestion has been made that
proteins in the plasma membrane act as the transducer of
environmental stimuli in microbial taxes.

Four main methods of locomotion are recognised among
micro-organisms.

Micro-organisms may possess *cilia* or *flagella* which act
as locomotion organelles. These structures are most
efficient and flagellated bacteria can travel as fast as
50 μm s⁻¹. Cilia and flagella cause movement of liquid
relative to the point of attachment of the organelle.
If the micro-organism is unable to move, this causes a
flow of liquid over the surface of the organism; if the
organism is free to move, it results in locomotion. These
organelles differ in that cilia typically cause movement
of liquid by beating at right angles to the long axis of
the cilium, while flagella cause movement along the length
of the flagellar axis (*Figure 3.7*). In practice, this

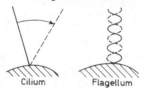

Cilium   Flagellum

*Figure 3.7 The different types of movement associated
with cilia and flagella*

distinction between ciliary and flagellar movement is not
always clear-cut. A considerable amount of work has been
reported on the mechanism of ciliary and flagellar
movement, and it is generally assumed that movement of
these organelles results from contractions of fibres
within the organelle and that these contractions are
propagated along the length of the cilium or flagellum.
A sequential clockwise contraction of each of the nine

fibres in the cilium or flagellum, or of protein subunits in the bacterial flagellum, could give rise to the beating patterns that have been observed with these organelles. There is clearly a similarity between fibril contraction in the cilium or flagellum and the contraction of muscle fibres although, as with muscle, the biochemistry of the process is far from clear. The enzyme ATPase has been detected in many isolated cilia and flagella but not in those from bacteria, and so it is possible that the mechanisms of fibre contraction may differ in these organelles.

The exotic if not bizarre *movements of spirochaetes* have been familiar to bacteriologists for many years. These organisms carry out a wide range of movements which include production of helical waves travelling along the length of the bacterium as well as bending, curling and lashing. The motor organelles in spirochaetes consist of a single fibril or bundle of fibrils, each about 10–20 nm in diameter, extending the length of the cell and attached at both ends. The fibrils usually lie under the surface layer of the bacterium. It has been suggested that these fibrils are analogous, perhaps homologous, to bacterial flagella.

Another type of movement which is found among micro-organisms is *gliding*. This is defined as the active movement of an organism in contact with a solid substratum when there is neither a visible organelle responsible for the movement nor a distinct change in shape of the organism. Gliding is carried out by a number of blue-green algae and myxobacteria as well as by certain diatoms and protozoa. Compared with movement caused by cilia and flagella, gliding is comparatively slow, and the organisms rarely move faster than 5 μm s^{-1}.

A fourth type of movement is characteristic of amoebae and is usually known as *amoeboid movement*. Amoebae appear to remain stuck to the substratum but the streaming flow of the cytoplasmic contents of the protozoon propels the organism along. During this movement, the organism under-goes a considerable change in shape and forms protuberances known as *pseudopodia*. The essential features of locomotion in unicellular organisms, such as *Amoeba proteus* and *Chaos chaos,* can be observed in the light microscope. Cytoplasmic streaming originates at the rear locomotory pole of the cell, known as the *uroid* (*Figure 3.8*), and is directed towards the front pole. There, the streaming cytoplasm, which is referred to as *endoplasm,* divides, turns to the periphery of the cell and for a

while becomes stationary in relation to the substratum.
This causes a gradual translocation of the stationary
cytoplasm or *ectoplasm* to the uroid where it again becomes
motile. During one cycle of these events, the cell
migrates over a distance corresponding to its length.
Several theories have been advanced to explain the
molecular basis of amoeboid movement, but none has yet
proved universally acceptable. Important in any theory
of amoeboid movement is the presence in these organisms
of thin filaments which, it is agreed, provide the
structural basis of contractility. These filaments are
sensitive to ATP, and resemble the actomyosin proteins
of contractile muscle. But exactly how the contractile
movements of these filaments contribute to the overall
movement of the amoeba remains a mystery.

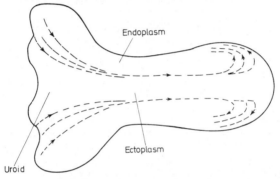

*Figure 3.8 Diagram showing the directions of cytoplasmic
streaming in an amoeba*

3.2.3 DIFFERENTIATION

Chemical and physical factors in the environment can
cause micro-organisms to differentiate, that is, to give
rise to structures morphologically different from
vegetative organisms. They can, for example, evoke the
formation of flagella, spores of various types, and
gametes. An account of the biochemical basis of certain
types of differentiation in micro-organisms can be found
in Chapter 10.

3.3   MODIFICATION TO THE ENVIRONMENT

As a result of microbial activity, the chemical
composition and physical state of the environment can
undergo considerable change.

3.3.1   CHANGES IN CHEMICAL COMPOSITION

The main changes in the chemical composition of the
environment following growth of a micro-organism are
accumulation of microbial cell material, removal of
nutrients from the medium, and excretion of products of
metabolism.   Either or both of the last two factors can
lead to cessation of growth of the micro-organism (*see*
page 380).

Substantial modifications to the chemical composition
of the environment are caused by secretion of enzymes by
micro-organisms.   Many micro-organisms secrete enzymes
that hydrolyse molecules which are too large to pass
across the plasma membrane.   These organisms can be
considered to begin metabolising these high molecular-
weight compounds before they enter the cell.   Quite often,
the amounts of hydrolytic enzyme secreted by a micro-
organism far exceed the amount required to hydrolyse
sufficient of the compound to make the products of
hydrolysis available to the organism in adequate amounts.
As a result, a large proportion, and sometimes all, of
the high molecular-weight compound in the environment is
hydrolysed, although only a small proportion of the
hydrolysis products is used by the micro-organism.

Many different types of extracellular hydrolytic
enzymes, as well as some non-hydrolytic enzymes, are
produced by micro-organisms.   By far the commonest are
those which catalyse hydrolysis of polymers such as
polysaccharides and proteins and molecules such as
triglycerides and phospholipids which tend to aggregate
in micelles.

Extracellular proteinases and peptidases are produced
by many different groups of micro-organisms and
especially by species of *Aspergillus, Bacillus,
Clostridium, Penicillium* and *Pseudomonas*.   Most of these
enzymes have a relatively low substrate-specificity.
The group includes type A clostridiopeptidases
(collagenases) produced by *Clostridium histolyticum,* and
the enzymes coagulase and staphylokinase which are
synthesised by certain staphylococci.

Probably the best studied examples of microbial extracellular polysaccharidases are the α-amylases which are
produced in abundance by many bacilli. Other enzymes
in this group include agarases, cellulases, chitinases,
dextrinases, glucanases and pectinesterases. Two rather
interesting microbial extracellular polysaccharidases are
neuraminidase, which is formed by certain clostridia and
probably catalyses cleavage of the terminal α-2,6 links
between N-acetylneuraminic acid and 2-acetylamino-2-
deoxy-D-galactose, and hyaluronate lyase, a product of
Bacillus subtilis and certain streptococci, which
catalyses cleavage of links between residues in various
mucopolysaccharides.

Microbial extracellular lipases have not been studied
to any great extent although they are known to be formed
by a number of micro-organisms, particularly moulds. In
common with plant and animal lipases, these microbial
enzymes usually have a very broad substrate specificity.
Somewhat greater specificity is found among microbial
extracellular phospholipases which catalyse hydrolysis of
phospholipids. This group includes the α-toxin of
Clostridium welchii which has been shown to hydrolyse
phosphatidylcholine.

Ribonucleases and deoxyribonucleases are not formed
extracellularly by many micro-organisms, although
representatives of this group of enzymes have been
detected in filtrates from cultures of certain
streptococci, staphylococci, aspergilli and penicillia.

Many of the enzymes which are produced extracellularly
in copious amounts by micro-organisms can also be
detected in the organisms, but examples have been
reported where this is not so. The extracellular
α-amylases produced by *B. subtilis* and *Pseudomonas
saccharophila,* for example, cannot be detected in these
bacteria although filtrates from cultures of the
organisms contain large amounts of the enzymes.
Moreover, not all 'extracellular' hydrolytic enzymes
produced by micro-organisms diffuse into the environment.
Some remain tightly bound to the micro-organism, either
within the cell wall or on the outside of the plasma
membrane. Strains of *Cytophaga* have been shown to
possess cell-bound cellulases, chitinases and agarases.

3.3.2 CHANGES IN PHYSICAL PROPERTIES

Some of the ATP used by micro-organisms is dissipated in the form of heat (*see* page 250) and this evolution of heat can cause a detectable rise in the temperature of a culture, particularly during the exponential phase of growth. Also, a few microbes are able to emit visible radiation under certain conditions. A particularly interesting change in the physical property of the environment is that caused by microbes that secrete copious amounts of polysaccharide. This can often accumulate to the extent that the culture becomes almost solid, even to the point at which it is nearly impossible to remove the cells from the culture by centrifugation.

FURTHER READING

ENVIRONMENTAL FACTORS

CHEMICAL FACTORS

Nutrients

BARTON-WRIGHT, E.C. (1952). *The Microbiological Assay of the Vitamin B-Complex and Amino Acids,* 179 pp. London; Pitman

CORRY, J.E.L. (1973). The water relations and heat resistance of micro-organisms. *Progress in Industrial Microbiology,* 12, 73-108

DUNDAS, I.E.D. (1976). Physiology of the halobacteria. *Advances in Microbial Physiology,* 14, in press

FRIES, N. (1965). The chemical environment for fungal growth. 3. Vitamins and other organic growth factors. In: *The Fungi,* Eds. G.C. Ainsworth and A.S. Sussman, Vol.2, pp.491-523. New York; Academic Press

GUIRARD, B.M. and SNELL, E.E. (1962). Nutritional requirements of micro-organisms. In: *The Bacteria,* Eds. I.C. Gunsalus and R.Y. Stanier, Vol.4, pp.33-93. New York; Academic Press

HUGHES, D.E. and WIMPENNY, J.W.T. (1969). Oxygen metabolism by micro-organisms. *Advances in Microbial Physiology,* 3, 197-232

HUTNER, S.H. (1972). Inorganic nutrition. *Annual Review of Microbiology,* 26, 313-346

KELLY, D.P. (1971). Autotrophy: concepts of lithotrophic bacteria and their organic metabolism. *Annual Review of Microbiology,* 25, 177-210

KOSER, S.A. (1968). *Vitamin Requirements of Bacteria and Yeasts*, 663 pp. Springfield, Illinois; Thomas

LARSEN, H. (1973). The halobacteria's confusion to biology. *Antonie van Leeuwenhoek*, 39, 383-396

LILLY, V.C. (1965). The chemical environment for fungal growth. 1. Media, macro-, and micronutrients. In: *The Fungi*, Eds. G.C. Ainsworth and A.S. Sussman, Vol.2, pp.465-478. New York; Academic Press

MORRIS, J.G. (1975). The physiology of obligate anaerobiosis. *Advances in Microbial Physiology*, 12, 169-245

PAYNE, J.W. (1975). Peptides and micro-organisms. *Advances in Microbial Physiology*, 13, 55-113

PERLMAN, D. (1965). The chemical environment for fungal growth. 2. Carbon sources. In: *The Fungi*, Eds. G.C. Ainsworth and A.S. Sussman, Vol.2, pp.479-489. New York; Academic Press

RITTENBERG, S.C. (1969). The role of exogenous organic matter in the physiology of chemolithotrophic bacteria. *Advances in Microbial Physiology*, 3, 159-196

SUOMALAINEN, H. and OURA, E. (1971). Yeast nutrition and solute uptake. In: *The Yeasts*, Eds. A.H. Rose and J.S. Harrison, Vol.2, pp.3-74. London; Academic Press

Antimicrobial compounds

FRANKLIN, T.J. and SNOW, G.A. (1975). Biochemistry of antimicrobial action. 2nd edit. 224 pp. London; Chapman and Hall

GALE, E.F., CUNDLIFFE, E., REYNOLDS, P.E., RICHMOND, M.H. and WARING, M.J. (1972). The molecular basis of antibiotic action. 456 pp. London; John Wiley and Sons

PHYSICAL FACTORS

BROCK, T.D. (1967). Life at high temperatures. *Science (New York)*, 158, 1012-1019

FARRELL, J. and ROSE, A.H. (1967). Temperature effects on micro-organisms. In: *Thermobiology*, Eds. A.H. Rose, pp.147-218. London; Academic Press

GOTTLIEB, S.F. (1971). Effect of hyperbaric oxygen on micro-organisms. *Annual Review of Microbiology*, 25, 111-152

HEDÉN, C.G. (1964). Effect of hydrostatic pressure on microbial systems. *Bacteriological Reviews*, 28, 14-29

LEGATOR, M.S. and FLAMM, W.G. (1973). Environmental mutagenesis and repair. *Annual Review of Biochemistry*, 42, 683-708

MORITA, R.Y. (1970). Application of hydrostatic pressure
  to microbial cultures. *Methods in Microbiology*, 2,
  243-257
MOSELEY, B.E.B. (1968). The repair of damaged DNA in
  irradiated bacteria. *Advances in Microbial Physiology*,
  2, 173-194
WEBB, J.L. (1963-1966). *Enzymes and Metabolic Inhibitors*,
  Vols. 1-4. New York; Academic Press

PHYSIOLOGICAL FACTORS

MORITA, R.Y. (1975). Psychrophilic bacteria. *Bacteriological Reviews*, 39, 144-167

RESPONSES TO THE ENVIRONMENT

GROWTH AND REPRODUCTION

ELLWOOD, D.C. and TEMPEST, D.W. (1972). Effects of
  environment on bacterial wall content and composition.
  Advances in Microbial Physiology, 7, 83-117

GROWTH INHIBITION AND DEATH

BENEVISTE, R. and DAVIES, J. (1974). Mechanisms of
  antibiotic resistance in bacteria. *Annual Review of
  Biochemistry*, 42, 471-506
GALE, E.F., CUNDLIFFE, E., REYNOLDS, P.E., RICHMOND, M.H.
  and WARING, M.J. (1972). The molecular basis of
  antibiotic action. 456 pp. London; John Wiley and
  Sons

MOVEMENT

HOLWILL, M.E.J. (1974). Some physical aspects of the
  motility of ciliated and flagellated micro-organisms.
  Science Progress, 61, 63-80
KOMNICK, H., STOCKEM, W. and WOHLFARTH-BOTTERMANN, K.E.
  (1973). Cell motility: mechanisms in protoplasmic
  streaming and amoeboid movement. *International Review
  of Cytology*, 34, 169-249

DIFFERENTIATION

ASHWORTH, J.M. and SMITH, J.E. (1973). Microbial
differentiation. *Symposium of the Society for General
Microbiology,* 23, 1-450

MODIFICATIONS TO THE ENVIRONMENT

DAVIES, R. (1963). Microbial extracellular enzymes,
their uses and some factors affecting their formation.
In: *Biochemistry of Industrial Micro-organisms,* Eds.
C. Rainbow and A.H. Rose, 708pp. London; Academic
Press

4

AN INTRODUCTION TO MICROBIAL
METABOLISM

Having considered the molecular architecture of micro-
organisms and their dependence on and response to the
chemical and physical factors in the environment, we
can now turn our attention to the central problem in
microbial physiology, namely, how the vast array of small
and large molecules which go to make up microbial cell
material is synthesised from the raw materials and sources
of energy available in the environment.

4.1 PRINCIPLES OF MICROBIAL METABOLISM

The cell wall and extramural layers of a micro-organism
are freely permeable to the majority of compounds present
in the environment, and the main barrier between a
nutrient and the metabolic machinery of a micro-organism
is the plasma membrane. Most solute molecules that pass
across the membrane do so through the mediation of
specific carrier or transport mechanisms. This initial
phase in the utilisation of a nutrient by a micro-organism
is discussed in Chapter 5 (page 159).

When inside the micro-organism, a nutrient is subjected
to a series of chemical modifications, that is, it is
metabolised. The sequence of reactions by which a
nutrient is metablised is referred to as a *metabolic
pathway*. On entering a pathway, a nutrient (*A*) is
converted into one or more compounds (*B-D*) known as

F

intermediates, each of which is further metabolised to give ultimately the end-product (*E*) of the pathway. Each of the reactions on a metabolic pathway

$$A \rightarrow B \rightarrow C \rightarrow D \rightarrow E$$

is catalysed by a specific enzyme. Another term which is often used when discussing cell metabolism is *precursor;* this is defined as any compound which is formed within the cell, or supplied by the environment, and which is metabolised to give some end-product.

A nutrient can enter one of two main types of metabolic pathway. Some, known as *anabolic pathways,* lead to an increase in the molecular complexity of the nutrient; others lead to a decrease in molecular complexity and are described as *catabolic pathways.* Anabolic pathways usually lead to synthesis of new cell constituents, while catabolic pathways result in the formation of low molecular-weight compounds, some of which are waste products of metabolism and are excreted while others serve as precursors in biosynthetic reactions and are channelled into anabolic pathways. Catabolic pathways also often furnish the cell with a supply of metabolic energy in the form of ATP or some other high-energy compound which can be used in biosynthetic reactions. Clearly, therefore, there is a very close link between anabolic and catabolic pathways of metabolism, although it is frequently convenient to consider them separately, especially when discussing the individual reactions involved.

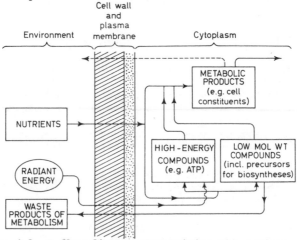

Figure 4.1   A flow diagram summarising the salient features of microbial metabolism

The terms 'anabolic' and 'catabolic' have also been used to refer to intermediates on various pathways, and even to the enzymes that catalyse reactions on these pathways. This use has led to some confusion since an intermediate or an enzyme can often quite justifiably be placed in both of these categories. An example is furnished by the intermediate acetolactate. This compound is produced during formation of acetylmethyl-carbinol which is an end-product of glucose catabolism in some micro-organisms (page 212), and it also acts as an intermediate in the biosynthesis of valine (page 273). The American microbiologist, Bernard Davis, suggested that the term *amphibolic* might be used to refer to intermediates of this nature, although he concedes that, since some are extremely difficult to classify, the term *diabolic* might be more appropriate!

4.2 STRATEGIES EMPLOYED IN STUDIES ON MICROBIAL METABOLISM

One of the main tasks of the microbial physiologist has been to discover the nature of each of the intermediates on various anabolic and catabolic pathways. This task of charting metabolic pathways in micro-organisms has progressed very considerably during the past two decades, but before we go on to discuss individual pathways and the ways in which they are regulated it is worth considering briefly how this very extensive body of knowledge has been obtained.

At the inaugural meeting of the Society for General Microbiology in February, 1945, that great pioneer of microbial physiology, Marjory Stephenson, drew a simple overall picture of the various strategies in research into the chemical activities of micro-organisms. She defined five levels at which this research should be carried out; these, together with two additional levels which have been recognised during the past ten years, are summarised in *Table 4.1*. The first three levels of investigation comprise a study of the nutritional requirements of a micro-organism which has been described in Chapter 3. At the fourth level of investigation the chemical composition of the environment is simplified so as to study the action of the micro-organism on individual (or a small number of) substrates under conditions in which growth cannot take place. For this, washed

Table 4.1   Levels of investigation in microbial physiology

Level	Type of investigation
1	Growth of mixed cultures of micro-organisms in complex 'natural' media
2	Growth of pure cultures of micro-organisms in complex non-defined media
3	Growth of pure cultures of micro-organisms in chemically defined media
4	Activities of cell suspensions of micro-organisms metabolising defined substrates, usually under conditions which prevent growth
5	Metabolism of individual substrates by cell-free preparations of micro-organisms using complete or fractionated cell-free extracts, or purified preparations of enzymes
6	Examination of the factors which regulate the synthesis and activity of enzymes on metabolic pathways
7	Movement and transport of molecules within micro-organisms

suspensions of micro-organisms harvested from cultures grown in a suitable medium are used.

Detailed analysis of a metabolic pathway requires ultimately the study of individual reactions and of the enzymes that catalyse these reactions. This is done using cell-free extracts of organisms, which are used either untreated or after being partially fractionated. At a later stage, it is often possible to isolate purified or partially purified preparations of a particular enzyme from these cell-free extracts, and to study the action of this enzyme on an individual substrate.

Since Marjory Stephenson formulated these levels of investigation in microbial physiology, two additional levels have been recognised. The first of these involves a study of mechanisms by which synthesis and activity of enzymes are controlled and regulated so that the micro-organism synthesises (or breaks down) just sufficient of a compound to satisfy its immediate needs, thereby conserving precursor compounds and metabolic energy. The study of the control of cell metabolism is still very much in its infancy, but certain regulatory mechanisms have already been shown to operate in micro-organisms and these are described in Chapter 8 (page 327).

Finally, there is the need to explain how molecules
that are formed as a result of the activity of metabolic
pathways are transported within the cell, either to be
excreted or to be assembled into various structures and
organelles. As yet, very little is known of this aspect
of cell metabolism, but the appreciation that metabolism
has not only magnitude but also direction in space is an
important prerequisite to studying microbial physiology
at this level. The British biochemist, Peter Mitchell,
coined the term *vectorial metabolism* (as compared with
scalar metabolism which has no direction in space) to
describe this process.

Marjory Stephenson emphasised the importance of
tackling any problem in microbial physiology at as many
of these levels as possible. The levels of investigation
represent the tactics of the attack on the problem. The
grand strategy involves a co-ordinated attack at several
different levels, shifting the main effort from one level
to another as dictated by the state of the problem and
the experimental armament available.

4.3   METHODS USED IN STUDIES ON MICROBIAL
METABOLISM

The availability of suitable experimental armament, that
is of techniques and methods that can be used to study
the chemical activities of microbes, has been a vitally
important factor in the growth and development of
microbial physiology. The more important of these
methods are briefly discussed below.

4.3.1 NUTRIENT BALANCES

Once methods had been devised for growing micro-organisms
in defined media, valuable information on the metabolism
of organisms was quickly obtained by analysing culture
fluids for disappearance of nutrients and appearance of
metabolic end-products. In this way it was possible to
construct a balance sheet showing the fate of each of
the major elements (carbon, nitrogen, phosphorus and
sulphur) involved in metabolism of the organism. By its
very nature, this type of experimental technique can
yield only preliminary information on metabolism of a
nutrient, but it enables the investigator to account for
all of the major metabolic products of an organism. It
was as a result of establishing nutrient balances that

ethanol and carbon dioxide were first recognised as the major products of glucose catabolism in yeast, while more recently carbon-balance experiments were instrumental in the discovery of phosphoketolase pathways for glucose breakdown in bacteria.

4.3.2   METABOLICALLY BLOCKED MICROBES

One of the most powerful tools available to the microbial physiologist for probing the nature of intermediates on metabolic pathways are microbes that are unable to catalyse individual reactions on various metabolic pathways. These metabolically blocked micro-organisms can be obtained in one of two main ways.

By far the more important is the production of *mutant strains* in which metabolic defects result from mutations in genes which code for specific enzymes. Mutant strains of micro-organisms were first used successfully by George Beadle and Edward Tatum in their experiments on biosynthetic pathways in *Neurospora crassa,* and the dictum that in almost all studies in microbial physiology the experimenter is better off with than without mutant strains of his experimental organisms is universally accepted. Many different types of mutant are used in experiments in microbial physiology. Among the most useful are *auxotrophic mutants* which require a specific nutrient or nutrients for growth, unlike the wild-type organism. Auxotrophic mutants can be used in two principal ways. It is possible to screen compounds for the ability to permit growth of the mutant, and so obtain information on the nature of the intermediates on a pathway beyond the blocked reaction. Metabolically blocked organisms also often accumulate and excrete the intermediate, metabolism of which has been blocked. Intermediates which are accumulated in this way are often growth factors for mutants with a block at an earlier reaction on the pathway, and this allows a ready detection of the accumulation using the technique of cross-feeding or *syntrophism.*

Individual metabolic reactions in a microbe can also be blocked by incorporating specific antimicrobial compounds in the growth medium. Inevitably, the method is far less specific and has yielded often equivocal data, as compared with the use of mutants. Nevertheless, providing that the antimicrobial compound used has a reasonably high specificity for a particular reaction, the method can be a valuable probe in studies on microbial metabolism.

4.3.3 RADIOACTIVELY LABELLED COMPOUNDS

Following the fate of individual atoms in molecules
during metabolism in growing cultures, cell suspensions
or cell-free extracts, is possible using nutrients and
substrates containing radioactive isotopes. The value
of this technique in studying the chemical activities of
micro-organisms can hardly be overstated and the
tremendous progress made during the past decade in
charting biosynthetic pathways in particular has been
due in no small measure to the use of radioactively
labelled substrates.

Over 30 years ago, Marjory Stephenson likened the
problem of the microbial physiologist to that of an
observer trying to gain an idea of life inside a
household by making first of all a careful scrutiny of
the food and supplies which arrive at the front and back
doors and by patiently examining the contents of the
dustbin. Since then, microbial physiologists have
devised some extremely ingenious and occasionally
diabolically cunning methods for studying life inside
the house, so much so that their efforts have been
contemptuously referred to as 'microbating' by their
colleagues who are less biologically inclined. However,
much still remains to be discovered about life inside
the house, and it is the pious hope of microbial
physiologists that methods will be devised for opening
the doors and paying a friendly visit rather than using
the somewhat brutal methods of investigation that are
now their stock in trade.

TRANSPORT OF COMPOUNDS INTO AND OUT OF MICRO-ORGANISMS

All micro-organisms possess a barrier which preserves the integrity of the organism and limits the entrance and exit of chemical compounds. This barrier, which has been termed the 'osmotic barrier', resides in the lipid-protein plasma membrane. Several pieces of evidence have led to this conclusion. For example, lipid-soluble compounds are often able to penetrate micro-organisms more quickly than lipophobic compounds. Moreover, the osmotic barrier can be broken by treating micro-organisms with lipid solvents, such as aqueous butanol or toluene, and this causes the release of low molecular-weight compounds from organisms. Microscopical observations on plasmolysable micro-organisms also indicate that the osmotic barrier resides in the plasma membrane.

The plasma membrane does not, however, constitute the only barrier to the passage of compounds from the environment into a micro-organism, for solutes must also pass across the cell wall and extramural layers. Capsules and slime layers are very loosely knit structures and probably have little restraining influence on the passage of most solutes. The cell wall, on the other hand, may provide a much more effective barrier to the diffusion of solutes. Estimates of the free space available for diffusion in packed organisms show that cell walls in many Gram-positive bacteria are not penetrated by dextrans of average molecular weight 10 000 daltons. This value may be an under-estimate for some bacteria. For example, polymers with a molecular weight below about 50 000 daltons are able to penetrate isolated walls of *Bacillus megaterium*.

It appears that fungal and yeast walls provide, on the whole, a greater restraining influence than bacterial walls. Experiments using inert polymers have shown that the threshold value for hyphal walls of *Neurospora crassa* is a molecular weight of about 4700 daltons, and for walls of *Saccharomyces cerevisiae,* 4500 daltons.

Factors other than molecular weight must be important in determining whether a compound penetrates a microbial cell wall. During genetic transformation, cell walls of competent bacteria are permeable to DNA with an average molecular weight of several million daltons. Very few data have been published on the diameters of pores in cell walls. One report claims that the hyphal walls of *N. crassa* have pores 4.0–7.0 nm in diameter, although it is possible that in these experiments, preparation of the walls for electron microscopy may have affected pore size. It is quite likely that the walls of some micro-organisms have a selective action in allowing the passage of some, but not all, low molecular-weight compounds. Certain cell-wall polymers, such as negatively charged teichoic acids, may act as ion-exchange resins and regulate the passage of positively charged ions through the wall.

5.1   SOLUTE-TRANSPORT PROCESSES

5.1.1   METHODS USED IN STUDYING SOLUTE-TRANSPORT

Passage of a compound from the environment into the interior of a micro-organism can be studied in one of two ways: (a) by following the disappearance of the compound from the environment, and (b) by analysing the organism for content of the transported solute.

A number of workers have measured the rate of solute uptake by suspending micro-organisms in buffer containing the solute, centrifuging or filtering portions at intervals, and assaying the supernatant liquid or filtrate for content of solute. Because of the relatively small amounts of solute taken up by micro-organisms, this method can only be used with suspensions containing large concentrations of organisms, and even then it places considerable demands on the accuracy of the analytical method employed. The technique has been used most extensively to study sugar uptake, particularly with yeasts.

Direct analysis of micro-organisms for the transported solute is generally preferred. It is usually difficult

to estimate the amount of solute that has been transported
if the solute is metabolised, and several techniques have
been devised for overcoming this problem. One way is to
study uptake of nutrients under conditions (e.g. in the
presence of poisons) which prevent metabolism of the
transported compound. Uptake of glucose by *Sacch.
cerevisiae* has been studied using organisms suspended in
buffer containing glucose and iodoacetate, the metabolic
poison being incorporated to prevent glycolysis of the
transported sugar. Galactose transport in *Escherichia
coli* is studied using mutant strains of the bacterium
that lack the ability to synthesise galactokinase and so
cannot phosphorylate the transported sugar. Another
technique uses non-metabolisable solutes that are
chemically related to the nutrient that is being studied.
Alpha-Methyl glucoside has been extensively used in
studies on glucose transport into bacteria; glucosamine,
galactose and sorbose have been used similarly with
yeasts. Non-metabolisable analogues have also been
employed in experiments on amino-acid transport; an
example is 2-amino*iso*butyrate which is an analogue of
alanine.

Uptake of solutes, either nutrients or non-metabolisable
analogues, is most easily studied using radioactively
labelled compounds. The rate of accumulation of radioactively labelled solutes can be studied at intervals as
short as 30 seconds by rapidly filtering portions through
membrane filters, washing the filters, and measuring the
radioactivity of the organisms by transferring the filter
to a vial containing scintillation fluid and placing the
vial in a liquid scintillation spectrometer.

Much useful information on the permeability properties
of the microbial plasma membrane has come from experiments
using structures other than intact cells. Studies using
protoplasts or sphaeroplasts can be difficult because of
the fragile nature of these structures, particularly when
they are subject to filtration on membrane filters. They
are, however, valuable in that any restraining or
interfering action of the cell wall on uptake of a solute
is eliminated.

Structures derived from bacterial protoplasts or
sphaeroplasts have recently, in the hands of H. Ronald
Kaback and his associates working at the Roche Institute
of Molecular Biology in New Jersey, U.S.A., yielded
valuable information on certain aspects of active
transport. Kaback's group have prepared *spherical
vesicular structures*, measuring 0.1-1.5 μm in diameter,

and bounded by the plasma membrane of the bacterium but enclosing little if any of the cytoplasmic contents. These vesicles are osmotically active, like protoplasts and sphaeroplasts. When prepared from sphaeroplasts of certain Gram-negative bacteria, they retain the lipopolysaccharide on the outside of the plasma membrane. The vesicles are made by submitting bacterial protoplasts or sphaeroplasts, prepared by growing bacteria in the presence of penicillin or by treating them with lysozyme and EDTA, to osmotic lysis in a suspension containing deoxyribonuclease and ribonuclease. When resuspended in a suitable buffer, the bacterial membranes reseal to give vesicular structures. Uptake of solute by these vesicles is studied in the same way as with intact cells. The advantage of using bacterial vesicles is that the availability of energy sources for the uptake process can be controlled, thereby providing a valuable means for characterising the nature of the molecules which furnish metabolic energy for the process.

Another structure which has been used to gain an insight into the mechanism of solute uptake in micro-organisms is the *liposome*. These are spherical structures in which a lipid bilayer encloses an aqueous solution. Liposomes can be prepared by shaking a portion of lipid dispersed in an aqueous solution. Not all lipids can be used to make liposomes. The mixture must include a polar lipid, and it is customary to include phosphatidic acid. Frequently, not one but several bilayers constitute the surface layer of the liposome. Experimentally, this is a disadvantage, one which it is possible to avoid by choosing optimal conditions for shaking. In addition, the population of liposomes formed is often heterogeneous in size, but a fraction more uniform in size can be obtained by gel filtration. Liposomes have been prepared from mixtures of microbial lipids, and used to study passage of solute molecules across the bilayer. The main disadvantage of liposomes is that proteins cannot be incorporated into the bilayer, which makes the data obtained using these structures of questionable value from a physiological standpoint.

5.1.2 CHARACTERISTICS OF SOLUTE-TRANSPORT PROCESSES

Types of transport process

A solute molecule can pass across a lipid-protein
membrane only if a driving force acts on the molecule
and there exists some means for the molecule to pass
across the membrane. Molecules may pass across membranes
by *passive diffusion* when the driving force is the
difference in concentration (with non-electrolytes),
or electrical potential (with ions), across the membrane.
Several observations, particularly the temperature
dependence of solute-transport processes, indicate that,
with the exception of water, few if any compounds pass
across the microbial plasma membrane by passive diffusion.

Most solutes which pass across the microbial membrane
do so as a result of the operation of specific *carrier* or
transport mechanisms. These are envisaged as being made
up of carrier or 'ferryboat' macromolecules which are
situated in the membrane, and which pick up solute
molecules and effect their transport to the opposite side
of the membrane.

There are three types of carrier-mediated transport
process. The simplest is *facilitated diffusion* in which
the only driving force is the difference in concentration
of the solute across the membrane. Transport is then a
'downhill' process. The membrane-bound carrier involved
in facilitated diffusion affects only the rate of movement
and not the final equilibrium. When metabolic energy is
coupled to facilitated diffusion, with the result that
solute is accumulated by a micro-organism, the process
is known as *active transport*. During active transport,
which is an 'uphill' process, the solute molecules remain
chemically unaltered during their passage from one side
of the membrane to the other. Expenditure of energy
during a transport process is usually demonstrated by
showing that uncoupling agents, (such as 2,4-dinitrophenol)
and reagents that prevent ATP synthesis, cause a decrease
in the rate of uptake. In some processes, such as
β-galactoside uptake by *Escherichia coli,* the energy-
coupling reactions can be dissociated from facilitated
diffusion, and each studied separately. The third type
of process is called *group translocation*. It differs
from both facilitated diffusion and active transport in
that the solute is modified chemically during the
transport process. Essentially, the carrier molecules

behave like enzymes in that they catalyse group-transfer reactions using the solute as substrate.

The process of solute accumulation can be regarded as being made up of an entrance process, and a process of exit or *efflux* which increases in velocity with the accumulation of intracellular solute. When the accumulation process has reached a steady state, the velocities of entrance and efflux are equal. Efflux usually involves downhill transport of solutes and, mainly because of its temperature dependence, it is generally considered to be a carrier-mediated process. Whether the carriers involved in efflux of a solute are the same as those that effect entry is not yet known. Several workers have measured the rate of efflux of solutes by adding radioactively labelled solute to suspensions of micro-organisms in which there is a steady-state concentration of accumulated non-radioactive solute. Since, under these conditions, the rate of efflux of solute from the organisms equals the rate of entrance, the rate at which the labelled solute is taken up will be a measure of the rate of efflux. It should be stressed that these experiments measure the rate of efflux under conditions in which the concentration of intracellular solute is very much higher than the concentrations attained during uptake of nutrients by growing organisms.

The physiological significance of carrier-mediated transport processes rests not only on the marked selectivity of these processes, but on the fact that they usually constitute the rate-limiting step in the metabolism of the available carbon and energy sources. This is shown by the finding that *E. coli,* on treatment with toluene to break the osmotic barrier, hydrolyses *o*-nitrophenyl-β-galactoside 15-20 times faster than organisms which have not been treated with toluene and in which, therefore, entry of the galactoside depends on a transport process. Since an increased rate of accumulation of a metabolisable sugar (or other carbon source) can increase the extent of catabolite repression of enzyme synthesis (*see* page 348), the rate of sugar transport may profoundly affect the metabolism of the entire organism.

Transport proteins

A series of very elegant experiments carried out by
Howard Rickenberg, George Cohen and their colleagues at
the Institut Pasteur in Paris, France, in the 1950s
provided the first evidence for the operation of a
carrier in the transport of β-galactosides into *E. coli*.
Their observations on the genetic control, inducibility
and specificity of the carrier led them to conclude that
it is a protein. At the same time, they coined the term
permease to describe transport proteins in general, a
term that has been subjected to some criticism but is
still nevertheless widely used. Since then, a great
deal of additional evidence has been obtained showing
that proteins are involved in the transport of solutes
across the microbial plasma membrane. Most of the data
are for bacteria; the properties of transport proteins
in eukaryotic micro-organisms are not so well understood.
  Indirect evidence for the protein nature of carriers
involved in transport came originally from the
observation that their synthesis could be induced and
repressed in the same way as synthesis of certain enzymes.
Synthesis of carriers can also be prevented by
concentrations of chloramphenicol that inhibit protein
synthesis in bacteria. *Para*-Fluorophenylalanine, which
is incorporated almost exclusively into proteins, also
prevents synthesis of carrier molecules in bacteria.
  Direct and more compelling evidence for the protein
nature of carriers has come as a result of the isolation
of certain carriers from bacteria. One of the first of
these proteins to be isolated was that which mediates
transport of β-galactosides into *E. coli*. This protein
is the product of the *y* gene in the *lac* operon in this
bacterium (*see* page 352), and it was isolated using a
rather ingenious method by Eugene Kennedy and his
colleagues at Harvard University in Boston, U.S.A.
They exploited the finding that the inhibitory effect
of N-ethylmaleimide on the transport protein can
be overcome by the compound thiodigalactoside which binds
very firmly to the solute-binding site on the protein.
Cells of, or isolated membranes from, *E. coli* in which
synthesis of the β-galactoside transport system had been
induced were treated with non-radioactive N-ethylmaleimide
in the presence of thiodigalactoside. N-Ethylmaleimide
forms a covalent linkage with sulphydryl groups and is
highly specific. This treatment therefore blocks all
sulphydryl groups on molecules in the cell or membrane

except those protected by thiodigalactoside. The cells
or membranes are then washed free of N-ethylmaleimide and
thiodigalactoside, and treated with radioactive
N-ethylmaleimide. As a result, the β-galactoside-transport
protein becomes uniquely labelled, which makes its
isolation by conventional protein isolation procedures
relatively easy.

However, most of the transport proteins which have been
isolated have been extracted from bacteria not by
conventional procedures, but using the technique of
osmotic shock. This technique, which was developed by
Leon Heppel at the National Institutes of Health in
Bethesda, Maryland, U.S.A., involves suspending cells in
a buffered hypertonic solution of sucrose containing
EDTA. The cells are then recovered by centrifugation or
filtration, and rapidly suspended in an ice-cold dilute
solution of magnesium chloride. When this suspension is
centrifuged, the supernatant liquid contains a number of
transport proteins. As yet, a reasonable physiological
explanation for the release of transport proteins by
osmotic shock has not been forthcoming. That the proteins
released are transport proteins was inferred by the
discovery that bacteria which had been subjected to
osmotic shock were unable to take up several compounds,
and that the affinity of the released proteins for various
solutes (as judged by measurements of apparent K_m values)
was of the same order as that of untreated bacteria.

A number of transport proteins, released by osmotic
shock, have been purified to homogeneity, and their
properties studied. They include proteins from *E. coli*
which bind arabinose, glucose, leucine together with
isoleucine and valine, and phosphate. Other transport
proteins have been examined from *Salmonella typhimurium*;
they include proteins which bind histidine and sulphate.
Transport proteins released by osmotic shock usually have
molecular weights in the range 30 000-50 000 daltons.
Estimates vary, but there are usually 1000-2000 molecules
of each type of transport protein in any one bacterium.

Kinetic considerations

The pioneer work at the Institut Pasteur in Paris, France,
on β-galactoside uptake by *E. coli* revealed the
catalytical properties of the transport process (and
hence prompted use of the term 'permease') and that the
process shows saturation kinetics in that the rate of

transport ceases to increase when the concentration of solute in the suspension is raised above a certain value. This similarity to an enzyme-catalysed reaction encouraged microbial physiologists to use equations derived during studies on enzyme kinetics to describe carrier-mediated transport processes. The most popular kinetic parameter derived for transport processes is the Michaelis constant, or K_m value. Values for K_m - or strictly speaking apparent K_m - are determined by plotting the reciprocal of the rate of solute accumulation (or efflux) against the reciprocal of the concentration of solute in the suspension. While apparent K_m values for transport processes are widely quoted, some physiologists have reservations about using these values because the concentration **dependence** of the uptake rate may not always follow the simple rectangular hyperbola characteristic of enzyme-catalysed reactions. For this reason, they prefer to quote a value (called the $S_{0.5}$ value) for the solute concentration that will give one half of the maximum rate of uptake.

Values for the apparent K_m and $S_{0.5}$ have been reported for accumulation of many different solutes by micro-organisms. Accumulation of a particular solute shows different values in various micro-organisms when measured under different conditions. For example, the K_m value for accumulation of glucose by *Sacch. cerevisiae* grown aerobically was reported to be 17.4 mM, whereas the value was 6.7 mM when the yeast was grown anaerobically. These data indicate that the affinity of glucose for the transport protein decreases when the yeast is grown aerobically rather than anaerobically, and are relevant in connection with the biochemical basis of the Pasteur effect (*see* page 343). Since biotin is usually present in the environment in very low concentrations, it is not surprising to find that the K_m value for uptake of this growth factor by *Lactobacillus plantarum* is as low as 3.2×10^{-8} M. As micro-organisms can often accumulate high concentrations of a solute, the K_m value for the entry process is assumed to be very much smaller than the value for efflux of the solute, i.e. that the affinity of the solute for the transport protein is greater during entry than during efflux.

Specificity

Microbial solute-transport processes show a wide range
of specificities. On the whole, proteins involved in
transport of ions are more specific in their action than
those which effect transport of organic compounds. With
transport of sugars and amino acids, the situation is
further complicated by the existence in many microbes of
more than one transport system for a particular solute.
E. coli, for instance, synthesises a transport protein
which is specific for leucine, isoleucine and valine.
In addition, this bacterium produces another transport
protein which is specific for just leucine, while
certain strains of *E. coli* can synthesise isoleucine-
specific and valine-specific transport proteins.
Likewise, *S. typhimurium* has four different systems for
uptake of histidine. One has a very high affinity for
the amino acid (with an apparent K_m value of 3×10^{-8} M),
while the others show a lower affinity. One of the low-
affinity group of histidine carriers is also the aromatic
amino-acid transport protein for this bacterium.
   A similar situation exists with the specificity of
transport proteins in *Sacch. cerevisiae.* This yeast has
at least three different transport proteins with affinity
for glucose. Moreover, it synthesises transport proteins
that are specific for, respectively, arginine, lysine,
methionine, histidine and for dicarboxylic amino acids.
In addition, the yeast synthesises a transport protein
that is capable of effecting transport of most if not
all amino acids. The significance of this wide range of
specificities for transport of organic compounds into
micro-organisms is quite unknown.

Effect of environmental factors

Reference has already been made to the effect of some
environmental factors on solute-transport processes in
micro-organisms, including the ability of certain solutes
to induce synthesis of the proteins required for their
transport, and the capacity of uncoupling agents to
decrease the rate of active transport. Compounds which
are related structurally to a solute can also affect the
rate of transport of a solute, the magnitude of the effect
depending on the concentration of the compound. For
example, the rate of accumulation of L-arginine by *Sacch.
cerevisiae* is decreased in the presence of L-canavanine,
L-lysine or D-arginine.

A number of metabolic poisons can also affect the operation of solute-transport processes in micro-organisms. In general, any reagent which reacts chemically with either protein or lipid in the membrane decreases the rate of solute uptake. Solute transport is usually inhibited by thiol reagents (such as *p*-chloromercuri-benzoate) in concentrations around mM, which indicates that free thiol groups are necessary for action of transport proteins. Heavy-metal ions also decrease the rate of uptake of many solutes by micro-organisms; uranyl ions completely prevent transport of glucose by *Sacch. cerevisiae*, while nickelous ions are only 75% effective. These heavy-metal ions, unlike copper, mercury and silver ions, do not penetrate the organisms and are thought to react with phospholipids in the plasma membrane.

The pH value of the suspending liquid influences the rate of solute uptake since H^+ and OH^- ions have free access to the outer surface of the membrane. Relatively few studies have been made on the effect of pH value on the action of individual solute-transport processes. It has been shown, however, that pH value can affect the ability of *Sacch. cerevisiae* transport proteins to discriminate between K^+ and Na^+; at low pH values, uptake of K^+ is preferred.

Solute uptake and efflux in micro-organisms are extremely sensitive to temperature. In the temperature range between the optimum for growth and $20^\circ C$, the temperature coefficient (Q_{10}) values for uptake and efflux of solutes are usually around 4-5, and in the range $15-5^\circ C$ may be as high as 10-12; these values are very much higher than those reported for metabolic processes and *in vitro* enzyme reactions in this temperature range.

5.1.3 MOLECULAR MECHANISMS OF SOLUTE TRANSPORT

The most intriguing, and at the same time the most puzzling, aspect of solute transport is the mechanism by which a solute molecule, once it has combined with a transport protein, is transferred from one side of the plasma membrane to the other, a distance of about 7.5 nm.

Evidence for a role for transport proteins in the passage of molecules across a membrane is now extensive, and all currently acceptable theories on the mechanism of solute transport invoke their activity. Membrane lipids are thought to anchor transport and related proteins in the membrane. Nevertheless, it is known that the

activities of certain proteins involved in solute
transport requires the presence of particular lipids.
There have been suggestions that membrane lipids are
involved metabolically in transport processes, although
this view is not now generally held. It was suggested,
for example, that lysylphosphatidylglycerol is involved
in the passage of lysine across the plasma membrane in
Staphylococcus aureus, but experimental evidence to
support the contention was not forthcoming.

Most students of solute-transport processes would
probably agree that, during facilitated diffusion, the
thermal energy and molecular deformation that result from
binding can account for the small amount of motion required
to effect downhill transport of the solute molecule. The
main efforts of biologists interested in solute transport
in recent years have been to explain how solutes pass
across a membrane by active transport and group translocation.

Active transport

Three main theories have been put forward to explain the
molecular basis of active transport of solutes in micro-
organisms.

Carrier conformational change. Probably the simplest
molecular explanation of active transport of solutes in
micro-organisms is that metabolic energy is required to
lower the affinity of the transport protein for the
solute at the inner membrane surface, conceivably by a
mechanism akin to the interactions between subunits in
allosteric proteins. A simplified schematic
representation of the carrier-conformational change
hypothesis is shown in *Figure 5.1.* Although experimental
evidence in support of the hypothesis is meagre it has
its devotees, notable among whom is Paul D. Boyer of the
University of California in Los Angeles, U.S.A.

Dehydrogenase-coupled active transport. The most
significant advance yet to be made as a result of
experiments on bacterial membrane vesicles has been the
discovery of transport of amino acids and sugars mediated
by the action of certain dehydrogenases. Experiments in
Kaback's laboratory established that addition of D-lactate

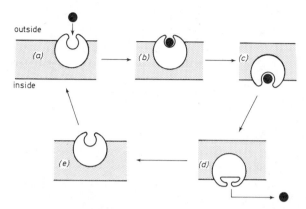

Figure 5.1 A diagram showing the sequence of molecular events which constitute the carrier-conformational change hypothesis for active transport of solutes. A solute molecule approaches a transport protein (a) and is bound to that protein (b). The protein-solute complex then moves to the opposite side of the membrane (c) where the protein undergoes a conformational change (d) which leads to release of the solute molecule. The transport protein then moves back to the opposite side of the membrane, where it reverts to its original conformation (e)

to a suspension of vesicles prepared from *E. coli* dramatically stimulates uptake of proline. Succinate can replace D-lactate but is less efficient. D-Lactate is converted by the bacterial membranes to pyruvate in a reaction catalysed by D-lactate dehydrogenase, and proline uptake is linked to the transfer of electrons in the reaction. Succinate is converted to fumarate through the activity of succinate dehydrogenase. Subsequent experiments provided evidence for uptake of several other amino acids and sugars by vesicles, each coupled to the activity of lactate dehydrogenase. Vesicles have since been prepared from several other bacteria, and their ability to transport amino acids and sugars demonstrated in the presence of D-lactate. Using vesicles prepared from membranes of *Staphylococcus aureus*, it has also been shown that uptake of amino acids is stimulated by dehydrogenation of L-α-glycerophosphate to give dihydroxyacetone phosphate.

Considerable thought has been given, mainly in Kaback's laboratory, to the mechanism of dehydrogenase-coupled

active transport. It is important to realise that
neither the generation nor utilisation of high-energy
phosphate bonds is involved in the process, which
operates solely as a result of electron transfer.
A model which summarises the present view of the mechanism
of dehydrogenase-coupled active transport is shown in
Figure 5.2. The essence of the model is that the carrier

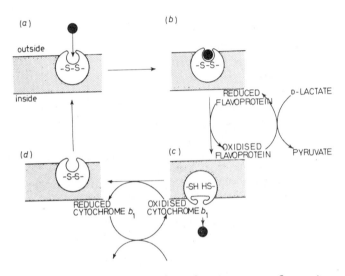

*Figure 5.2 A diagram showing the sequence of events
during transport of a solute molecule across a membrane
by a process linked to oxidation of D-lactate*

or transport proteins undergo cyclic reduction and
re-oxidation in the form of sulphydryl–disulphide
interconversions, and that only the oxidised form of the
carrier has a high affinity for the solute. Active
uptake of an amino acid or sugar is associated with
reduction of the appropriate carrier by D-lactate
dehydrogenase via a flavoprotein. Reduction of the
oxidised form of the carrier is accompanied by a
conformational change in the carrier molecule that
results in translocation of the bound solute from the
outer to the inner surface of the membrane. The reduced
or low-affinity form of the carrier releases the solute
and is itself then re-oxidised, probably by a part of the
electron-transport chain from cytochrome b_1 and beyond, to
form the oxidised form.

The discovery of dehydrogenase-linked active transport has been a major achievement in the advance towards an understanding of the molecular mechanism of active transport in micro-organisms. However, evidence has yet to be forthcoming to show unequivocally that the process operates in intact bacteria. It would also be interesting to know whether there is a similar process operating in eukaryotic micro-organisms.

Ion-gradient-mediated active transport. The British biochemist, Peter Mitchell, suggested many years ago that the link between solute transport and metabolism is effected by a gradient of protons. Mitchell has continued to advocate this view, and in recent years he has been eloquently supported by Franklin Harold working in Denver, Colorado, U.S.A. The gist of this hypothesis, the salient points of which are depicted diagrammatically in *Figure 5.3,* is that the activity of the respiratory chain (*Figure 5.3a*), or of a Mg^{2+}-dependent ATPase (*Figure 5.3b*), or in anaerobically growing organisms oxidation-reduction of pairs of substrates (e.g. by the Stickland reaction; *see* page 215) through an oxidation chain arranged across the membrane, leads to the extrusion of protons, and hence to the generation of a difference in pH value and of electrical potential across the membrane. This asymmetric distribution of protons gives rise to a proton-motive force contributed by the difference in pH value and the difference in electrical potential across the membrane. The difference in pH value across a bacterial membrane may be of the order of one unit (the interior being alkaline), with a membrane potential of around -150 mV. The influx of protons made possible by the formation of a gradient of pH value and electrical potential is thought to be accompanied, on an appropriate carrier, by a simultaneous influx of neutral molecules (such as sugars and amino acids; *Figure 5.3c*) or of anions (such as phosphate; *Figure 5.3d*) each riding on the same carrier as the inwardly flowing protons. The influx of neutral molecules and anions is referred to as *symport*. Experimental evidence for a proton-solute symport is not extensive, but it has been shown to be involved in lactose uptake in *E. coli* and in uptake of amino acids by *Saccharomyces carlsbergensis*. The existence of a

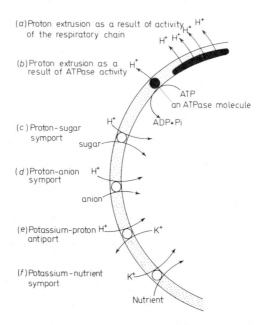

(a) Proton extrusion as a result of activity of the respiratory chain

(b) Proton extrusion as a result of ATPase activity

ATP
an ATPase molecule
ADP+Pi

(c) Proton-sugar symport

sugar

(d) Proton-anion symport

anion

(e) Potassium-proton antiport

(f) Potassium-nutrient symport

Nutrient

Figure 5.3   The ways in which circulation of protons and cations across the microbial plasma membrane may be coupled to transport of solutes

steady-state electrical potential across the plasma membrane would cause external cations (such as K^+) to accumulate in cells, which in turn would tend to collapse the potential difference. In order that a significant potential difference may be maintained across the plasma membrane, it is postulated that *antiporters* exist in the membrane with the task of translocating cations, such as K^+, outwards (*Figure 5.3e* and *f*).

Group translocation

The discovery that some sugars are accumulated by bacteria not in the free form but as phosphorylated derivatives led to the discovery in Saul Roseman's laboratory in Johns Hopkins University in Baltimore, Maryland, U.S.A. of a transport process now referred to as *group translocation* or *vectorial phosphorylation*. The process operates during

sugar transport in many bacteria including *E. coli,*
S. typhimurium, Staph. aureus, Staph. lactis, Aerobacter
aerogenes, Bacillus subtilis and *Lactobacillus plantarum;*
it has been studied most intensively in the first three
of these bacteria. Group translocation is not thought
to be involved in amino-acid uptake but it has been
implicated recently in uptake of adenine and possibly
other purines by *E. coli*. Group translocation of sugars
has also been demonstrated using bacterial membrane
vesicles. Although some yeasts are known to accumulate
certain sugars as phosphorylated derivatives, it has yet
to be shown that this involves vectorial phosphorylation
similar to that which operates in bacteria.

The overall equation for vectorial phosphorylation is:

$$\text{sugar + phosphoenolpyruvate} \xrightarrow{Mg^{2+}} \text{sugar-phosphate + pyruvate}$$

The reaction is specific for phosphoenolpyruvate (PEP)
although many different sugars can serve as the acceptor.
For this reason, the system is often referred to as the
phosphoenolpyruvate-glycose phosphotransferase system,
or *PEP transferase* system for short. Most of the
monosaccharide phosphates are the 6-phosphate esters.
The one exception is fructose which is transported into
E. coli as fructose 1-phosphate.

The PEP transferase system is quite complicated.
Basically, it consists of the result of two separate
reactions. In the first reaction, catalysed by Enzyme I,
a phosphoryl group is transferred from phosphoenolpyruvate
to a relatively heat-stable protein (HPr). The phosphoryl
group is attached to N-1 in a histidine residue in the
protein. The equation for the reaction is:

$$\text{phosphoenolpyruvate + HPr} \xrightarrow[\text{Enzyme I}]{Mg^{2+}} \text{phospho-HPr + pyruvate}$$

A number of HPr proteins have been examined in detail.
All have molecular weights around 9000 daltons. Those
from *E. coli* and *S. typhimurium* have similar amino-acid
compositions, but they differ in this property from the
HPr protein of *Staph. aureus*. Neither Enzyme I nor HPr
protein is bound to the plasma membrane, and both would
appear to be cytosol constituents.

Whereas the mechanism of phosphorylation of HPr
proteins in reactions catalysed by Enzyme I is the same
or very similar in all bacteria, the second reaction of
the PEP transferase system:

$$\text{phospho-HPr + sugar} \xrightarrow[\text{Factor III}]{\text{Enzyme II}} \text{sugar-phosphate + HPr}$$

varies in different bacteria and for individual sugars. Unlike Enzyme I, Enzymes II are firmly bound to the plasma membrane. Several of the Enzymes II involved in sugar transport into *E. coli* have been separated into two protein fractions, designated Enzyme II-A and Enzyme II-B. As a result of reconstitution experiments, it has been found that each II-A protein is specific for a particular sugar, and is designated accordingly by a superscript (e.g. Enzyme II-A^{glu}). On the other hand, the II-B protein, which has a molecular weight of around 36 000 daltons, is a general component involved in transport of glucose, fructose and mannose. It can account for up to 10% of the membrane protein. In *E. coli*, the Enzymes II-A require for activity phosphatidylglycerol, which is quantitatively a minor component of the plasma membrane in this bacterium.

The second of the reactions that go to make up the group translocation system for active transport of lactose in *Staph. aureus* has also been studied in detail, and it differs from the corresponding reaction for monosaccharide transport in *E. coli*. The principal difference is that, in addition to Enzyme II^{lac}, which is membrane-bound, the reaction also involves a soluble protein, Factor III^{lac}. The manner in which this factor is involved in the reaction is shown in the following equations:

$$3\,\text{phospho-HPr + Factor III}^{lac} \rightleftharpoons \text{triphospho-Factor III}^{lac} + 3\,\text{HPr}$$

$$\text{triphospho-Factor III}^{lac} + 3\,\text{lactose} \xrightarrow{\text{Enzyme II}^{lac}} 3\,\text{lactose phosphate + Factor III}^{lac}$$

The binding of phosphoryl groups to Factor III^{lac} would also appear to involve histidine residues in the protein. It will be interesting to see whether research on group translocation processes in other bacteria will lead to the discovery of still greater complexity in the processes, or whether some order will emerge, such as, for instance, a functional analogy between Enzyme II-A in *E. coli* and Factor III in *Staph. aureus*.

The fact that it is now possible to discuss molecular mechanisms involved in active transport and group translocation across the microbial plasma membrane represents a spectacular advance in our understanding of solute transport in micro-organisms. A decade ago such a discussion would not have been possible. At the same time, much remains to be discovered about the molecular basis of solute transport in microbes. Entirely new mechanisms of solute transport may well be discovered. Also, the microbial physiologist wants to know whether the various phospholipids and sterols in the microbial plasma membrane are involved in solute transport by being mandatorily associated with the activity of a transport protein, or whether they merely provide an appropriate menstrum in which the proteins are embedded. Few microbial physiologists would demur when it is suggested that the molecular events at the plasma membrane constitute probably their greatest challenge in the forthcoming decade.

5.2   ENDOCYTOSIS

Many protozoa, particularly amoebae, can transport solutes and particles across the plasma membrane by a process in which vesicles are pinched off the membrane after the formation of a small invagination (*Figure 5.4*). The vesicles and their contents pass into the cell and are digested by intracellular enzymes. The process is often known by the blanket term *endocytosis,* although some workers prefer to use the terms *pinocytosis* (Gr. 'drinking by cells') for uptake of liquids and *phagocytosis* (Gr. 'eating by cells') for uptake of solid particles. Compared with solute uptake by carrier-mediated transport processes, endocytosis is a relatively non-specific process. The process is, in fact, often studied using small (0.1 μm diameter) polystyrene latex spheres.

Very little indeed is known about the molecular events involved in endocytosis. The main unanswered questions concern the ways in which the changes in membrane behaviour that take place during endocytosis are triggered off, and whether or not the process involves growth or

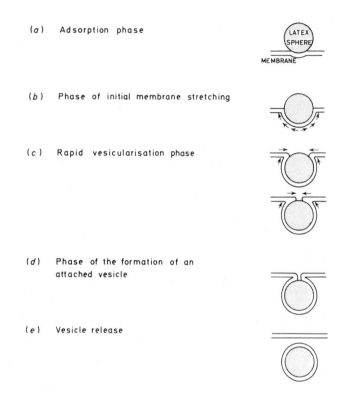

(*a*)   Adsorption phase

(*b*)   Phase of initial membrane stretching

(*c*)   Rapid vesicularisation phase

(*d*)   Phase of the formation of an attached vesicle

(*e*)   Vesicle release

Figure 5.4   The proposed phases during endocytosis of a polystyrene latex sphere at a microbial plasma membrane

just stretching of the plasma membrane. The speed at which endocytosis takes place argues in favour of a stretching process rather than membrane growth. It has been calculated that it may be necessary for a localised area of plasma membrane to stretch to an area four times greater than that of the original membrane. Unfortunately, nothing is known about the molecular events that take place during membrane stretching.

FURTHER READING

SOLUTE-TRANSPORT PROCESSES

HAMILTON, W.A. (1975). Energy coupling in microbial transport. *Advances in Microbial Physiology*, 12, 1-53

HAROLD, F.M. (1972). Conservation and transformation of energy by bacterial membranes. *Bacteriological Reviews*, 36, 172-230

HENDERSON, P.J.F. (1971). Ion transport by energy-conserving biological membranes. *Annual Review of Microbiology*, 25, 393-428

KABACK, H.R. and HONG, J.S. (1973). Membranes and transport. *Critical Reviews in Microbiology*, 2, 333-376

KOCH, A.L. (1972). The adaptive responses of *Escherichia coli* to a feast and famine existence. *Advances in Microbial Physiology*, 6, 147-217

SIMONI, R.D. (1972). Macromolecular characterization of bacterial transport systems. In: *Membrane Molecular Biology*, Eds. C.F. Fox and A.D. Keith, pp.289-322. Stamford, Connecticut; Simon Associates Inc.

ENDOCYTOSIS

HOLTER, H. (1965). Passage of particles and macro-molecules through cell membranes. *Symposium of the Society for General Microbiology*, 15, 89-114

6

ENERGY-YIELDING METABOLISM

All of the free energy reaching this planet originates
in the nuclear fusion reactions that take place in the
Sun. These reactions emit hard gamma radiation which,
after extensive absorption and re-emission, is radiated
from the surface of the Sun. It is from this reservoir
of free energy that all living organisms (with the
exception of the chemolithotrophic bacteria) directly or
indirectly draw in order to exist.

6.1 PRINCIPLES OF BIOENERGETICS

The account of the molecular architecture of micro-
organisms given in Chapter 2 shows how exquisitely
ordered and structured a microbial cell is. All the
same, we know that there is a perpetual tendency towards
randomness and disorder in the universe, or in
thermodynamical terms an increase in entropy. The
famous Israeli biophysicist, the late Aharon Katchalsky,
once stated that life is a constant struggle against the
tendency to produce entropy. To understand how a
microbial cell - indeed any cell - can exist and multiply
in the face of this tendency, we need to look more
closely at the laws which govern the ways in which
chemical reactions proceed.

Not all chemical reactions proceed spontaneously, and
whether a reaction will or will not so proceed depends
on the value for the *standard free energy change* or ΔG°
value for the reaction. The ΔG° value for any reaction

is equal to the difference in the standard free energies of the reactants and the standard free energies of the products, due account being taken of the stoicheometry of the equation:

$$\Delta G^{\circ} \; = \; \Sigma G^{\circ}_{\text{products}} \; - \; \Sigma G^{\circ}_{\text{reactants}}$$

The standard free energy of a chemical compound is equal to the total amount of energy it can yield on complete decomposition. In bioenergetics, pH 7.0 is designated as the reference state rather than a pH value of zero as is the practice in physical chemistry. The standard free energy change for a reaction, at pH 7.0, indicated by the ΔG°' value, can be calculated using the equation:

$$\Delta G^{\circ}{}' \; = \; - \; RT \; \ln K'_{eq}$$

where K'_{eq} is equal to the thermodynamic equilibrium constant for the reaction. Values for K'_{eq} can be calculated by determining the activities of the reaction components at equilibrium. Thus, for the reaction:

$$A + B \rightleftharpoons C + D$$

the thermodynamic equilibrium constant equals:

$$K'_{eq} \; = \; \frac{(C)\,(D)}{(A)\,(B)}$$

where (A), (B), (C) and (D) are the activities at equilibrium of, respectively, A, B, C and D. By determining the equilibrium constant for a reaction, it is thus possible to calculate the ΔG°' for the reaction. Reactions which have a negative ΔG°' value proceed spontaneously in the direction in which they are written when all of the components are at a concentration of one molar. These are termed *exergonic reactions*. When the ΔG°' value for a reaction is positive, the reaction does not proceed spontaneously in the direction written, under the same conditions; these are called *endergonic reactions*.

Biochemists have shown many times that living cells, including micro-organisms, carry out reactions which, it can be calculated, have a large positive ΔG°' value at pH 7.0. This is particularly true of many reactions on biosynthetic pathways. This seemingly paradoxical situation can be resolved by either or both of two explanations. Firstly, the reactions may take place in the microbial cell under conditions of reactant concentration and pH value such that the actual free energy change is negative, although the calculated value is positive. Secondly, the reaction that occurs in the cell may be a modified form of that which is thought to take place, with the modified form having a negative rather than a positive ΔG^{o}' value. There is a wealth of evidence in favour of the second explanation, although the possibility that some reactions take place in the cell for the reasons adduced for the first explanation cannot be dismissed.

Living cells can carry out a seemingly endergonic reaction by coupling it with an exergonic reaction, the ΔG^{o}' value for which, under the same conditions, is more negative than is the value for the endergonic reaction positive. The coupling is possible because, as already stated, the ΔG^{o}' value for a reaction is the difference between the sum of the standard free energies of the products and the reactants. Although the number of compounds that are reactants in the endergonic component of a coupled reaction is very large indeed, the reactants that form the exergonic component are relatively small in number. The main ones are nucleoside 5'-triphosphates (such as ATP, GTP and UTP). Also included are acyl phosphates, such as acetyl phosphate and phosphoenol-pyruvate, and acyl thio-esters such as acetyl-CoA. Despite the protestations of some thermodynamicists, these have come to be known as *high-energy compounds*. Their distinctive property is that they all have relatively high ΔG^{o}' values for hydrolysis. The data in *Table 6.1* show that these ΔG^{o}' values are higher than those for hydrolysis of esters such as glycerol 1-phosphate and glucose 1-phosphate, which are known as *low-energy compounds*.

The most important of the high-energy compounds as far as cellular metabolism is concerned is ATP. The ATP molecule is uniquely designed for its role in energy metabolism. In the intact cell, the molecule is highly charged; at pH 7.0 each of the phosphate groups is completely ionised. ATP exists in the cell mainly as a complex with Mg^{2+}. The biochemically utilisable energy in the molecule is stored in the two terminal high-energy phosphate bonds (designated by the symbol $\sim$). The energy of each of these bonds is 7.3 kcal. The prime importance of ATP may seem at first sight strange, since the ΔG^{o}' value for hydrolysis of ATP, to yield ADP and inorganic phosphate, is well below the values for hydrolysis of

G

Table 6.1   Standard free energy values of hydrolysis ($\Delta G^{\circ}{}'$) at pH 7.0 for some biologically important compounds

Compound	$\Delta G^{\circ}{}'$ Value (kcal per mole)
Phosphoenolpyruvate	-14.8
1,3-Diphosphoglycerate	-11.8
Acetyl phosphate	-10.1
Acetyl-CoA	-7.7
Adenosine triphosphate (→ ADP + Pi)	-7.3
Glucose 1-phosphate	-5.0
Glucose 6-phosphate	-3.3
Glycerol 1-phosphate	-2.2

phosphoenolpyruvate and 1,3-diphosphoglycerate among other compounds. The unique property of ATP is that the $\Delta G^{\circ}{}'$ value for its hydrolysis is in the middle of the scale shown in *Table 6.1*. It can therefore function as a carrier of phosphate groups, which originate from compounds higher on the scale than ATP (such as phosphoenolpyruvate), to acceptor molecules that form low-energy phosphate compounds that are below ATP on the scale.

The way in which ATP, with its high negative $\Delta G^{\circ}{}'$ value of hydrolysis, can be coupled to an endergonic reaction to convert it into an exergonic one is well illustrated by the first reaction of the Embden-Meyerhof-Parnas scheme, namely that leading to the formation of glucose 6-phosphate from glucose (*see* page 189). The reaction is:

glucose + Pi ⟶ glucose 6-phosphate + H_2O + 3.30 kcal

Since the reaction is endergonic, it cannot proceed from left to right spontaneously. However, when coupled with the reaction:

$$ATP \longrightarrow ADP + Pi - 7.3 \text{ kcal}$$

which is exergonic, to give the reaction:

$$glucose + ATP \longrightarrow glucose \ 6\text{-phosphate} + ADP$$

phosphorylation of glucose to give glucose 6-phosphate does proceed spontaneously because the $\Delta G^{o\prime}$ value of the coupled reaction is -4.00 kcal (i.e. +3.30 + [-7.3 kcal]). It must be emphasised that, although almost all reactions that take place during microbial metabolism are catalysed by enzymes, the $\Delta G^{o\prime}$ value for the reaction is in no way affected by the action of the enzyme.

Because of its ability to react with other high-energy compounds in the cell, ATP has aptly been termed the 'currency' of the energy metabolism of cells. The sum of the concentrations of AMP, ADP and ATP in a micro-organism is around 10-20 μmoles per gram dry weight, the concentration of ATP usually exceeding those of AMP and ADP combined. The microbial cell can conveniently be considered as a battery, in that the status of the cell is a measure of the extent to which the ATP-ADP-AMP system is 'filled' with high-energy phosphate bonds. Daniel Atkinson, working in the University of California in Los Angeles, U.S.A., was the first to express this quantitatively in the form of the *energy charge* or *adenylate charge* of a micro-organism, which can be calculated using the formula:

$$\text{Energy charge} = \frac{0.5 \ [ADP] + 2[ATP]}{[AMP] + [ADP] + [ATP]}$$

Using this formula, values for the energy charge of a micro-organism fall in the range 0-1.0. We shall return to a consideration of energy charge when we discuss regulation of energy production by micro-organisms (Chapter 8).

This chapter deals with ways in which micro-organisms obtain a supply of ATP from the great variety of energy sources which can be placed at their disposal. Although the metabolism associated with the channelling of the energy of chemical compounds or of visible radiation into

ATP is often extensive, basically there are just two
types of process involved. The first of these is termed
substrate-level phosphorylation, which involves reactions
in which ADP reacts with a high-energy compound that has
a $\Delta G^{\circ}{}'$ value for hydrolysis greater than that of ATP.
Mention has already been made of this property of the
ADP-ATP system which indeed makes ATP such an ideal
carrier of phosphate groups.

The second process is of much greater importance in
most micro-organisms, and is known as *oxidative
phosphorylation.* As the name implies, it is a process
in which synthesis of ATP from ADP is associated with
oxidation-reduction reactions. Any oxidation-reduction
reaction involves two pairs of compounds (called redox
couples) which differ in their affinity for electrons.
The redox couple that has the greater affinity for
electrons will be the oxidising couple, and that with
the smaller affinity will be the reducing couple. For
example, in the reaction:

$$\text{NAD}_{rd} + \text{flavoprotein}_{ox} \quad \text{NAD}_{ox} + \text{flavoprotein}_{rd}$$

where 'rd' signifies 'reduced' and 'ox' signifies 'oxidised',
the flavoprotein_{ox} - flavoprotein_{rd} couple has a
greater affinity for electrons than the NAD_{ox} - NAD_{rd}
couple, so that electrons are accepted by oxidised
flavoprotein which itself then becomes reduced.
The electron affinities of couples can be measured and
quoted as an electron-motive force or potential. In
bioenergetics, this potential is usually quoted as the
standard reduction potential or E_0' value. Redox couples
can therefore be arranged in a table showing the
increasing electron affinity (or decreasing electron
pressures) of the couples. *Figure 6.1* shows data for E_0'
values of couples of interest to the student of
bioenergetics. In any oxidation-reduction reaction
there is a decline in free energy which can be related
to the difference in the E_0' values of the participating
couples by the equation:

$$\Delta G^{\circ}{}' = -nF\Delta E_0'$$

where n is the number of electrons involved and F the
Faraday (23.04 kcal volt^{-1}). Providing that the E_0'
values of redox couples in biological systems differ by
about 170 mV or more, the $\Delta G^{\circ}{}'$ value of an oxidation-
reduction reaction can be of sufficient magnitude (i.e.

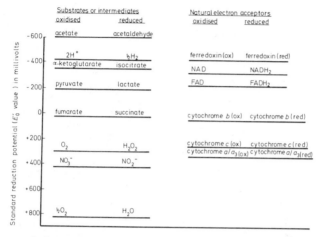

Figure 6.1   Standard reduction potentials of some redox couples

greater than 7.3 kcal) to lead to synthesis of ATP from ADP.

The coupling of ATP synthesis to the change in $\Delta G^{o\,\prime}$ values of oxidation-reduction reactions forms the basis of oxidative phosphorylation. The data in *Figure 6.1* indicate how certain biological redox couples could take part in oxidation-reduction reactions which, because the couples differ sufficiently in $E_0^\prime$ values, will have a change in Gibbs free energy sufficient to lead to synthesis of ATP from ADP. The redox couples listed in *Figure 6.1* are selected to illustrate the point because, as we shall see on page 202, they are the couples which are actually involved in many microbial oxidative phosphorylation chains.

6.2   ENERGY FROM ORGANIC COMPOUNDS

A wide range of oxidisable organic compounds can be used as energy sources by micro-organisms. The catabolism involved in the oxidation of organic compounds leads also to the formation of many different two-, three- and four-carbon compounds which, as we shall see in the next chapter, provide raw materials for biosynthetic reactions.

In any discussion on the energy-yielding metabolism of chemo-organotrophic micro-organisms, one soon encounters

terms *fermentation* and *respiration*. Unfortunately
se terms have come to have several different meanings,
and it is as well to state at this juncture the way in
which they are now used by microbial physiologists. The
term *fermentation* should be strictly reserved for those
energy-yielding pathways in which organic compounds act
as both the electron donor and the electron acceptor.
Some micro-organisms obtain energy from oxidations in
which inorganic compounds act as final electron acceptors,
and this process is referred to as *anaerobic respiration*
to distinguish it from *aerobic respiration* in which
oxygen acts as the final electron acceptor.

6.2.1 CARBOHYDRATES

Carbohydrates are frequently used as energy sources for
micro-organisms in laboratory media, so that it is not
surprising that the mechanisms by which these compounds
are made to yield energy have been closely studied.
Several different metabolic pathways for carbohydrate
breakdown are now known, and it will be seen in the
account that follows that pyruvate occupies a key
position in these pathways since it stands at an
intersection from which radiate several different
terminal pathways (*Figures 6.2 to 6.6*).

Pathways of carbohydrate breakdown

Despite the fact that research on catabolism of
carbohydrates by micro-organisms has been going on for
well over three-quarters of a century, new reactions and
schemes are still being discovered. In the following
pages, four major interconnected pathways of carbohydrate
breakdown by micro-organisms are described. It is
probably most instructive to see these four pathways as
the highways or motorways in carbohydrate breakdown.
The reactions which permit other carbohydrates and
related compounds to enter these major catabolism
highways are discussed on page 195, and these reactions
can be considered as forming link or approach roads in
the catabolism highway system.

Embden-Meyerhof-Parnas (EMP) pathway. Studies on carbohydrate breakdown by micro-organisms began in 1897 when the Buchners accidentally discovered that cell-free extracts of yeast can carry out an alcoholic fermentation of sucrose. During the first half of the present century, biochemists gradually unravelled the dozen or so reactions by which sugars are fermented to ethanol and carbon dioxide and, in honour of three of the more illustrious investigators, the series of reactions making up this glycolytic pathway has come to be known as the *Embden-Meyerhof-Parnas* (EMP) *scheme*. The discovery of the intermediates on this pathway was a notable contribution in the development of biochemistry in that the experimental

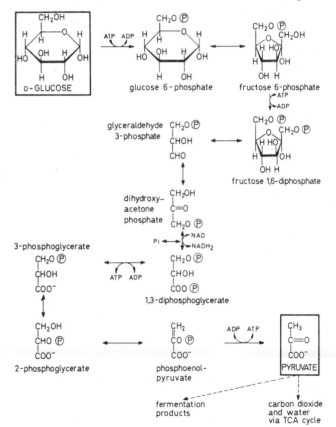

Figure 6.2   The Embden-Meyerhof-Parnas scheme

approach used set a pattern for subsequent studies on
other biochemical pathways. The reactions of the EMP
scheme as far as pyruvate are shown in *Figure 6.2*. With
three exceptions – the phosphorylation of glucose
catalysed by hexokinase, of fructose 6-phosphate catalysed
by phosphofructokinase and the conversion of
phosphoenolpyruvate to pyruvate catalysed by pyruvate
kinase – each of the reactions in the scheme is readily
reversible.

The characteristic reaction of the EMP scheme is the
splitting of fructose 1,6-diphosphate by the enzyme
aldolase to give a mixture of triose phosphates, which
explains why the scheme is sometimes referred to as the
hexose diphosphate scheme. Each of the triose phosphates
is then converted to pyruvate. During this conversion,
two molecules of ATP are formed from each triose phosphate
molecule, one during the dephosphorylation of 1,3-diphos-
phoglycerate and another during the conversion of
phosphoenolpyruvate into pyruvate. These are substrate-
level phosphorylations.

Since two molecules of ATP were used for priming the
glucose substrate to form fructose diphosphate, the
overall gain in the conversion of one molecule of glucose
to two molecules of pyruvate is two molecules of ATP.
The $\Delta G^{o\prime}$ value for the reaction:

$$\text{glucose} \longrightarrow 2 \text{ pyruvate}$$

is around -47 kcal. Assuming that synthesis of each
mole of ATP from a mole of ADP accounts for 7.3 kcal,
the EMP pathway is seen to be only just over 30%
efficient as an energy-yielding process.

Hexose monophosphate pathway. For many years, the EMP
scheme was the only known pathway for the breakdown of
carbohydrates by micro-organisms. However there had for
some time been a suspicion that other pathways must
exist, mainly because the EMP scheme does not explain
how ribose, which is required for RNA synthesis, can be
formed, nor does it explain how pentoses and other sugars
can be used as energy sources by micro-organisms. During
the 1940s and 1950s, work in a number of laboratories,
notably those of Otto Warburg in Berlin, W. Germany, and
Frank Dickens in London, and in the U.S.A. of Bernard
Horecker and Ephraim Racker, established the existence
of a pathway for carbohydrate breakdown which involves
the formation of hexose monophosphates and pentose

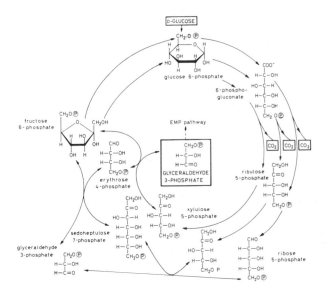

Figure 6.3 The hexose monophosphate pathway

phosphates, together with some reactions of the EMP
pathway. This pathway can therefore be viewed as a
'shunt' or 'loop' pathway of the EMP scheme. It is
usually referred to as the *hexose monophosphate* (HMP),
pentose phosphate or *Warburg-Dickens* pathway.

Figure 6.3 shows a cycle of reactions that make up one
version of the HMP pathway. The net result of these
reactions is the conversion of a molecule of glucose
6-phosphate to one of glyceraldehyde 3-phosphate and
three molecules of carbon dioxide. The glyceraldehyde
3-phosphate can then enter the EMP pathway, and as a
consequence there would be a net yield of one ATP
molecule from the entry of a glucose molecule into the
HMP pathway.

The first reaction on the pathway, that catalysed by
glucose 6-phosphate dehydrogenase, is:

glucose 6-phosphate + NADP $\rightleftharpoons$ 6-phosphogluconate + NADPH$_2$

This is followed by a second reaction which also yields
NADPH$_2$:

6-phosphogluconate + NAD $\rightleftharpoons$ ribose 5-phosphate + NADPH$_2$

Production of reducing power in these reactions is an important function of the HMP pathway in most microbes.

The characteristic reactions of the HMP pathway are those catalysed by *transaldolase* and *transketolase*. The first of these enzymes catalyses transfer of a dihydroxyacetone moiety, from sedoheptulose 7-phosphate or fructose 6-phosphate, to a suitable acceptor (glyceraldehyde 3-phosphate or erythrose 4-phosphate). Transketolase is also a transferase, and catalyses transfer of a ketol group. This enzyme catalyses two different reactions in the scheme shown in *Figure 6.3*. These are the reactions leading to production of sedoheptulose 7-phosphate and glyceraldehyde 3-phosphate, and of fructose 6-phosphate and glyceraldehyde 3-phosphate.

It is generally believed that the HMP glycolytic pathway is not a major energy-yielding pathway in most microbes, since it is only half as efficient as the EMP pathway, but that it is used principally to provide pentose phosphates for nucleotide synthesis and $NADPH_2$ as a source of reducing power.

Phosphoketolase pathways.   Some bacteria catabolise glucose using a pathway which involves the action of a *phosphoketolase*, an enzyme which cleaves acetyl phosphate from a phosphorylated C_5 or C_6 compound. This type of pathway was discovered in *Leuconostoc mesenteroides* which ferments glucose according to the following overall equation:

$$\text{glucose} \longrightarrow \text{lactate} + \text{ethanol} + CO_2$$

At first sight, it might appear that this bacterium catabolises glucose by the EMP pathway with some of the pyruvate being converted to ethanol and carbon dioxide, and the remainder being reduced to lactate. That this supposition is incorrect emerged when it was shown that the bacterium is unable to synthesise phosphofructokinase, aldolase, or triose phosphate isomerase. Further work revealed the reactions that go to make up the pathway by which *Leuconostoc mesenteroides* breaks down glucose, and these are shown in *Figure 6.4*. The key reaction on this pathway is that catalysed by the *phosphoketolase*, namely the cleavage of xylulose 5-phosphate to acetyl phosphate and glyceraldehyde 3-phosphate. As a means of synthesising ATP, the pathway is again less efficient than the EMP pathway, in that catabolism of one molecule of glucose leads to synthesis of only one molecule of ATP.

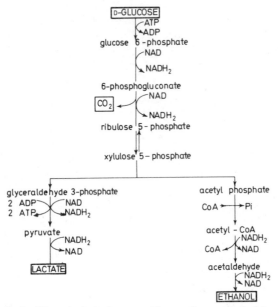

Figure 6.4 *Phosphoketolase pathway for catabolism of glucose by* Leuconostoc mesenteroides

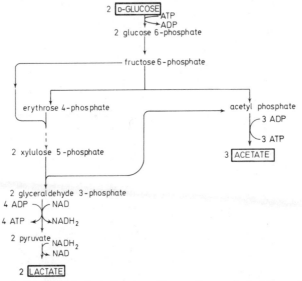

Figure 6.5 *Phosphoketolase pathway for catabolism of glucose by* Bifidobacterium bifidus

A more involved phosphoketolase pathway operates during glucose catabolism in *Bifidobacterium bifidus* (once called *Lactobacillus bifidus*). On this pathway, which is depicted in *Figure 6.5,* there are two reactions catalysed by a phosphoketolase. The first of these is the cleavage of fructose 6-phosphate to give erythrose 4-phosphate and acetyl phosphate, and the second is the breakdown of xylulose 5-phosphate to glyceraldehyde 3-phosphate and acetyl phosphate. The ATP yield from this pathway is **five molecules of ATP from each two molecules of glucose.**

Entner-Doudoroff pathway. A fourth pathway for carbohydrate breakdown in micro-organisms was first shown to operate in *Pseudomonas saccharophila,* and is usually referred to as the *Entner-Doudoroff* pathway (*Figure 6.6*). Like the

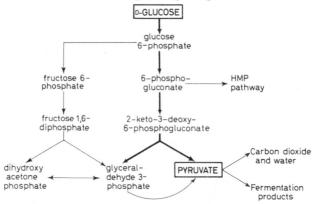

Figure 6.6   The Entner-Doudoroff pathway and its relationship to other glycolytic pathways

HMP pathway, the Entner-Doudoroff pathway produces only one molecule of ATP from each molecule of glucose. However, unlike the other major pathways, it is used by only a very restricted number of micro-organisms, mostly bacteria of the genus *Pseudomonas*. Indeed, the distribution of this pathway among micro-organisms is so restricted that it is thought it may be a separate evolutionary line.

Micro-organisms differ considerably in the extent to which they use the two major pathways (EMP and HMP) for carbohydrate breakdown. During glucose fermentation by *Sacch. cerevisiae,* almost all of the glucose is metabolised by the EMP pathway but, when the yeast is grown aerobically, as much as 30% of the glucose is channelled into the HMP scheme. The EMP pathway is also the major route of glucose catabolism in *Propionibacterium arabinosum* and the homofermentative bacteria. In *Penicillium chrysogenum,* however, two-thirds of the glucose go through the HMP pathway. Other microbes which favour the HMP pathway for glucose catabolism are *Brucella abortus* and species of *Acetobacter.*

Catabolism of other carbohydrates. Many other carbohydrates in addition to glucose can serve as energy sources for micro-organisms, and there are several ways in which these carbohydrates can be introduced into the main catabolic pathways. With monosaccharides, the first step involves a phosphorylation, as with glucose, and the ability of a monosaccharide to be used as a source of energy by a micro-organism often depends upon the availability and specificity of the appropriate kinase.

The first step in D-galactose utilisation involves phosphorylation to give D-galactose 1-phosphate, a reaction catalysed by galactokinase:

D-galactose + ATP $\longrightarrow$ D-galactose 1-phosphate + ADP

D-Galactose 1-phosphate is then converted to glucose 1-phosphate in a reaction catalysed by glucose phosphate uridylyl transferase:

D-galactose 1-phosphate + UDPG $\longrightarrow$ D-glucose 1-phosphate + UDPGal

UDP-Glucose is regenerated by the epimerisation of UDP-galactose, a reaction catalysed by UDP-glucose epimerase:

UDP-glucose   UDP-galactose

Finally, glucose 1-phosphate is converted to glucose 6-phosphate by phosphoglucomutase.

D-Fructose, after conversion to fructose 6-phosphate, can be incorporated into the EMP pathway. An alternative pathway for catabolism of D-fructose operates in *Aerobacter aerogenes.* The sugar is converted into

fructose l-phosphate during group translocation involving
the phosphoenolpyruvate phosphotransferase transport
system (*see* page 174), after which phosphorylation of
fructose l-phosphate to fructose 1,6-diphosphate is
catalysed by a specific kinase, synthesis of which is
induced by D-fructose. A similar coupling with the
phosphoenolpyruvate phosphotransferase group-
translocation system operates in *Salmonella typhimurium*
during mannitol catabolism. D-Mannitol l-phosphate,
which is the product of the transport process, is
converted to D-fructose 6-phosphate in a reaction
catalysed by mannitol l-phosphate dehydrogenase.

Entry of pentoses into the main highways of
carbohydrate catabolism is effected by similar types of
reaction. D-Ribose can be utilised by some micro-
organisms after conversion to ribose 5-phosphate. Xylose,
on the other hand, may be converted to xylulose before
being phosphorylated. D-Arabinose is catabolised in at
least three different ways by bacteria. In *Aerobacter
aerogenes*, it is first isomerised to yield D-ribulose
which in turn is phosphorylated at C-5 and epimerised at
C-3 to give xylulose 5-phosphate, which is an intermediate
on the major pathways. D-Ribulose is also the first
product of catabolism of D-arabinose in *E. coli*. However,
this intermediate is then converted into D-ribulose
l-phosphate which, in turn, is cleaved to dihydroxyacetone
phosphate (an intermediate on the EMP pathway) and
glycolaldehyde. The latter product is converted to
glycollate which can enter the dicarboxylic acid cycle
(*see* page 226). Finally, in *Pseudomonas saccharophila*,
the initial product of D-arabinose catabolism is
D-arabinolactone which is converted to D-arabonic acid,
2-keto-3-deoxy-D-arabonic acid and finally , by
dehydrogenation, to two compounds (pyruvate and glycollate)
which can enter other catabolic highways. The existence
of these different pathways for catabolism of D-arabinose
by bacteria illustrates vividly the diversity which
exists in microbial carbohydrate catabolism.

Utilisation of tri- and oligosaccharides is usually
preceded by a hydrolytic breakdown to monosaccharides.
Some organisms, however, possess mechanisms for directly
phosphorylating disaccharides. Certain lactobacilli and
Neisseria meningitidis, for example, contain an enzyme,
maltose phosphorylase, which catalyses the reaction:

$$\text{maltose} + PO_4^{3-} \longrightarrow \text{glucose l-phosphate} + \text{glucose}$$

As a result of this reaction a molecule of phosphorylated glucose is produced without expenditure of ATP.

Aerobic respiration

The $\Delta G°'$ value for the reaction by which glucose is completely oxidised to carbon dioxide and water is -686 kcal. When they catabolise glucose only as far as pyruvate, micro-organisms make available to themselves only 47 of the 686 kcal. Many microbes are able to use some of the energy in the pyruvate molecule to make ATP by carrying out a complete oxidation of an acetyl residue formed from pyruvate.

Tricarboxylic acid cycle. Oxidation of acetyl residues involves the operation of the *tricarboxylic acid* (TCA) *cycle (Figure 6.7).* Enzymes which catalyse reactions of the TCA cycle have been found in extracts of a wide range of micro-organisms, and there is no doubt that the cycle represents the main pathway of terminal respiration in micro-organisms. Operation of the TCA cycle also provides the micro-organism with precursors for biosynthetic reactions (*see* page 250).

Before entering the cycle, pyruvate is oxidised to acetyl-CoA by a complex reaction, the overall equation for which is:

$$CH_3-CO-COOH + CoA-SH + NAD \longrightarrow CH_3-CO-S-CoA + NADH_2 + CO_2$$

pyruvate   acetyl-CoA

In addition to NAD and CoA, three other cofactors, namely thiamine pyrophosphate (TPP), lipoic acid and magnesium ions, participate in this reaction.

With the exception of the complex reaction in which α-ketoglutarate is oxidised to succinyl-CoA, a reaction which because of its requirements for TPP and lipoic acid

bears some resemblance to the oxidation of pyruvate to
acetyl-CoA, each of the ten reactions of the TCA cycle is
readily reversible. The carbon atoms of the acetate (in
the form of acetyl-CoA) are liberated as carbon dioxide,
while the hydrogen atoms are removed by four different
dehydrogenases. In three of these four oxidations, the
hydrogen atoms are accepted by NAD (or NADP); with
succinate dehydrogenase, however, they are transferred
directly to a flavoprotein (FAD).

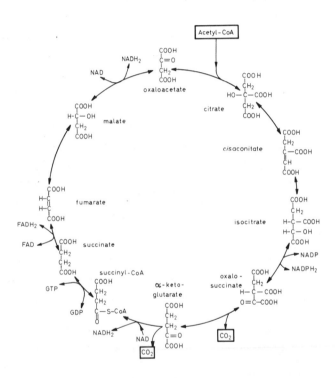

Figure 6.7   The tricarboxylic acid cycle

Electron-transport chains. During aerobic respiration, hydrogen ions from the reduced carriers are ultimately combined with molecular oxygen to produce water, thereby effecting complete oxidation of the acetyl group. However, the hydrogen ions are not transferred directly to oxygen. Instead, they and the accompanying electrons are shuttled along a chain of carrier molecules that go to make up the *electron-transport* or *respiratory chain,* and only in the last reaction in the chain, that catalysed by cytochrome oxidase, are the hydrogen ions combined with the ultimate electron acceptor.

This shuttling of electrons along a chain of carriers would be a cumbersome process if the aim was simply to allow them to be accepted finally by molecular oxygen. In fact, it serves another function since the transfer is coupled or linked with production of high-energy phosphate bonds as a result of *oxidative phosphorylation* (*see* page 201). This is the process whereby some of the energy in the acetyl group is made available to micro-organisms.

The carriers and reactions involved in electron-transport chains have been studied in a number of micro-organisms. Although many of the reactions are basically similar in prokaryotic and eukaryotic microbes, important differences nevertheless exist in the manner in which these two groups of micro-organisms carry out oxidative phosphorylation.

ELECTRON TRANSPORT IN EUKARYOTES. In eukaryotic micro-organisms, the electron-transport chain is located in mitochondria, organelles which are often dubbed the *power houses of the cell.* As described in Chapter 2 (page 72) each mitochondrion contains the enzymes of the TCA cycle, together with the components of the electron-transport chain and proteins that regulate the coupling of electron flow to ATP synthesis. The TCA-cycle enzymes are in the matrix of the mitochondrion, and the carriers of the electron-transport chain mainly in the inner mitochondrial membrane. Moreover, the carriers are present in amounts that are in simple ratio to one another, and it is thought that they are situated on the inner membrane in the precise sequence in which they act. These units of the electron-transport chain are referred to as *respiratory assemblies,* and each mitochondrion is believed to contain several thousand such assemblies.

The carriers which go to make up the electron-transport chain in mitochondria, and the sequence in which they function, are shown in *Figure 6.8.* Information about these

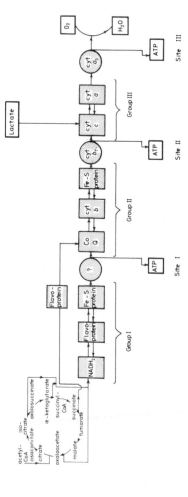

Figure 6.8 Diagrammatic representation of a eukaryotic-type electron-transport chain and its relationship to the TCA cycle, showing the sites at which ATP is synthesised

carriers and their positions in the chain came very
largely from research on mitochondria isolated from
animal cells, particularly from beef and pigeon hearts.
Much less work has been done with mitochondria from
eukaryotic microbes, but in those organisms whose
mitochondria have been subjected to investigation,
particularly the yeasts *Candida utilis* and *Sacch.
cerevisiae*, it is generally held that the components of
the electron-transport chain and the sequence in which
they function are similar, although not identical, with
those in animal-cell mitochondria.

When a pair of electrons passes along the chain from
$NADH_2$ to molecular oxygen, there are three steps,
usually referred to as Sites I, II and III, where passage
of the electrons yields sufficient energy to cause
synthesis of ATP from ADP. These sites are indicated
on *Figure 6.8*. This discovery came originally from the
experimental observation that the number of moles of
inorganic phosphate recovered in the organic form as
each gram atom of oxygen is consumed, when isolated
mitochondria oxidise $NADH_2$, equals 3.0; this is known as
the *P:O ratio*. Other substrates yield different P:O
ratios; with succinate, for instance, the value is 2.0,
which is explained by the fact that succinate enters the
chain at coenzyme Q and so by-passes the first
ATP-yielding site. Similarly, when lactate is used as
substrate, the P:O ratio is 1.0, because both Sites I
and II are by-passed as lactate enters the chain at
cytochrome *c* (*Figure 6.8*).

Knowing the P:O ratios for oxidation of $NADH_2$ and of
succinate, and with the knowledge that oxidation of
pyruvate to acetyl-CoA also gives a P:O ratio of 3.0,
a calculation can be made of the total number of high-
energy bonds produced as a result of the complete
oxidation of pyruvate. This number equals 15, which is
made up of yields of three ATP molecules as a result of
the activities of pyruvate, isocitrate, α-ketoglutarate
and malate dehydrogenases, two from oxidation of
succinate, and one from substrate-level phosphorylation
during the conversion of succinyl-CoA into succinate.
A yield of 15 high-energy phosphate bonds represents
112.5 kcal of the 280 kcal available from complete
oxidation of pyruvate, an efficiency of just over 40%.

The carriers in the electron-transport chain are thought
to exist as three separate groups, now referred to as
Groups I, II and III. The carriers in any one group all

have very similar standard redox potentials, and together
they are thought to serve as a 'redox potential buffer
pool'. Only when electrons are transferred from a
carrier in one group to a carrier in another group, or
from a carrier in Group III to molecular oxygen, is
sufficient energy made available (that is about 180 mV)
to effect synthesis of ATP from ADP. Despite many years
of work on the electron-transport chain in laboratories
throughout the world, it is still far from clear how a
carrier in one group interacts with a carrier in another
group to achieve this energy coupling. Britton Chance,
whose work at the University of Pennsylvania in
Philadelphia, U.S.A., has done much to unravel the
complexities of the electron-transport chain, believes
that carriers with alternating redox potentials couple a
carrier in one group with a carrier in another, or a
carrier in Group III with molecular oxygen. The nature
of the coupling carrier between Groups I and II, that is
at Site I, is not known, but that between Group II and
Group III is thought to be a variant of cytochrome b,
known as cytochrome b_T. Cytochrome a_3 is thought to
couple Group III and molecular oxygen (*Figure 6.9*).

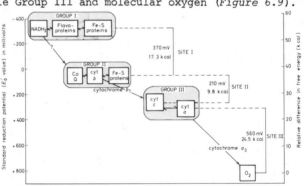

*Figure 6.9 Decline in free energy as pairs of electrons
flow down a eukaryotic electron-transport chain from* $NADH_2$
*to molecular oxygen. Values for the energy released at
Sites I, II and III are calculated from published values
for the standard reduction potentials of NAD, cytochrome
b, cytochrome c, cytochrome a and the* $\frac{1}{2}O_2/H_2O$ *couple
(see Figure 6.1)*

For a long time, even greater mystery surrounded the
molecular mechanism by which ATP is synthesised during
electron transport. Originally, the *chemical coupling
hypothesis* for explaining this process held sway. This

hypothesis postulates that high-energy bonds are generated in carriers during the passage of electrons from one group of carriers to the next, and that the high-energy bonds so formed are the precursors of the high-energy bonds in ATP. In recent years, the *chemiosmotic hypothesis* has come to be favoured. This hypothesis, which was developed mainly by the British biochemist Peter Mitchell, states that the energy needed to synthesise ATP from ADP comes from the energy made available when a gradient of protons is set up across the inner mitochondrial membrane. The reverse of this process, that is the use by the cell of the energy of ATP to create a gradient of protons across the plasma membrane, forms the basis of the proton-gradient hypothesis to explain active transport of solutes into microbes (*see* page 173).

Several antimicrobial compounds act by interfering with the electron-transport chain in mitochondria. The barbiturate amobarbital (sometimes known as amytal) and the highly toxic insecticide rotenone act at Site I. Antimycin A_1 and hydroxyquinoline-N-oxide act at Site II, while carbon monoxide and cyanide act at Site III.

amobarbital

antimycin A_1

ELECTRON TRANSPORT IN PROKARYOTES. The carriers and enzymes involved in electron transport in prokaryotic microbes are located in the plasma membrane. Electron-transport chains have been studied in detail in a number of bacteria, and they are known to differ in several respects from electron-transport chains in eukaryotic microbes. Cytochromes and quinones are synthesised by bacteria, but these carriers often differ, albeit slightly, from their counterparts in eukaryotes. For example, bacteria synthesise cytochromes of the *c* type, but these carriers often have different properties from those synthesised in eukaryotic microbes, and so are designated accordingly (e.g. cytochrome c_5). Another important difference between the electron-transport chains in bacteria and eukaryotes is that, in the former, the chains are often branched, with electrons able to enter the chain by any one of several different dehydrogenases and to leave by one of a number of different oxidases. The multiplicity of terminal oxidases is a particularly interesting feature. These oxidases are often recognised by their capacity to bind carbon monoxide, and they include a separate class of cytochromes, known as cytochromes *o*.

The branched electron-transport chain in *Azotobacter vinelandii* is shown in *Figure 6.10*. The three terminal branches of the chain are used to different extents depending on the conditions under which the bacterium is grown. Highly aerobic conditions for growth promote electron transport by the non-phosphorylating branch which terminates in cytochrome a_2. But when *A. vinelandii* is grown under oxygen-limiting conditions, the 'minor' branches, which terminate in cytochromes a_1 and *o*, are used in preference.

Incomplete oxidations. As a result of the operation of the TCA cycle, carbohydrates are completely oxidised to carbon dioxide and water. With some micro-organisms, however, there is an incomplete oxidation of carbohydrate resulting in accumulation of oxidisable organic compounds in the culture medium. Incomplete oxidations are often encountered when abnormally high concentrations of carbohydrate substrate are made available to an organism. Probably the best known examples of organisms which carry out incomplete oxidations are the acetic acid bacteria - members of the genus *Acetobacter* - which characteristically oxidise ethanol to acetic acid. The oxidising activities of these bacteria extend to other alcohols including aliphatic polyols. A study of these activities, in so far as they pertain to the oxidation of sugars, led to the formulation of the Bertrand-Hudson rule which states that oxidation of a secondary alcohol group to a ketone group occurs only if the secondary alcohol group is situated in between a primary and another secondary alcohol group, and is *cis* to the secondary alcohol group. In accordance with this rule, D-gluconate is oxidised to 5-keto-D-gluconate and mannitol to D-fructose. It is likely that at least some of these incomplete oxidations lead to the production of high-energy phosphate bonds following electron-transport.

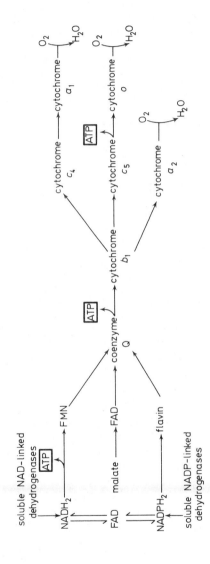

Figure 6.10 The electron-transport chain in Azotobacter vinelandii, indicating the sites of ATP synthesis. It should be noted that the standard reduction potentials of many of the carriers in the chain almost certainly differ from those listed in Figure 6.1

Microbial luminescence. Visible light is emitted by a
number of living organisms including bacteria, fungi and
protozoa. Luminescence caused by micro-organisms,
especially certain dinoflagellates (e.g. *Pyrodinium
bahamense*) which are responsible for the glow sometimes
seen in the wake of a ship, and fungi such as *Panus
stipticus,* has been known for centuries. Luminescence
is often encountered with bacteria growing on meat, and
it has been with these luminescent bacteria, especially
Achromobacter fischeri, that most of the experimental
work on the mechanism of luminescence has been carried
out.

The light-emitting process is linked to the respiratory
chain and the details of the process are different in
bacteria and fungi. The reactions involved in bacterial
luminescence are:

$$NADH_2 + FMN \xrightarrow{\text{transdehydrogenase}} NAD + FMNH_2$$

$$FMNH_2 + O_2 \xrightarrow[\text{aldehyde}]{\text{luciferase}} FMN + light$$

The first reaction, catalysed by a transdehydrogenase,
leads to formation of $FMNH_2$ which, in the presence of an
enzyme known as *luciferase,* a long-chain aldehyde and
molecular oxygen, is oxidised with emission of light.
Details of the second reaction, and in particular of the
role of the long-chain aldehyde (which is often
palmitaldehyde), are poorly understood. It is not known,
for instance, whether the aldehyde acts catalytically or
whether it is oxidised to the corresponding fatty acid.
Light-emitting bacteria contain as much as 2-5% of the
soluble protein luciferase. The enzyme has a molecular
weight of approximately 60 000 daltons, and it is
relatively rich in acidic amino acids. The wavelength
of the radiation emitted by bacteria is around 490 nm.

The reactions which lead to light emission in fungi
differ, as does the wavelength of the radiation (about
530 nm), from those that take place in bacteria. Again,
a luciferase is involved but this enzyme, together with
the emitter molecule known as *luciferin,* are enclosed in
a membrane-bound structure. Reduced nicotinamide
adenine dinucleotide reduces an intermediate, the nature
of which is not as yet understood. The reduced
intermediate in turn reduces luciferin in a reaction
which involves luciferase and molecular oxygen, and leads
to emission of light.

Microbial luminescence is usually viewed as a wasteful
process since it leads to a lowering of the efficiency
of the respiratory chain. It has been suggested that it
may be a vestigial system in organic evolution which was
used originally for detoxifying small amounts of molecular
oxygen.

Anaerobic respiration

Micro-organisms which grow aerobically can often also **grow
under anaerobic conditions using inorganic compounds other**
than molecular oxygen as the final electron acceptor.
For example, many aerobic bacteria can grow anaerobically
using nitrate as a final electron acceptor - a process
known as *nitrate respiration*. The products of nitrate
reduction include nitrite and various other reduced forms
of nitrogen. Some *Bacillus* spp. carry out anaerobic
respirations in which nitrate is converted to ammonia.
If dinitrogen, nitric oxide or nitrous oxide is the
product of nitrate respiration, the process is called
denitrification. Nitrate respiration differs from
assimilatory nitrate reduction (page 264) in that the
reduction products are not utilised further and are
usually excreted by the micro-organism. With many micro-
organisms that carry out nitrate respiration (e.g. *E. coli*),
the electron-transport chain would seem to be the same as
under aerobic conditions with nitrate reductase
substituting for cytochrome oxidase.
  Sulphate (SO_4^{2-}) can act as a final electron acceptor
for species of *Desulfovibrio*; thiosulphate ($S_2O_3^{2-}$) or
sulphite (SO_3^{2-}) can substitute for sulphate. The
product of the reduction is hydrogen sulphide. Further
details of this process of *dissimilatory sulphate
reduction* are given on page 269. Anaerobic respiration
in which carbon dioxide acts as the ultimate electron
acceptor is also known. *Methanobacterium omelianskii*,
for example, oxidises ethanol according to the following
overall equation:

$$2C_2H_5OH + CO_2 \longrightarrow 2CH_3COOH + CH$$

In *Clostridium aceticum*, hydrogen is oxidised with carbon
dioxide as the electron acceptor:

$$4H_2 + 2CO_2 \longrightarrow CH_3COOH + 2H_2O$$

Because the standard reduction potential ($E_0^!$ value) of the redox couple acting as the terminal electron acceptor in anaerobic respiration may be less than the value (+820 mV) for the couple (O_2/H_2O) which acts during aerobic respiration, it is possible that the decline in free energy that occurs as a result of passage of a pair of electrons from the penultimate cytochrome on the chain to the terminal acceptor may be less than is required for synthesis of an ATP from and ADP molecule.

Carbohydrate fermentations

The amount of NAD in a micro-organism is limited, and it must continually be regenerated if metabolism is to continue. Under anaerobic conditions $NADH_2$ formed during glycolysis, cannot be oxidised through an electron-transport system with oxygen as the final electron acceptor. In some micro-organisms it can instead be regenerated by oxidation-reduction reactions involving pyruvate or some compound derived from pyruvate. These further reactions of pyruvate vary considerably among different micro-organisms and, especially among bacteria, lead to the formation of characteristic fermentation products such as alcohols and acids. In some organisms, this further metabolism of pyruvate can also lead to the production of high-energy phosphate bonds. The inability of algae to grow heterotrophically under anaerobic conditions is probably caused by a lack of mechanisms for regenerating NAD.

Alcoholic fermentation. Brewers and wine-makers have for centuries exploited the ability of certain yeasts (particularly strains of *Sacch. cerevisiae*) to carry out an alcoholic fermentation in which sugars are broken down to give almost quantitative yields of ethanol and carbon dioxide. The conversion of pyruvate to these end-products involves two reactions. In the first, the enzyme pyruvate decarboxylase catalyses a decarboxylation of pyruvate to acetaldehyde, a reaction requiring TPP. In the second reaction, acetaldehyde is reduced to ethanol at the expense of $NADH_2$, the enzyme involved being alcohol dehydrogenase. Alcoholic fermentation is carried out by many yeasts, by moulds such as *Fusarium* spp., and by members of the Mucorales, but curiously enough by very few bacteria.

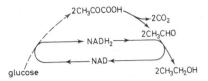

During his pioneer studies on the mechanism of alcoholic fermentation in yeast, Carl Neuberg showed that, after adding bisulphite or alkali to the yeast culture, glycerol appeared as the major end-product of fermentation. These modified fermentations have come to be known as Neuberg's second and third forms of fermentation respectively, the normal 'unsteered' fermentation being the first form. The explanation for this effect of adding bisulphite or alkali is that, in their presence, acetaldehyde is no longer able to act as an acceptor for the hydrogen of $NADH_2$ which is instead accepted by dihydroxyacetone phosphate to give glycerol 3-phosphate; glycerol is then formed by dephosphorylation. The inability of acetaldehyde to accept the hydrogen of $NADH_2$ in the presence of these steering agents is the result of its forming an addition compound with the bisulphite and by its undergoing a dismutation reaction under alkaline conditions to give acetate and ethanol.

Lactic acid fermentation. Lactic acid is another common product of carbohydrate fermentations, especially among the lactic acid bacteria and *Rhizopus* spp. Lactic acid-producing micro-organisms can be subdivided into *homolactic fermenters* which produce lactate in almost

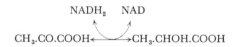

quantitative yields from carbohydrate substrates, and *heterolactic fermenters* which produce, in addition to lactate, considerable amounts of other fermentation products such as ethanol, acetate and carbon dioxide. In the homolactic fermentation, pyruvate is reduced directly to lactate by lactate dehydrogenase working in reverse. The optical activity of the lactate formed varies with the organism. It depends upon the stereospecificity of the lactate dehydrogenase and upon whether the organism possesses lactate racemase, an enzyme which reversibly converts D-lactate to the L-form.

Heterolactic fermenters produce lactate by a **phospho-ketolase** pathway. The final step, the reduction of pyruvate, is the same as in the homolactic fermentation, but pyruvate is formed by the breakdown of glucose by the HMP pathway, and a cleavage of xylulose 5-phosphate to give glyceraldehyde 3-phosphate and acetyl phosphate (*see Figure 6.4*).

Propionic acid fermentation. Propionic acid is a product of carbohydrate fermentation by the propionic acid bacteria (*Propionibacterium* spp.). The acid is formed together with acetate and carbon dioxide, and an examination of the relative amounts of these fermentation products indicates that two molecules of pyruvate are converted to propionate for every one that is oxidised to acetate and carbon dioxide. The pathway for the conversion of pyruvate to propionate is shown in *Figure 6.11*.

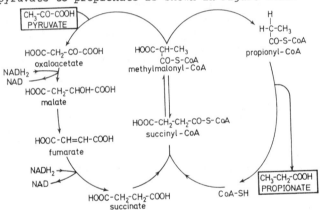

Figure 6.11 Pathway for the conversion of pyruvate into propionate

Succinyl-CoA, which is formed from pyruvate in the TCA cycle, is isomerised to methylmalonyl-CoA. In a reaction with pyruvate, catalysed by the biotin-dependent methylmalonyl-CoA carboxyltransferase, methylmalonyl-CoA gives rise to propionyl-CoA and oxaloacetate. Propionate is then formed from propionyl-CoA. Propionate production in other organisms is probably by a similar mechanism, although *Clostridium propionicum* seems to be an exception in that lactate and not succinate is an intermediate.

Formic acid fermentations. A number of bacteria, especially members of the Enterobacteriaceae, can carry out a *formic acid fermentation*. In this, metabolism of pyruvate leads to a number of different products the nature of which varies with each organism. However, all of these organisms produce formate which either accumulates or, under acid conditions, is converted by the action of 'formic hydrogenlyase' (a result of the combined action of formate dehydrogenase and hydrogenase) to molecular hydrogen and carbon dioxide. Bacteria carrying out a formic fermentation fall into one of three groups.

$$HCOOH \xrightarrow{\text{'formic hydrogenlyase'}} H_2 + CO_2$$

The first of these, *the mixed acid type,* is represented by *E. coli*. In this organism, for every four molecules of pyruvate formed, two are reduced to lactate by lactate dehydrogenase, and the remaining two are metabolised to give acetate (as acetyl phosphate) and formate in a reaction known as the *phosphoroclastic split:*

$$CH_3-CO-COOH + H_3PO_4 \rightleftharpoons CH_3-CO-O-\overset{\displaystyle O}{\underset{\displaystyle OH}{\overset{\|}{\underset{|}{P}}}}-OH + H-COOH$$

This is a complex reaction, variants of which are found in different micro-organisms. In the *E. coli* reaction, TPP and CoA are involved as cofactors. One molecule of the resulting acetate, possibly as acetyl-CoA, is reduced in a two-stage process to ethanol; this involves oxidation of two molecules of $NADH_2$. Since acetyl phosphate can donate a phosphate group to ADP, the phosphoroclastic splitting of a C—C bond may lead to ATP synthesis.

In representatives of the second group of formic fermentation bacteria some of the pyruvate is metabolised as in the mixed acid fermentation, but the bulk of it is condensed to give acetylmethylcarbinol or acetoin which may then be reduced to 2,3-butanediol (*Figure 6.12*). This type of fermentation is characteristic of *Aerobacter, Serratia* and *Bacillus* spp., and is known as the *butanediol type.*

Aerobacter aerogenes and *Escherichia coli* are species of bacteria which are in many respects alike, but which particularly in water analysis need to be distinguished since the presence of the latter, but not the former, is indicative of possible faecal contamination. The species are identifiable by using a series of physiological

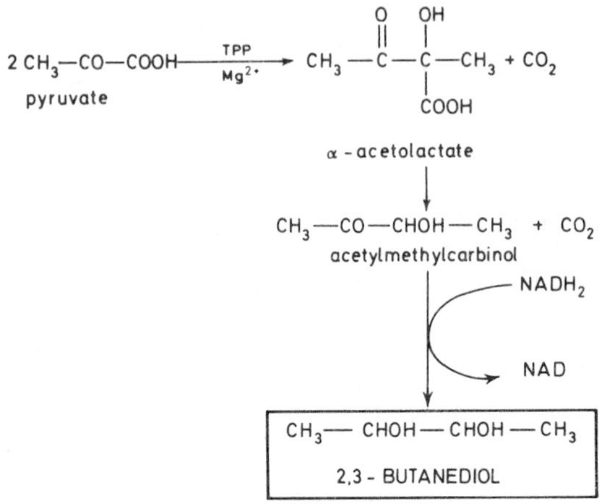

$$2\ CH_3-CO-COOH \xrightarrow[Mg^{2+}]{TPP} CH_3-\overset{\displaystyle O}{\overset{\displaystyle \|}{C}}-\underset{\displaystyle COOH}{\overset{\displaystyle OH}{\underset{\displaystyle |}{\overset{\displaystyle |}{C}}}}-CH_3 + CO_2$$

pyruvate

α - acetolactate

$$CH_3-CO-CHOH-CH_3 \ + \ CO_2$$
acetylmethylcarbinol

NADH$_2$

NAD

$$CH_3-CHOH-CHOH-CH_3$$
2,3 - BUTANEDIOL

Figure 6.12   Pathway for the conversion of pyruvate into 2,3-butanediol in bacteria

reactions often referred to as the IMViC tests, two of which are based on the different formic acid fermentations carried out by the two species of bacteria.

Since *E. coli* carries out a mixed acid fermentation leading to the formation of lactate and formate, while the butanediol fermentation of *A. aerogenes* leads to the production of less of these acids, cultures of *E. coli* are distinguished from those of *A. aerogenes* by their ability to give an acid reaction (a red colour) with the indicator methyl red, showing that the pH value of the culture has fallen below 4.5. This test is the M component of the IMViC series.

The ability of *A. aerogenes,* but not *E. coli,* to synthesise acetylmethylcarbinol is the basis of another diagnostic test, namely the *Voges-Proskauer test.* This is a test for the presence of acetylmethylcarbinol in a culture. The culture is made alkaline with potassium hydroxide, which oxidises acetylmethylcarbinol to diacetyl ($CH_3-CO-CO-CH_3$). Diacetyl in turn reacts with compounds in the peptone component of the medium containing guanidine residues to give a pink-coloured compound. The reaction can be made more sensitive by adding α-naphthol and creatine to the culture medium.

The two remaining tests in the IMViC series are the capacity of *E. coli,* but not *A. aerogenes,* to produce indole (I) from tryptophan, a capacity which is based on the ability of the bacterium to synthesise the enzyme tryptophanase, and the presence in *A. aerogenes* but not in *E. coli* of a transport protein for citrate (C) which enables the former bacterium to utilise this tricarboxylic acid.

The formic acid fermentation of *E. coli* leads to the formation in cultures of this bacterium of equimolar amounts of carbon dioxide and molecular hydrogen, whereas *A. aerogenes* produces an excess of carbon dioxide. This difference in behaviour can also be used as a test for distinguishing the two species of bacteria.

A third group, characteristic of *Clostridium* spp., carry out a *butyric type* fermentation. This leads to a variety of products, but butyrate is always formed either as an end-product or as an intermediate in the formation of *n*-butanol. The intermediates formed during butyric fermentations are shown in *Figure 6.13*. Some clostridia

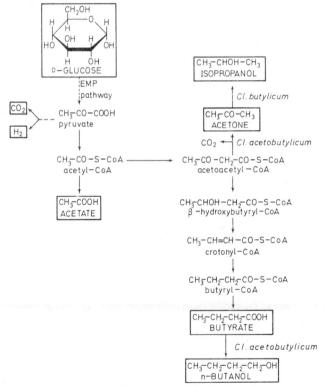

Figure 6.13 Overall pathways in the fermentation of glucose by clostridia

of the butyric type (e.g. *Cl. butyricum*) produce acetate, butyrate, carbon dioxide and molecular hydrogen. Others, typified by *Cl. acetobutylicum,* produce acetone and butanol in addition to the above products, while in *Cl. butylicum,* isopropanol is formed at the expense of acetone. During the course of these butyric fermentations there is a production of additional high-energ phosphate bonds, for instance in the conversion of acetyl-CoA to acetate.

6.2.2 NITROGENOUS COMPOUNDS

Among organic nitrogenous compounds, amino acids, purines and pyrimidines are the ones most frequently available as substrates for micro-organisms. Since many of these compounds are at an oxidation level lower than that of carbohydrates, they are potentially useful as sources of energy for both aerobic and anaerobic microbes.

Amino acids

The initial reaction in catabolism of amino acids usually involves removal of either the amino group or the carboxyl group. Deamination can take any one of several different forms, depending on the nature of the microbe involved and sometimes of the conditions under which it has been grown. The product of the deamination reaction is either a compound which can be channelled directly into one or other of the energy-yielding pathways already described, or one which can readily be metabolised to produce a potential source of energy. For example, α-keto*iso*caproate, which is a product of deamination of leucine, yields on further catabolism acetoacetate and acetyl-CoA, which is oxidised by the TCA cycle.

Deamination reactions were among the first metabolic reactions to be demonstrated in microbes, largely as a result of the pioneer work of Marjory Stephenson and Ernest Gale in the University of Cambridge, England. One of the commonest deamination reactions is *oxidative deamination,* in which an amino acid is converted to the corresponding keto acid via an imino acid:

$$\underset{NH_2}{\overset{COOH}{R-\overset{|}{\underset{|}{C}}H}} \xrightarrow[H_2O]{\tfrac{1}{2}O_2} \underset{NH}{\overset{COOH}{R-\overset{|}{\underset{\parallel}{C}}}} \longrightarrow \underset{O}{\overset{COOH}{R-\overset{|}{\underset{\parallel}{C}}}} + NH_3$$

An example is provided by the catabolism of glycine to
give glyoxylic acid in *E. coli.* Glyoxylic acid can then
be catabolised to yield energy, as shown on page 226.
Both L- and D-amino acid oxidases are widely distributed
in micro-organisms. Occasionally, the reaction leads to
the formation not of water but of hydrogen peroxide,
which is one of the reasons why some microbes synthesise
the enzyme catalase which catalyses the breakdown of
hydrogen peroxide. In *Sacch. cerevisiae,* and probably in
most other eukaryotic microbes, catalase is located in
intracellular organelles known as *peroxisomes.*

Reductive deamination, which is common among
anaerobically grown micro-organisms, gives rise to a
saturated fatty acid:

$$R-CH(NH_2)-COOH \longrightarrow R-CH_2-COOH + NH_3$$

Clostridia, for instance, deaminate glycine to yield
acetate.

Many clostridia growing in media containing mixtures
of amino acids obtain most of their energy not from
individual amino acids but by a coupled oxidation-
reduction reaction between suitable pairs of acids.
This energy-yielding mechanism was first described over
35 years ago and after the name of its discoverer, the
British microbial physiologist Lawrence Stickland, is
known as the *Stickland mechanism.* The reaction appears
to be confined to clostridia, although suggestions have
been made that it also operates in certain other micro-
organisms.

From the point of view of participation in the
Stickland mechanism amino acids can be divided into two
groups: (a) hydrogen acceptors, including glycine, proline
and ornithine; and (b) hydrogen donors such as alanine,
isoleucine, leucine and valine. The reaction involves
oxidation of the donor amino acid to a keto acid which
is then oxidised further to a fatty acid. The $NADH_2$
formed is used to reduce another amino acid - one of the
hydrogen-acceptor group - or occasionally a non-
nitrogenous compound. Glycine, for example, is reduced
to acetate and ammonia, proline to δ-aminovalerate. Not

H

all of the amino acids in each group are used by every clostridium that carries out the Stickland reaction.

Deamination to produce an unsaturated acid is characteristic of *desaturative deamination:*

$$R-CH_2-CH(NH_2)-COOH \longrightarrow R-CH=CH-COOH + NH_3$$

E. coli catabolises histidine to give urocanic acid, the reaction being catalysed by the enzyme histidinase:

histidine   urocanic acid

Transamination reactions are frequently carried out during amino-acid catabolism by all micro-organisms. These reactions lead to the production of another amino acid in a reaction which also involves keto acids. A particularly common transamination reaction is:

alanine + α-ketoglutarate $\longrightarrow$ pyruvate + glutamate

Some micro-organisms possess special mechanisms for obtaining energy from breakdown of amino acids. For example, certain microbes can obtain energy from arginine by the *arginine dihydrolase* system, which was the name given originally to the mechanism by which arginine is broken down, in the mistaken belief that it was effected by a single enzyme. There are, in fact, three enzymes involved in this conversion of arginine to ornithine and ATP. The first of these, arginine deiminase, catalyses the deamination of arginine to citrulline, which is then converted to ornithine in a reaction coupled with ATP synthesis:

citrulline + Pi + ADP $\rightleftharpoons$ ornithine + CO_2 + NH_3 + ATP

This reaction takes place in two stages, with carbamoyl phosphate, produced in the first stage, being broken down to carbon dioxide and ammonia in the second (*see* page 269) for further discussion of the interrelationships between arginine, citrulline and ornithine).

Decarboxylation of amino acids is carried out by a wide range of micro-organisms, the product of the reaction. being an amine:

$$R-CH(NH_2)-COOH \longrightarrow R-CH_2-NH_2 + CO_2$$

Many of these amines have unpleasant smells, and are responsible for the noisome odours which are characteristic of putrefaction. Cadaverine, so called because its odour resembles that of corpses, is formed by decarboxylation of lysine. Other amines produced by microbial decarboxylation reactions are pharmacologically active. Examples include tyramine formed from tyrosine, and histamine from histidine.

Purines and pyrimidines

These compounds can become available to micro-organisms as a result of hydrolytic breakdown of nucleic acids and their component nucleotides and nucleosides. The degradation reactions lead to production of compounds that can be channelled into energy-yielding pathways. An example is provided by catabolism of uracil by a mycobacterium, which gives barbituric acid and then urea and malonic acid:

uracil   barbituric acid

Malonic acid can then be converted into acetyl-CoA which in turn is oxidised by the TCA cycle.

6.2.3 TRIGLYCERIDES AND FATTY ACIDS

Triglycerides, which are glycerol esters of long-chain fatty acids, and waxes (mono-alcohol esters of fatty acids) are among the most highly reduced substrates available to micro-organisms and, as such, are potential sources of energy. The initial step in the catabolism of these fatty-acid esters involves hydrolysis to yield glycerol (or a monohydric alcohol) and free fatty acids. This hydrolysis is catalysed by intracellular or extracellular lipases (*see* page 145). After phosphorylation glycerol can be incorporated into the EMP pathway and so be made to yield ATP. Catabolism of the fatty acid involves a sequence of β-oxidations. The overall scheme for this oxidation was outlined many years ago by Knoop although details of the individual reactions were elucidated much later. The first step (*Figure 6.14*)

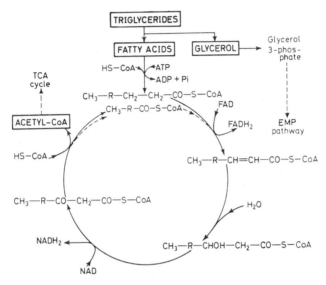

Figure 6.14 Cyclic pathway for β-oxidation of long-chain fatty acids

involves formation of the CoA ester of the fatty acid which is then oxidised to a β-unsaturated derivative in a reaction catalysed by flavoprotein acyl dehydrogenases. Then follows a hydration of the unsaturated compound to give a β-hydroxyacyl-CoA derivative. In a second dehydrogenation, catalysed by β-hydroxyacyl dehydrogenases, a β-ketoacyl derivative is formed; these dehydrogenases are NAD- or NADP-linked. Finally, the ketoacyl derivative is cleaved to give acetyl-CoA (which enters the TCA cycle) and the CoA ester of a fatty acid containing two carbon atoms fewer than the initial substrate. Energy is generated by oxidative phosphorylation. It is probable that unsaturated fatty acids and hydroxy fatty acids are oxidised via this pathway after conversion to their CoA esters.

In many bacteria, the catabolic pathways followed by the lower (C_2-C_{10}) fatty acids are thought to differ from those followed by the higher acids (C_{12}-C_{18}) since it is possible to inhibit oxidation of longer-chain acids without affecting oxidation of the shorter-chain acids. This difference in metabolism of short- and long-chain fatty acids by bacteria is probably a reflection of differences in the properties of one or more of the enzymes concerned in the oxidation scheme. However, little if anything is known of the nature of these differences.

Another type of variation in fatty-acid catabolism is seen among species of *Aspergillus* and *Penicillium* which carry out an incomplete oxidation of certain fatty acids to methyl ketones. Methyl ethyl ketone, for example, is formed from valeric acid, and methyl undecyl ketone from myristic acid. The mechanisms of these incomplete oxidations are unclear although, since β-unsaturated and β-hydroxy fatty acids can also act as substrates, it is possible too that they are intermediates.

6.2.4 HYDROCARBONS

Some micro-organisms can oxidise hydrocarbons and use these compounds as sources of energy. The ability to grow on hydrocarbons is found among many of the major groups of microbes, including bacteria (species of *Corynebacterium, Mycobacterium* and *Pseudomonas*), yeasts (species of *Candida, Debaryomyces, Hansenula* and *Pichia*), and filamentous fungi (*Cladosporium, Fusarium* and *Homodendrum* spp.). Moreover, the variety of hydrocarbons which can be used is wide, although detailed information on the reactions involved in oxidation processes is confined to the simpler aliphatic hydrocarbons and to the alicyclic hydrocarbon camphor.

Alkanes

Oxidation of *n*-alkanes is usually accomplished by oxidation of a terminal methyl group to form the corresponding primary alcohol. This product is then oxidised to a fatty acid which in turn is oxidised by the β-oxidation cycle (*see* page 218) leading to formation of ATP from oxidation of acetyl-CoA (*Figure 6.15*).

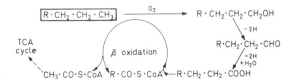

Figure 6.15   Pathway for the monoterminal oxidation of alkanes

   Diterminal oxidation of *n*-alkanes is also known to occur.   A *Corynebacterium* sp., for instance, has been shown to oxidise C_{10}-C_{14} alkanes to, eventually, the corresponding dicarboxylic acids.   Also, species of *Pichia* can carry out a diterminal oxidation of *n*-undecane. Oxidation of a methylene group in the alkane chain, that is *subterminal oxidation,* also occurs in some microbes. Undecan-2-one was shown by isotopic methods to have arisen from *n*-undecane in cultures of *Mycobacterium rhodocrous:*

$$CH_3 - (CH_2)_9 - CH_3 \longrightarrow CH_3 - (CH_2)_8 - \overset{\overset{\displaystyle O}{\|}}{C} - CH_3$$

Oxidation at both C-1 and C-2 in an alkane has been shown to take place in some microbes, such as *Mycobacterium smegmatis.*

   Microbial oxidation of *n*-alkanes requires molecular oxygen, and the enzymes which catalyse the reactions are known as *oxygenases*.   Some oxygenases catalyse the insertion of both atoms of molecular oxygen into the substrate, whereas others which are referred to as *hydroxylases* or *mixed-function oxygenases* catalyse the insertion of only one atom of oxygen.   Oxygenases and hydroxylases are organised into short chains along with other electron carriers including cytochromes, but movement of electrons along these chains is not accompanied by formation of ATP.   Thermodynamically, therefore, reactions catalysed by oxygenases and hydroxylases are prodigal of energy since there is no device for trapping the energy evolved.

   However, little information is available on the nature of the carriers associated with the activity of these hydroxylases.   Experiments on methyl-group oxidation of alkanes by pseudomonads have indicated that the chain includes rubredoxins, which are carriers couples of which have standard reduction potentials of

0.0 to -100 mV. The following scheme has been put
forward for methyl-group oxidation of *n*-alkanes in
Pseudomonas oleovorans:

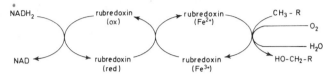

It is possible, too, that hydroxylases catalyse
subterminal oxidation of *n*-alkanes by micro-organisms.

Alkenes

Although alkenes can be oxidised at the terminal methyl
group, of special interest is the way in which they are
oxidised at the double bond at the opposite end of the
molecule. Oxidation at the double bond involves molecular
oxygen and the formation of a diol. An example is
provided by *Candida lipolytica* which oxidises hexadec-1-ene
to hexadec-1,2-diol:

$$CH_3 - (CH_2)_{13} - CH{=}CH \xrightarrow{O_2} CH_3 - (CH_2)_{13} - \overset{OH}{\underset{}{CH}} - \overset{OH}{\underset{}{CH}}$$

Pseudomonads use an alternative method for oxidising
alkenes at the double bond, which leads to production of
1,2-epoxides. Thus, oct-1-ene is oxidised to octene-1,2-
epoxide. In general, however, bacteria prefer to oxidise
the saturated end of an alkene molecule. For instance,
hept-1-ene is oxidised to hept-1-ene-6-carboxylate. It
is unclear whether both oxygenases and hydroxylases are
involved in oxidation of the double bond in alkenes.
The formation of epoxides suggests that the reaction is
catalysed by a hydroxylase, although evidence for the
action of an oxygenase has come from experiments on
oxidation of hexadec-1-ene by *Candida lipolytica*.

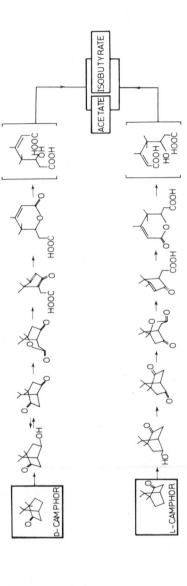

Figure 6.16 Pathways for catabolism of D- and L-camphor to acetate and isobutyrate by Pseudomonas putida

Alicyclic hydrocarbons

Microbes are known to utilise a wide variety of other
types of hydrocarbon but, with the exception of aromatic
hydrocarbons, information on the pathways by which these
compounds are oxidised is confined to just one compound,
namely the alicyclic hydrocarbon camphor, due largely to
work in the laboratory of Irwin Gunsalus in the
University of Illinois, U.S.A.

Catabolism of D- and L-camphor has been studied with
two bacteria, namely *Pseudomonas putida* and *Mycobacterium
rhodocrous*. In the pseudomonad, the catabolic fate of
both enantiomers of camphor has been traced to the point
at which the intermediates enter the major energy-
yielding pathways in the bacterium. Data on camphor
catabolism in the mycobacterium are less extensive
although it is known that the catabolic routes followed
differ from those in the pseudomonad. *Figure 6.16* shows
the intermediates formed during catabolism of D- and
L-camphor by *Ps. putida*. Introduction of additional
oxygen atoms into the camphor molecule takes place in
the first three reactions, each of which is catalysed by
a specific hydroxylase. In the fourth reaction on the
pathway the bicyclic bornane ring is cleaved during a
reaction in which the 5-keto-1,2-campholide gives rise
to the appropriate cyclopentanone acetic acid. The
cyclopentanone ring is then cleaved, giving rise finally
to three molecules of acetate and one of isobutyrate.
Acetate can be used as an energy source after conversion
to acetyl-CoA, while isobutyrate is converted to succinyl-
CoA which too can give rise to ATP.

6.2.5 AROMATIC COMPOUNDS

Compounds which contain aromatic rings can be oxidised
by a number of micro-organisms. Usually, this takes
place under aerobic conditions, although some bacteria
can catabolise aromatic compounds in the absence of
molecular oxygen. Under aerobic conditions, metabolism
of an aromatic compound leads ultimately to one of a
small number of diphenolic intermediates which include
catechol, protocatechuate, homoprotocatechuate,
gentisate and homogentisate. Highly specific oxygenases
then catalyse opening of the aromatic ring in these
diphenolic intermediates. The cleavage may occur between

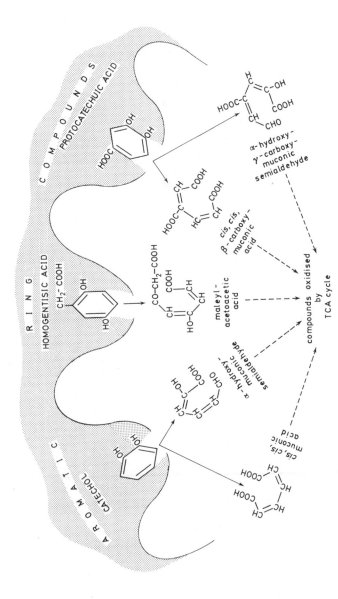

Figure 6.17 Intermediates formed during the ring opening of certain aromatic compounds by micro-organisms

two carbon atoms each of which is linked to an hydroxyl group; this is referred to as *ortho fission*. The reaction initiates a pathway that generally leads to formation of 3-ketoadipate which in turn is metabolised to give compounds such as acetyl-CoA and succinate that are oxidised in the TCA cycle (*Figure 6.17*). Alternatively, cleavage of the aromatic ring may take place between an hydroxylated carbon atom and a non-hydroxylated carbon atom, which constitutes *meta fission*. This reaction initiates a pathway that usually results in formation of pyruvate.

Information on the pathways followed during anaerobic catabolism of aromatic compounds is restricted to breakdown of benzoate by *Rhodopseudomonas palustris* growing photosynthetically. In this bacterium, benzoate is reduced to cyclohex-1-ene-1-carboxylate which is then metabolised by reactions similar to those employed in β-oxidation of fatty acids, namely addition of water to give 2-hydroxycyclohexanecarboxylate, and dehydrogenation to 2-ketocyclohexanecarboxylate. Ring cleavage of the last compound yields pimelate which can be oxidised to yield ATP (*Figure 6.18*).

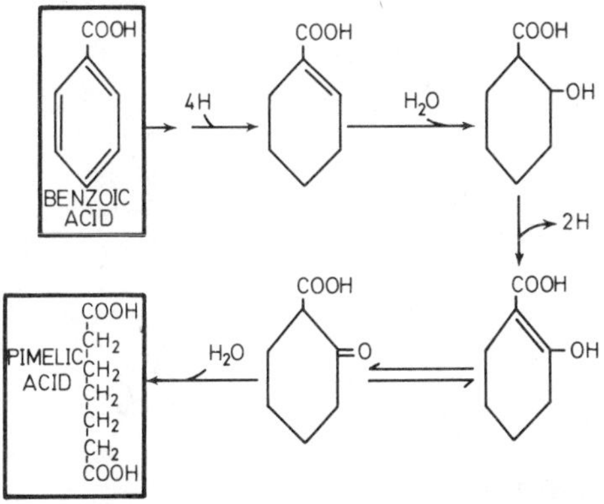

Figure 6.18 Pathway for catabolism of benzoic acid by Rhodopseudomonas palustris *growing anaerobically*

6.2.6 TWO-CARBON COMPOUNDS

The mechanisms of oxidation of two-carbon compounds by
micro-organisms present some very interesting features.
Acetate, and compounds (e.g. ethanol) that can give rise
to acetate, enter the TCA cycle directly (as acetyl-CoA)
and can then be oxidised completely. Two-carbon compounds
which are more highly oxidised than acetate (e.g.
glycollate) cannot enter the cycle directly. In many
micro-organisms, these substrates are first oxidised to
glyoxylate which is then oxidised completely by a
dicarboxylic acid cycle, details of which came largely
from work by the British biochemist Hans Kornberg
(*Figure 6.19*). In this cycle, glyoxylate reacts with

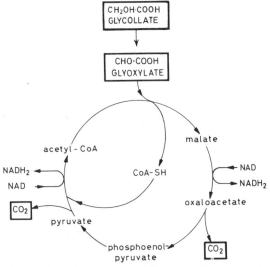

*Figure 6.19 Dicarboxylic acid cycle for oxidation of
two-carbon compounds*

acetyl-CoA to give malate in a reaction catalysed by
malate synthase. Malate is oxidised to oxaloacetate, and
then to pyruvate and carbon dioxide via phosphoenolpyruvate;
pyruvate then gives rise to acetyl-CoA. The reduced
nicotinamide nucleotides formed as a result of these
oxidations are re-oxidised by molecular oxygen with
concomitant formation of ATP.

Another pathway for glyoxylate catabolism operates in
Micrococcus denitrificans; this is known as the

β-*hydroxyaspartate pathway.* Glyoxylate formed by this bacterium is transaminated to give glycine which then condenses with a further molecule of glyoxylate to form erythro-β-hydroxyaspartate. This compound then undergoes a dehydratase reaction to yield oxaloacetate and ammonia.

The ability to use oxalate, the most highly oxidised two-carbon compound, as a sole carbon source is restricted to a small number of micro-organisms (e.g. *Pseudomonas oxalaticus*). The pathway involves oxidation to formate by the following series of reactions:

oxalate + succinyl-CoA→ oxalyl-CoA + succinate

$$\text{oxalyl-CoA} \xrightarrow{\text{TPP}} \text{formyl-CoA} + CO_2$$

formyl-CoA + succinate→ formate + succinyl-CoA

The main energy-yielding reaction in these organisms is the oxidation of formate by formate dehydrogenase.

6.2.7 ONE-CARBON COMPOUNDS

Although the ability to use one-carbon compounds as a source of energy is possessed by a sizeable number of microbes, with one or two exceptions the reactions involved in the energy-yielding metabolism of these organisms are cloaked in mystery. We have already noted that formate can be oxidised to carbon dioxide in a reaction catalysed by formate dehydrogenase, and which can be linked to NAD reduction.

In recent years, interest has centred on the mechanisms by which methane and methanol are used as sources of ATP, not least because these substrates are being contemplated as sources of energy for growing bacteria such as species of *Methylobacter* and *Methylococcus* to provide single-cell protein. For a long time it was thought that methane was oxidised directly to methanol, which in turn was oxidised to formaldehyde and thence to formate. But recent work in the Edinburgh laboratory of John Wilkinson has suggested the strong possibility that the sequence of reactions involved in oxidation of methane by bacteria is:

$$CH_4 \longrightarrow CH_3-O-CH_3 \longrightarrow CH_3-O-CH_2OH \longrightarrow CH_3-O-CHO$$

methane   dimethyl ether   methoxymethanol   methyl formate

$$CO_2 + H_2 \longleftarrow H-COOH \longleftarrow H-CHO \longleftarrow CH_3OH$$

        formic acid   formaldehyde   methanol

Evidence for this pathway has come largely from the
discovery that, by suitably manipulating the conditions
under which the bacteria are cultured, it is possible to
detect the presence of intermediates in the culture.
Because so far it has not been possible to prepare
extracts of methane-oxidising bacteria with demonstrable
ability to oxidise methane or dimethyl ether, nothing is
known of the properties of the enzymes which might
catalyse reactions on the first part of the pathway.
More is known about the later reactions, and methanol,
formaldehyde and formate dehydrogenases have been detected
in extracts of several bacteria that have been grown on a
one-carbon substrate. The methanol dehydrogenase in a
pseudomonad has been shown to be associated with a
pteridine, the function of which has yet to be elucidated.
   Another one-carbon compound which can be oxidised by
some micro-organisms is carbon monoxide. Although
microbial oxidation of carbon monoxide is thought to be
very important in the biosphere, very little is known of
the reactions involved.

6.2.8   ENDOGENOUS RESERVE POLYMERS

The majority of micro-organisms are able to survive,
often for long periods of time, when totally deprived of
exogenous nutrients. Organisms existing under these
conditions frequently remain viable and may show other
evidence of active metabolism such as motility and the
ability to respire. Clearly, these organisms are
producing sufficient energy for at least the maintenance
of life (turnover of nucleic acids and protein, osmotic
regulation) and this must be derived from oxidation of
intracellular constituents.
   Compounds, such as polysaccharides and poly-β-
hydroxybutyrate, which are laid down as energy stores in
micro-organisms, are described in Chapter 2, p. 80.
In the absence of an exogenous source of energy, micro-
organisms including both heterotrophs and autotrophs
rapidly break down these intracellular energy reserves
if available and, when these are depleted, often proceed
to use cellular protein and RNA. Intracellular energy
reserves are first broken down by depolymerising enzymes
to monomers (sugars, fatty acids, β-hydroxybutyrate)
which then enter one of the energy-yielding pathways
already described for exogenous substrates. It is worth
noting that the energy yield from the metabolism of

intracellular glycogen and starch via the EMP pathway, namely three molecules of ATP per molecule glucose equivalent, is greater than the yield from a glucose molecule entering the pathway (two molecules of ATP). This is because phosphorolysis of these polymers yields glucose 1-phosphate which is converted to glucose 6-phosphate before entering the EMP pathway. Therefore, ATP is not required for producing glucose 6-phosphate and the overall energy yield is correspondingly greater. Polysaccharides and lipids are hydrolysed by intracellular polysaccharidases and lipases. Information on the pathway for degradation of poly-β-hydroxybutyrate is most complete for *Azotobacter beijerinckii,* and the reactions involved are shown in *Figure 6.20.* Activity of the depolymerising enzyme gives rise to D(-)-3-hydroxy-butyrate, which is then converted to acetoacetate in a reaction catalysed by an NAD-specific D(-)-3-hydroxy-butyrate dehydrogenase. Acetoacetate is then converted to acetoacetyl-CoA in a reaction catalysed by a thiophorase. Finally, acetoacetyl-CoA is converted to acetyl-CoA which can be a source of ATP after oxidation in the TCA cycle.

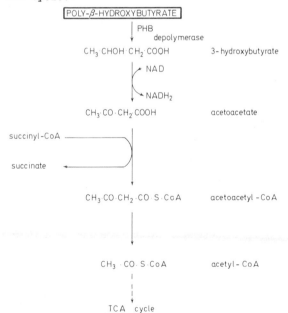

Figure 6.20 Pathway for the degradation of poly-β-hydroxybutyrate in Azotobacter beijerinckii

Micro-organisms are often able to utilise more than
one endogenous substrate as an energy source and
frequently show a preference for one substrate over
others. Yeasts (strains of *Sacch. cerevisiae*) can
ferment and respire endogenous substrates, glycogen
being metabolised most rapidly followed by trehalose.
In some organisms, such as *A. aerogenes* and *E. coli,*
protein and RNA are degraded once the polysaccharide
reserves have been depleted. Lipids are the first
endogenous substrates to be metabolised by mycobacteria
whereas, in *Pseudomonas aeruginosa,* intracellular protein
is the preferred substrate.

6.3   ENERGY FROM INORGANIC COMPOUNDS

The ability to obtain energy by oxidation of inorganic
compounds is confined to a relatively small group of
microbes, namely the chemolithotrophic bacteria. The
energy-yielding metabolism of these bacteria has
attracted far less attention than that of the chemo-
organotrophs, although the subject has at one time or
another interested some of the great pioneers in
microbial physiology including Sergius Winogradsky,
Martinus Willem Beijerinck, Jan Kluyver and Cornelius
van Niel. There are two main reasons for the comparative
unpopularity of this subject. Firstly, chemolithotrophic
bacteria are not of any medical importance. Secondly,
they are on the whole rather more difficult to cultivate
in the laboratory compared with chemo-organotrophic
bacteria. Nevertheless, information on the energy-
yielding metabolism of chemolithotrophs has gradually
accumulated over the years, and this is discussed in the
following paragraphs.

6.3.1   ELECTRON TRANSPORT IN CHEMOLITHOTROPHS

The inorganic substrates oxidised by chemolithotrophic
bacteria are, in general, rather specific for a given
organism. *Table 6.2* lists some of the better known
chemolithotrophic bacteria, together with the inorganic
substrates (electron donors) oxidised and the electron
acceptors used. In addition to those listed in *Table 6.2*
several other inorganic substrates, including thiocyanate
and Mn^{2+}, can be oxidised by certain of these organisms.
Both aerobic and anaerobic chemolithotrophs are known.

Table 6.2 Electron donors and acceptors used by some chemolithotrophic bacteria

Organism	Energy source (electron donor)	Electron acceptor	End-product of oxidation from electron donor	End-product of oxidation from electron acceptor
Beggiatoa spp.	Hydrogen sulphide	Oxygen	Sulphur	Water
Desulfovibrio desulfuricans	Hydrogen	Sulphate	Water	Hydrogen sulphide
Hydrogenomonas spp.	Hydrogen	Oxygen	Water	Water
Nitrobacter spp.	Nitrite	Oxygen	Nitrate	Water
Nitrosomonas spp.	Ammonium ion	Oxygen	Nitrite	Water
Thiobacillus denitrificans	Sulphur, thiosulphate and other inorganic sulphur compounds	Nitrate	Sulphate	Nitrogen
Thiobacillus ferro-oxidans	Ferrous ion, thiosulphate	Oxygen	Ferric ion	Water
Thiobacillus thio-oxidans	Sulphur, thiosulphate and other inorganic sulphur compounds	Oxygen	Sulphate	Water

Some, such as the hydrogen bacteria, are facultative autotrophs, whereas others like the thiobacilli are obligate autotrophs.

Cytochromes have been detected in extracts of many chemolithotrophic bacteria and, on general biochemical grounds, this is taken as evidence for the occurrence of oxidative phosphorylation. This conclusion is supported by reports that extracts of many of these bacteria can carry out oxidative phosphorylation. With some electron donors, it is easy to understand how oxidation can be coupled to oxidative phosphorylation. For example, the standard reduction potential (E'_0 value) of the $H_2/2H^+$ couple is -420 mV (*see Figure 6.1*), which makes it easy to explain how ATP synthesis in hydrogen bacteria can be coupled to electron flow knowing the standard reduction potentials of the cytochromes that have been detected in these bacteria. But this is not so with many other chemolithotrophic bacteria. For instance, the standard reduction potential of the nitrate/nitrite couple is around +420 mV which is appreciably higher than the standard reduction potential of the cytochrome c that has been detected in *Nitrobacter* species that oxidise nitrite, namely +250 mV. It has been suggested that, in order to lower the standard reduction potential of the substrate to a value more compatible with its reduction of the cytochrome, nitrite may be converted into adenyl nitrite. Similar difficulties are encountered in attempting to explain how bacteria oxidise Fe^{2+} with formation of ATP. The standard reduction potential of the Fe^{2+}/Fe^{3+} couple is +770 mV which is so close to that of the O_2/H_2O couple as to make ATP formation very unlikely. Again it has been suggested that the inorganic substrate is chemically modified to give rise to a more electronegative redox couple.

The initial steps in the metabolism of many of the electron donors are well established. With facultatively autotrophic hydrogen bacteria, such as *Hydrogenomonas ruhlandii,* the substrate is metabolised in a reaction catalysed by the enzyme hydrogenase:

$$H_2 \rightleftharpoons 2H^+ + 2e$$

Thiocyanate, which is oxidised by *Thiobacillus thioparus,* is first hydrolysed to cyanate and sulphide:

$$CNS^- + H_2O \longrightarrow HCNO + SH^-$$

Cyanate is then further hydrolysed to ammonia and carbon dioxide. Sulphide and ammonia may both function as electron donors. Oxidation of NH_4^+ by *Nitrosomonas* spp. is thought to involve the following reactions:

$$NH_4^+ + H_2O \rightarrow NH_2OH + 3H^+ + 2e$$

$$NH_2OH \rightarrow NO^- + 3H^+ + 2e$$

$$NO^- + H_2O \rightarrow NO_2 + 2H^+ + 2e$$

The second of these reactions is almost certainly the one used for cytochrome reduction by these bacteria. Little is known of the metabolism of nitrite, although as already mentioned it has been suggested that it is initially converted into adenyl nitrite.

Knowledge of the reactions undergone by the electron acceptors is also incomplete. Cytochrome oxidases presumably operate in those bacteria that use molecular oxygen as the electron acceptor. It is assumed too that nitrate reductases function in those organisms which use nitrate instead of oxygen. The reactions involved in dissimilatory sulphate reduction by *Desulfovibrio* spp. are comparatively involved. The first reaction, catalysed by ATP-sulphurylase, leads to formation of adenosine 5'-phosphosulphate (APS):

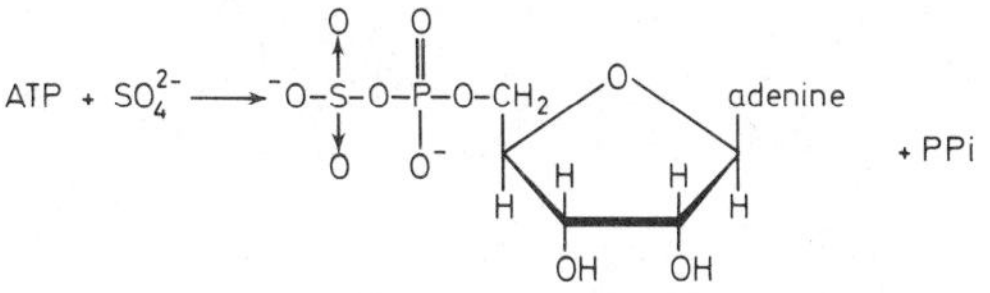

adenosine 5′-phosphosulphate

In a reaction catalysed by APS reductase, APS is converted to sulphite and AMP, and the sulphite is then further reduced to sulphide. The intermediates formed *en route* to sulphide are unknown; the suggestion has been made that they are unstable and remain bound to enzymes.

6.3.2 PRODUCTION OF REDUCING POWER IN CHEMOLITHOTROPHS

Organic cell constituents are, in general, much more reduced than carbon dioxide and correspond approximately to the empirical formula CH_2O or C_2H_3O. When growing autotrophically, chemolithotrophic bacteria must, therefore, produce not only ATP to drive the synthetic reactions, but

also reduced nicotinamide nucleotides to effect reduction
of the carbon dioxide substrate. The metabolic pathways
by which carbon dioxide is converted into organic cell
material by these bacteria are discussed on page 255.
In hydrogen bacteria, formation of reduced nicotinamide
nucleotides is coupled to the action of the enzyme
hydrogenase. With *Nitrobacter* spp. and the iron and
sulphur bacteria, very little is known of the way in
which reduced nucleotides might be formed. With some of
these bacteria, the standard reduction potentials are
such as to make direct formation of reduced nicotinamide
nucleotides highly improbable, and it is possible that
they have other mechanisms for generating $NADH_2$. One
which it has been suggested operates in species of
Ferrobacillus, *Nitrobacter* and *Nitrosomonas* is an
ATP-dependent reduction of NAD.
    Much thought has been given to the physiological basis
of obligate autotrophy in chemolithotrophic bacteria.
Two main explanations have been proposed. Some physiologists believe that bacteria are obligately autotrophic
because they are unable to synthesise certain enzymes,
including some that catalyse reactions in the TCA cycle
as well as $NADH_2$ oxidase. An alternative explanation is
that obligately autotrophic bacteria are unable to transport organic compounds into the cell.

6.4 ENERGY FROM VISIBLE RADIATION

Algae, along with green-sulphur, and purple-sulphur and
·non-sulphur - bacteria and the phytoflagellates, resemble
green plants in being able to convert the energy of
visible radiation directly into the high-energy bonds of
ATP. This they do by a process of *photosynthesis*.
A great deal has been written in the past concerning
microbial and green plant photosynthesis particularly
with regard to the nature of the end-products. It is now
clear, however, that the only product common to the
process in all of these organisms is ATP, formed by a
process of *photophosphorylation*. The use that is made
of this ATP is determined by the chemical environment and
the enzymes synthesised by the organisms.
    In photolithotrophic microbes (algae, green sulphur
bacteria and many purple sulphur bacteria) most of the
ATP is used in synthesis of cell constituents from carbon
dioxide, and this process requires a source of reducing
power. In algae, blue-green algae, phytoflagellates and green

plants, the source of this hydrogen is water, which is
thereby oxidised to molecular oxygen. However, molecular
oxygen is never evolved during bacterial photosynthesis,
and in photolithotrophic bacteria an alternative
reductant, usually hydrogen sulphide, is required. The
non-sulphur purple bacteria, on the other hand, are
organotrophs and use organic compounds as raw materials
in the photosynthesis of cell material. When these
organic compounds are at an oxidation level similar to
that of cell constituents, the need for an exogenous
inorganic reductant disappears. An example of direct
photo-assimilation of an organic substrate is the
conversion by *Rhodospirillum rubrum* of 3-hydroxybutyrate
to the polymer poly-β-hydroxybutyrate which occurs
without photoreduction or any uptake or formation of
carbon dioxide. When the organic substrate is more
oxidised than cell material, there is an oxidation of
part of the substrate to carbon dioxide and this
oxidation provides reducing power for synthesis of cell
material from other molecules of substrate. With organic
substrates which are more reduced than the cell material,
there is a partial oxidation of the substrate coupled
with a reduction and assimilation of carbon dioxide. In
the dark, some of these photo-organotrophic bacteria
oxidise organic substrates via the TCA cycle.

The ability of phototrophic micro-organisms to utilise
the energy of visible radiation calls for the presence in
these micro-organisms of pigments that are capable of
absorbing this energy. These photosynthetic pigments are
contained, along with other pigments and associated
electron-transport carriers and enzymes, in membraneous
structures, which are described in Chapter 2 (*see* page
75). In algae, the structures are chloroplasts, but in
bacteria and blue-green algae the photosynthetic apparatus
is present in a variety of different morphological forms.

6.4.1 PHOTOSYNTHESIS IN EUKARYOTES AND BLUE-GREEN ALGAE

Experiments using monochromatic light led to the conclusion
that isolated algal chloroplasts possess two functionally
distinct light-trapping systems, now referred to as
Photosystem I and *Photosystem II* (*Figure 6.21*). In both
photosystems, conversion of light energy into chemical
energy occurs through the co-operation of a large number
of light-absorbing chlorophyll molecules which transfer
this energy by inductive resonance to a smaller number of

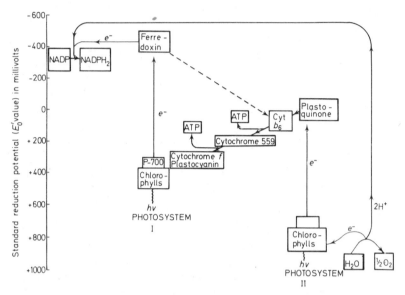

Figure 6.21   A diagrammatic representation of non-cyclic photophosphorylation in chloroplasts.   The broken line indicates the path that may be taken by electrons in cyclic photophosphorylation

photochemically active molecules.   The complex comprising the photosynthetically active pigments and associated molecules is known as the *reaction centre* in the photosystem.   The reactive pigment in Photosystem I is a single-electron carrier with a standard reduction potential of around +450 mV, and called pigment P-700. Although firm evidence is lacking, it is generally believed that there is also a similar reaction centre in Photosystem II.

Absorption of a quantum of light by the reaction centre in Photosystem I or II results ultimately in the expulsion of an electron at a high energy potential, the energy being derived from the quantum of light as in a photoelectric cell.   As a result of this expulsion, a molecule in the photosystem is positively charged.   Electrons expelled from Photosystem I are accepted by ferredoxin molecules which are one-electron carriers (*Figure 6.21*). They have a molecular weight of about 12 000 daltons and contain two iron and two labile sulphur atoms in each molecule.   The standard reduction potential of these ferredoxin molecules is greater than about -400 mV, so that the electrons accepted by the molecules are capable of yielding a considerable amount of energy. There is some evidence for an even more highly electronegative carrier accepting electrons expelled from the reaction centre in Photosystem I, but little is known about its properties. By contrast, electrons expelled from Photosystem II are accepted by plastoquinone, which has a much lower standard reduction potential of around zero millivolts.

Operation of both photosystems is required to raise the electrons from the level of the H_2O/O_2 couple (+820 mV) to the level of $NADPH_2$ (-340 mV). Electrons expelled from Photosystem I and accepted by ferredoxin are used, together with hydrogen ions obtained from splitting or photolysis of water, to reduce NADP to $NADPH_2$ (*Figure 6.21*). Electrons expelled from Photosystem II, after being accepted by plastoquinone, are passed down an electron-transport chain and, as a result, ATP is synthesised from ADP. Three different cytochromes are localised in the chloroplasts of eukaryotic cells, namely cytochrome *f* which is a *c*-type cytochrome, and two cytochromes *b* designated cytochrome b_6 and cytochrome 559. Also in the electron-transport chain is plastocyanin, which is an acidic protein containing two atoms of copper in each molecule. Although several viewpoints prevail, the concensus of opinion is that phosphorylation of ADP to ATP takes place during transfer of electrons from cytochrome b_6 to cytochrome 559 and from cytochrome 559 to plastocyanin. At the end of the electron-transport chain, the electrons enter Photosystem I and so restore electroneutrality to the pigment molecules, which explains why the overall process is known as *non-cyclic photophosphorylation*. Meanwhile, the positive charge on the molecules in Photosystem II is neutralised by electrons derived from photolysis of water (*Figure 6.21*).

Some workers on photosynthesis believe that there is an alternative fate for electrons that are expelled by Photosystem I and accepted by ferredoxin, in that they might be transferred directly to cytochrome b_6 in the electron-transport chain, and so be made to yield ATP. This process is referred to as *cyclic photophosphorylation*. However, it has to be admitted that the operation of this process is viewed with not a little scepticism by many authorities in the field, who opine that electrons expelled from Photosystem I are used only to reduce NADP.

Several antimicrobial compounds inhibit photophosphorylation in chloroplasts. The herbicide 3-(3,4-dichlorophenyl)
-1,1-dimethylurea (DCMU) inhibits reduction of cytochrome
559, while antimycin A_1, which prevents Site II activity
in oxidative phosphorylation (*see* page 203), stops
reduction of cytochrome *f*.

Overall, the photosynthetic reactions which take place
in blue-green algae are the same as those in chloroplasts,
and include non-cyclic and possibly cyclic photophosphorylation, and reduction of NADP with water as the
proton donor. Cytochromes of the *c*-type are found in
blue-green algae; one of them, cytochrome c_{553}, functions
in place of the cytochrome *f* found in chloroplasts.

6.4.2   BACTERIAL PHOTOSYNTHESIS

Photosynthetic bacteria do not possess an accessory
pigment system, and it is generally believed that the
photosynthetic apparatus in these microbes has only one
type of photosystem. Light is absorbed by chlorophyll
molecules in the unit, and the energy is passed on to a
reaction-centre pigment often referred to as pigment
P-890. In bacterial photophosphorylation, a cyclic
process, electrons expelled from the centre are accepted
by an electronegative carrier, which may or may not be a
ferredoxin (*Figure 6.22*). The ferredoxins found in
photosynthetic bacteria differ in properties from those
of green and blue-green algae, and are of the
characteristic bacterial type. They are blackish-brown
in colour with absorption maxima around 390 nm, and

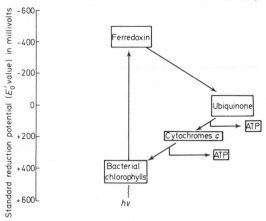

*Figure 6.22   A diagrammatic representation of photo-
phosphorylation in bacteria*

contain between four and six atoms of iron in each molecule. Ferredoxins from green bacteria resemble those from clostridia in molecular weight (about 6000 daltons) and amino-acid composition. Ferredoxins from *Chromatium* spp., on the other hand, have a molecular weight nearer 10 000 daltons and an exceptionally low standard reduction potential of about -490 mV.

Electrons accepted by the initial carrier are transferred to ubiquinone, and then proceed to pass down an electron-transport chain in which there are probably two phosphorylation sites (*Figure 6.22*). A wide range of cytochromes, mainly of the c-type, have been detected in photosynthetic bacteria. Purple bacteria contain a high-potential cytochrome c_2, equivalent in potential to the cytochrome f of green algae, as well as another c-type cytochrome which contains two haem groups (called 'cytochrome cc'). *Chlorobium thiosulfatophilum* contains three c-type cytochromes, none of which has a high standard reduction potential equivalent to that of cytochrome f. At the end of the electron-transport chain the electrons return to the photosystem, thereby restoring electroneutrality.

Some researchers on bacterial photosynthesis hold that non-cyclic photophosphorylation also operates in photosynthetic bacteria. Growing autotrophically, these microbes need to synthesise $NADPH_2$ from NADP, and one possible way in which to do this is for the bacteria to use an exogenous electron donor (molecular hydrogen, hydrogen sulphide, thiosulphate) as a source of hydrogen ions, and to effect synthesis of $NADPH_2$ in a manner similar to that which operates in chloroplasts (*Figure 6.22*). Other workers in the field, championed by Howard Gest at Indiana University, U.S.A., believe that non-cyclic photophosphorylation does not operate in photosynthetic bacteria, and that the bacteria obtain $NADPH_2$ by a process of reverse electron flow using, for instance, succinate and the relatively high standard reduction potential of the succinate-fumarate couple (*see Figure 6.1*).

Under certain conditions, photosynthetic bacteria and some algae evolve molecular hydrogen when illuminated in the presence of an oxidisable substrate. The process is thought to involve a reduction of hydrogen ions by reduced nicotinamide nucleotides. It has been suggested that this 'hydrogen valve' may be important as a mechanism for draining electrons away from the photo-

phosphorylation process when the organism has a
diminished requirement for ATP. Yet another possible
fate for the reduced nicotinamide nucleotides is in
nitrogen fixation (*see* page 264), which is carried out
by both purple and green bacteria when growing
anaerobically in light.

FURTHER READING

PRINCIPLES OF BIOENERGETICS

LEHNINGER, A.L. (1971). *Bioenergetics,* 2nd ed., 270 pp.
Menlo Park, California; W.A. Benjamin Inc.
MORRIS, J.G. (1974). *A Biologist's Physical Chemistry,*
2nd edit. 408 pp. London; Edward Arnold Ltd.

ENERGY FROM ORGANIC COMPOUNDS

Carbohydrates

ANDERSON, R.L. and WOOD, W.A. (1969). Carbohydrate
metabolism in microorganisms. *Annual Review of
Microbiology,* 23, 539-578
CHANCE, B. (1972). The nature of electron transfer and
energy coupling factors. *Federation of European
Biochemical Societies Letters,* 23, 3-20
DE LEY, J. and KERSTERS, K. (1964). Oxidation of
aliphatic glycols by acetic acid bacteria.
Bacteriological Reviews, 28, 164-180
FRAENKEL, D.G. and VINOPAL, R.T. (1973). Carbohydrate
metabolism in bacteria. *Annual Review of Microbiology,*
27, 69-100
HASTINGS, J.W. (1968). Bioluminescence. *Annual Review
of Biochemistry,* 37, 597-630
HORIO, T. and KAMEN, M.D. (1970). Bacterial cytochromes.
II. Functional aspects. *Annual Review of Microbiology,*
24, 399-428
JURTSHUK, P., ACORD, W.C. and MUELLER, T.J. (1974).
Bacterial terminal oxidases. *Critical Reviews in
Microbiology,* 3, 399-468
KAMEN, M.D. and HORIO, T. (1970). Bacterial cytochromes.
I. Structural aspects. *Annual Review of Biochemistry,*
39, 673-700
MORRIS, J.G. (1975). The physiology of obligate
anaerobiosis. *Advances in Microbial Physiology,* 12,
169-245

MORTLOCK, R.P. (1975). Pathways of catabolism of
unnatural carbohydrates by micro-organisms. *Advances
in Microbial Physiology,* 13, 1-53
SOKATCH, J.R. (1969). *Bacterial Physiology and
Metabolism,* 443 pp. London; Academic Press
WHITE, D.C. and SINCLAIR, P.B. (1971). Branched electron-
transport systems in bacteria. *Advances in Microbial
Physiology,* 5, 173-211

Nitrogenous compounds

THIMANN, K.V. (1963). *The Life of Bacteria,* 2nd edn.
pp.306-363. New York; The Macmillan Company

Triglycerides and fatty acids

WAKIL, S.J. (1970). Fatty acid metabolism. In:
Lipid Metabolism, Ed. S.J. Wakil, pp.1-48. New York;
Academic Press

Hydrocarbons

GUNSALUS, I.C. and MARSHALL, V.P. (1972). Monoterpene
dissimilation: chemical and genetic models. *Critical
Reviews in Microbiology,* 1, 291-310
KLUG, M.J. and MARKOVETZ, A.J. (1971). Utilization of
aliphatic hydrocarbons by micro-organisms. *Advances
in Microbial Physiology,* 5, 1-43
MARKOVETZ, A.J. (1972). Subterminal oxidation of
hydrocarbons by micro-organisms. *Critical Reviews in
Microbiology,* 1, 225-237

Aromatic compounds

DAGLEY, S. (1971). Catabolism of aromatic compounds by
micro-organisms. *Advances in Microbial Physiology,* 6,
1-46
GIBSON, D.T. (1972). The microbial oxidation of aromatic
hydrocarbons. *Critical Reviews in Microbiology,* 1,
199-223

242 *Energy-yielding metabolism*

Two-carbon compounds

KORNBERG, H.L. (1966). Anaplerotic sequences and their
role in metabolism. *Essays in Biochemistry*, 2, 1-31

One-carbon compounds

QUAYLE, J.R. (1972). The metabolism of one-carbon
compounds by micro-organisms. *Advances in Microbial
Physiology*, 7, 119-203
RIBBONS, D.W., HARRISON, J.E. and WADZINSKI, A. (1970).
Metabolism of single carbon compounds. *Annual Review
of Microbiology*, 24, 135-158

Endogenous reserve polymers

DAWES, E.A. and SENIOR, P.J. (1973). The role and
regulation of energy reserve polymers in micro-
organisms. *Advances in Microbial Physiology*, 10,
135-266

ENERGY FROM INORGANIC COMPOUNDS

HORIO, T. and KAMEN, M.D. (1970). Bacterial cytochromes.
II. Functional aspects. *Annual Review of Microbiology*,
24, 399-428
KELLY, D.P. (1971). Autotrophy: concepts of
lithotrophic bacteria and their organic metabolism.
Annual Review of Microbiology, 25, 177-210
RITTENBERG, S.C. (1969). The roles of exogenous organic
matter in the physiology of chemolithotrophic bacteria.
Advances in Microbial Physiology, 3, 159-196
SUZUKI, I. (1974). Mechanisms of inorganic oxidation
and energy coupling. *Annual Review of Microbiology*,
28, 85-101

ENERGY FROM VISIBLE RADIATION

BISHOP N.I. (1971). Photosynthesis: the electron-
transport system of green plants. *Annual Review of
Biochemistry*, 40, 197-226
EVANS, M.C.W. and WHATLEY, F.R. (1970). Photosynthetic
mechanisms in prokaryotes and eukaryotes. *Symposium
of the Society for General Microbiology*, 20, 203-220

GEST, H. (1972). Energy conversion and generation of
reducing power in bacterial photosynthesis. *Advances
in Microbial Physiology,* 7, 243-282
HALL, D.O. and RAO, K.K. (1972). *Photosynthesis,* 68 pp.
London; Edward Arnold Ltd.
PARSON, W.W. (1974). Bacterial photosynthesis. *Annual
Review of Microbiology,* 28, 41-59

7

ENERGY EXPENDITURE: BIOSYNTHESES

In a growing culture of micro-organisms, some of the ATP
formed by the pathways described in the previous chapter
is used in the synthesis of cell components. The present
chapter describes the pathways which microbes use for
synthesising cell components. But before considering
these biosynthetic pathways, we must look in a little
more detail at the ways in which ATP formed in energy-
yielding pathways is channelled into other metabolic
processes that utilise ATP. In other words, we begin
this chapter with an inspection of the ATP balance sheet
in micro-organisms.

7.1   THE ATP BALANCE SHEET

We have already noted that some of the ATP produced in
the energy-yielding pathways used by micro-organisms is
consumed in the synthesis of new cell constituents. When
discussing this aspect of metabolism, microbial
physiologists refer to the tight *coupling* between
catabolic and anabolic reactions in a population of
growing micro-organisms. All microbes contain a small
quantity of free ATP, often referred to as the *ATP pool*.
However, the amount of ATP in a micro-organism is
usually around 10 µmoles per gram dry weight of organism,
which is quite small when one considers the amount of ATP
consumed in producing a gram dry weight of microbe. This
latter value can be calculated for certain organisms,
such as anaerobically growing bacteria (*see* page 246), and

for *Aerobacter cloacae* growing anaerobically on glucose
it equals about 15 mmoles.   In other words, a microbe
may contain, in the free form, about one-thousandth of
the amount of ATP needed to produce that microbe.

In the following paragraphs there is a discussion of the
main debit items in the ATP balance sheet of micro-
organisms.   It is perhaps rather surprising that, after
almost a century of research on the energy-yielding
metabolism of microbes and well over half a century of
work on the ways in which microbes carry out biosynthetic
reactions, microbial physiologists are a long way off
being able to submit a balance sheet that would satisfy
any self-respecting accountant.

7.1.1   ENERGY EXPENDED IN BIOSYNTHESES

Over the years, several attempts have been made to
calculate the amount of microbial cell material
synthesised from a known amount of ATP produced by the
microbes.   These calculations can easily be made, and
with some confidence, when organisms are grown under
conditions where the amount of ATP produced from
consumption of a known amount of energy-yielding
substrate can be calculated accurately.   Since the yield
of ATP from oxidative phosphorylation in microbes, and
particularly in bacteria, is very difficult to assess
because of our present ignorance of the efficiency
(that is, the P:O ratio) of the process, such
calculations have so far been confined largely to
microbes growing anaerobically on substrates from which
the ATP yield, by substrate-level phosphorylation, can
be calculated.   Values for this relationship are
expressed as the number of grams dry weight of cell
material produced from one mole of ATP.   Conventionally,
they are referred to as Y_{ATP} values or, in deference to
Sydney Elsden and Tom Bauchop, two British microbial
physiologists who pioneered this area of study, as
Bauchop-Elsden values.

The first Bauchop-Elsden values to be determined were
for anaerobic growth of *Streptococcus faecalis,*
Saccharomyces cerevisiae and *Pseudomonas lindneri*
(now known as *Zymononas mobilis*), and the average value
for these microbes was 10.5 (that is 10.5 grams dry
weight of organisms produced from each mole of ATP).
Since then, many further determinations have been made

and, with very few exceptions, they all fall near the
value of 10.5, which indicates, somewhat surprisingly
perhaps, that the processes of biosynthesis have
apparently about the same efficiency in different
micro-organisms.

As we shall see later in this chapter, knowledge about
the individual reactions on pathways that lead to
synthesis of microbial cell constituents is now extensive,
and the belief that microbial biosynthetic pathways have
been almost completely charted has tempted microbial
physiologists to try to calculate the amount of ATP
required to synthesise unit mass (usually one gram dry
weight) of microbe, knowing the requirement for ATP on
each of the pathways and assuming that all monomers are
available preformed. *Table 7.1* gives these data for a
typical bacterium. They show that, from one mole of
ATP, it should be possible to produce about 31 g dry
weight of bacterium. The data in *Table 7.1* make several
assumptions – for instance that the amount of ATP
expended in synthesis of mRNA is the same as with other
species of RNA – but even making allowances for these,
the calculation for the amount of microbial cell material
that can be made from one mole of ATP is always well
above the experimentally determined value of 10.5. This
discrepancy is taken to indicate that a microbe expends
a sizeable proportion of the ATP that it makes in
processes other than those which lead to the formation
of cell material. Microbial physiologists have
appreciated for a long time that a microbe uses ATP for
processes other than biosynthesis; the surprise when this
calculation is made is that the proportion is so large.

7.1.2 MAINTENANCE ENERGY

All microbes expend some ATP to maintain the integrity
of the organism, principally in the synthesis of
macromolecules, such as proteins and nucleic acids, that
are continually being degraded and resynthesised. The
turnover of cell components in micro-organisms,has been
studied mainly in relation to breakdown and synthesis of
proteins, although it is probably just as extensive with
many other cell macromolecules. The rate of turnover of
cell protein varies with the environmental conditions in
which a microbe finds itself. In a population of growing
microbes, the rate of protein turnover rarely exceeds

Table 7.1 Calculation of the amount of energy expended in synthesis of one bacterium from preformed monomers

Component	% dry weight of cell	Approx. mol. wt. (daltons $\times 10^{-6}$)	No. of molecules per cell	No. of monomers per molecule	Total no. of monomers per cell	No. of ATP molecules needed for addition of one monomer	No. of ATP molecules needed to synthesise one cell ($\times 10^{-6}$)	% total ATP expended
DNA	3	2000	1	6.5×10^6	6.5×10^6	2	13	0.27
RNA	10	1.0	15 000	3.1×10^3	47×10^6	2	94	1.94
Protein	62	0.06	1 600 000	550	880×10^6	5	4 400	91.0
Lipid	10	0.001	15 000 000	1	15×10^6	7	105	2.17
Polysaccharide	15	0.2	113 000	1 000	113×10^6	2	226	4.67

4838×10^6 Molecules of ATP are required to synthesise one bacterium (2.5×10^{-13} g dry weight); 6.02×10^{23} molecules (the number in one mole) should therefore synthesise 31 g dry weight of cells.

2-3% h^{-1}, but when the cells are subjected to starvation conditions the rate rises appreciably.

Several attempts have been made to estimate the amount of ATP needed by a microbe for maintenance energy. Earlier studies aimed to calculate the amount of energy source that must be continually added to a population of non-growing microbes in order just to prevent loss of viability. But these studies were vitiated by the strong possibility of restricted growth of some cells in the population using compounds released from dying or dead cells, a phenomenon known as *cryptic growth* (*see* page 393). Undoubtedly the most elegant method for calculating the maintenance energy requirement is that devised by John Pirt and described on page 389. This entails plotting the reciprocal of the yield value against the reciprocal of the growth rate for continuously growing cultures of micro-organisms. The slope of the straight line obtained gives a value for the *maintenance coefficient*. The values for this coefficient are predictably greater for microbes grown anaerobically because of the lower efficiency of energy-yielding metabolism under anaerobic conditions. Thus, the maintenance coefficient for aerobic growth of *Aerobacter cloacae* is 0.09 grams of glucose per gram dry weight bacterium, and for anaerobic growth 0.47 grams glucose per gram dry weight. Using the latter value, it is possible to calculate the amount of ATP required to maintain one gram dry weight bacterium, which is about one-twentieth of the amount of ATP needed to produce one gram dry weight of bacteria.

7.1.3 ENERGY EXPENDED IN SOLUTE UPTAKE

Energy is expended in uptake of solutes during active transport and group translocation (*see* page 159). However, it is extremely difficult if not impossible to calculate the ATP expenditure because of our ignorance of the molecular mechanisms involved and of the proportion of molecules which are transported into a microbe by active transport or group translocation, as distinct from the energy-independent process of facilitated diffusion, under any particular set of environmental conditions. Chancing one's arm, it might be guessed that the amount of ATP expended in solute uptake is an appreciable proportion of the total ATP used by the cell.

7.1.4 HEAT PRODUCTION

Some ATP is squandered by micro-organisms, and liberated
as heat. One way in which heat is evolved is in
biosynthetic reactions which require ATP but not all of
the energy in one high-energy phosphate bond. Thus,
only about 3 kcal are required to form an amide or ester
bond in a molecule, although in reactions leading to
formation of these bonds one complete high-energy bond
(i.e. 7.3 kcal) is actually used. The bulk of the energy
in this bond is given off in the form of heat. Heat
production by a microbial culture can be determined using
sensitive microcalorimetry techniques, and calculations
show that it does not represent a large proportion of the
total ATP produced. Heat production, which is a problem
in many very large-scale cultures used in the fermentation
industries, is greater in microbes than in higher
organisms. It has been calculated that heat is produced
at the rate of one kcal per hour per gram dry weight of
Escherichia coli; comparable values for Drosophila are
0.1 kcal and for man 0.01 kcal. Excessive rates of heat
production by microbes may indicate a lower efficiency
of ATP coupling in biosynthetic reactions than in higher
organisms.

7.1.5 ENERGY EXPENDED IN MOVEMENT

Energy is also consumed in the movement of cilia and
flagella by motile organisms. It has been calculated
that as much as 10% of the total ATP expended by certain
algae is used in flagellar locomotion. Energy expenditure
by flagella is, on the whole, a very inefficient process.
Hydrodynamic calculations show that only about 10% of the
energy expended by the flagellum is used to propel *Polytoma
uvella,* the remainder being necessary to sustain the
transverse oscillations in the viscous medium.

7.2 RAW MATERIALS FOR BIOSYNTHESES

The first requirement in the biosynthesis of cell
constituents is a suitable supply of low molecular-weight
compounds (e.g. sugars, amino acids) which serve as
precursors or raw materials in biosyntheses. These
compounds may be present in the environment and the

organism can then channel them directly into biosynthetic
pathways. The environment is never able to supply all of
the low molecular-weight compounds needed in the
biosynthetic reactions of a micro-organism, so that at
least some, and often all, of these compounds need to be
manufactured by the organism from available nutrients.

Among chemo-organotrophic micro-organisms, many of
these biosynthetic raw materials are formed during the
breakdown of organic compounds to yield ATP. A wide
variety of two-, three-, four- and five-carbon compounds
are furnished by catabolic reactions, and these compounds
can often be tapped off and used in biosyntheses.

7.2.1 ANAPLEROTIC SEQUENCES

There is an especially heavy drain on intermediates of
the TCA cycle, particularly in micro-organisms
synthesising amino acids, certain of which are formed
from pyruvate, α-ketoglutarate or fumarate. Another
TCA-cycle intermediate, succinyl-CoA, is used in
tetrapyrrole synthesis, while acetyl-CoA is consumed in
the biosynthesis of fatty acids. Theoretically the TCA
cycle can continue to operate using one molecule of
oxaloacetate, providing that this molecule is continually
regenerated. If TCA-cycle intermediates are tapped off
as biosynthetic raw materials, the cycle can continue to
operate only if this deficit is made good. Micro-
organisms have been shown to possess several metabolic
devices for ensuring that adequate supplies of TCA-cycle
intermediates are available. These metabolic routes
have been described as *anaplerotic sequences* (Gr.
'filling up') by the British biochemist Hans Kornberg.
Anaplerotic sequences, the principal ones of which are
described below, are virtually irreversible under
physiological conditions.

Many organisms, especially those in which the catabolic
pathways follow the latter part of the EMP scheme, are
able to replenish certain TCA-cycle intermediates by
reactions involving carbon-dioxide fixation. These
reactions can be divided into two categories, namely
heterotrophic and autotrophic. Heterotrophic carbon-
dioxide fixation is carried out by probably all micro-
organisms and involves the addition of carbon dioxide to
organic acceptor molecules; but it does not result in
the total synthesis of these acceptor molecules from
carbon dioxide. Autotrophic carbon-dioxide fixation is

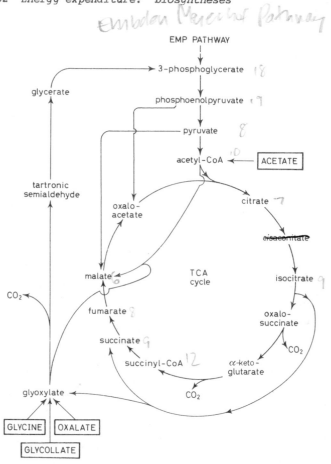

Figure 7.1   Reactions leading to the replenishing of TCA-cycle intermediates in micro-organisms.   See text for explanation

characteristic of lithotrophic micro-organisms and is dealt with later in this chapter.

Several types of carbon dioxide-fixing reactions leading to the synthesis of TCA-cycle intermediates have been shown to operate in micro-organisms; these are shown in *Figure 7.1.* The reaction between phosphoenolpyruvate and carbon dioxide to yield oxaloacetate has been demonstrated using cell-free extracts of several micro-organisms. At least three different enzymes have been shown to catalyse this reaction:

$$\text{phosphoenolpyruvate} + CO_2 \xrightarrow{\substack{\text{phosphoenol-}\\ \text{pyruvate}\\ \text{carboxylase}}} \text{oxaloacetate} + Pi$$

$$\text{phosphoenolpyruvate} + CO_2 + GDP \xrightarrow[\text{(ADP)carboxykinase}]{\substack{\text{phosphoenol-}\\ \text{pyruvate}}} \begin{array}{l}\text{oxaloacetate}\\ +\\ GTP\ (ADP)\end{array}$$

$$\text{phosphoenolpyruvate} + CO_2 + Pi \xrightarrow[\substack{\text{carboxytrans-}\\ \text{phosphorylase}}]{\substack{\text{phosphoenol-}\\ \text{pyruvate}}} \begin{array}{l}\text{oxaloacetate}\\ +\\ PPi\end{array}$$

Carboxylation of pyruvate to give malate, a reaction catalysed by the malate enzyme, could also serve as a means of replenishing TCA-cycle intermediates. In most micro-organisms, however, the equilibrium in this reaction appears to favour pyruvate formation, and it is unlikely that the reaction contributes appreciably to the synthesis of malate. In some micro-organisms, pyruvate can be carboxylated by the reaction:

$$\text{pyruvate} + CO_2 + ATP \rightleftharpoons \text{oxaloacetate} + ADP + Pi$$

The enzyme catalysing this reaction – pyruvate carboxylase is one of a group of carbon dioxide-fixing enzymes that require biotin as a prosthetic group.

7.2.2 METABOLISM OF TWO-CARBON COMPOUNDS

In order that carbon dioxide-fixing reactions can operate in replenishing TCA-cycle intermediates, it is essential that pyruvate or phosphoenolpyruvate be available as acceptor molecules. In the strict anaerobe, *Clostridium kluyveri*, pyruvate can be formed by carboxylation of acetyl-CoA:

$$CH_3CO.S.CoA + CO_2 + 2H \rightarrow CH_3.CO.COOH + CoA.SH$$

This reaction, which only takes place under strongly reducing conditions such as obtain in *Cl. kluyveri,* involves the participation of ferredoxin.

Aerobic micro-organisms, however, cannot carboxylate two-carbon compounds in this way. When growing on acetate as the sole carbon source, these organisms synthesise TCA-cycle intermediates by reactions of the *glyoxylate cycle* which is also known as the glyoxylate by-pass of the TCA cycle (*Figure 7.1*). The key reactions in this cycle are those which involve the breakdown of isocitrate to succinate and glyoxylate, catalysed by isocitrate lyase, and the formation of malate from glyoxylate and acetyl-CoA by malate synthase. As a result of these reactions, the oxidation steps of the TCA cycle between isocitrate and succinate — those reactions that result in the evolution of carbon dioxide — are by-passed. Although the TCA and glyoxylate cycles share certain intermediates, their functions are quite different. The TCA cycle leads to the combustion of acetate with release of energy, while the glyoxylate cycle results in the synthesis of four-carbon dicarboxylic acids.

With other two-carbon compounds such as glycine, glycollate and oxalate, some of which are more highly oxidised than acetate, quite different mechanisms operate for replenishing TCA-cycle intermediates. These substrates are first converted to glyoxylate, glycollate and glycine by oxidation and oxalate by reduction. Glyoxylate then enters the *glycerate pathway.* Glyoxylate carboligase catalyses the condensation of two molecules of glyoxylate to yield carbon dioxide and tartronic semialdehyde:

$$2CHO.COOH \rightarrow CHO.CHOH.COOH + CO_2$$

Tartronic semialdehyde is then reduced to glycerate in a reaction catalysed by tartronic semialdehyde reductase (*Figure 7.1*). Glycerate, after phosphorylation, is metabolised to pyruvate by the EMP pathway. Carbon dioxide-fixing reactions using phosphoenolpyruvate or pyruvate as acceptor molecules can then operate. The major reactions of the glycerate pathway have been demonstrated using extracts from many different micro-organisms.

7.2.3 METABOLISM OF ONE-CARBON COMPOUNDS

We saw in Chapter 3 (page 100) that a variety of one-carbon compounds, ranging in oxidation state from carbon dioxide to methane, can be used as sole carbon sources by many microbes. The unique property of these microbes is their capacity to synthesise three-carbon compounds from one-carbon compounds. The C_3 compounds formed then enter the biosynthetic pathways used in all other organisms, and which are described later in this chapter.

Assimilation of carbon dioxide

The ability to use carbon dioxide as a sole carbon source is the property of autotrophic microbes. The main pathway by which carbon dioxide is incorporated into organic compounds in autotrophs was charted as a result of work by Melvin Calvin and his colleagues in the University of California at Berkeley, U.S.A., and in recognition of this work is known as the *Calvin cycle*.

Calvin and his colleagues began their work on the mechanism of carbon-dioxide fixation in 1946 using two unicellular algae, *Chlorella pyrenoidosa* and *Scenedesmus obliquus*, as experimental organisms. The methods used by this group in following the path of carbon-dioxide fixation involved adding $^{14}CO_2$, in the form of bicarbonate, to an illuminated suspension of algae and, after a few seconds or minutes depending on the experiment, killing the organisms by rapid immersion in ethanol. Extracts of the algae were then prepared and the radioactive-labelled intermediates in the extracts identified by a combination of paper chromatography and autoradiography. Later, when the major steps in the pathway had been discovered, confirmation of the scheme was obtained by demonstrating the ability of cell-free extracts of the algae to carry out the individual reactions and later still by purifying the enzymes concerned.

The Calvin cycle differs fundamentally from other carbon dioxide-fixing mechanisms in that it can result in the total synthesis of hexose from carbon dioxide. Two of the enzymes involved in the cycle are unique in that they are not found on other pathways. These are phosphoribulokinase, which phosphorylates ribulose 5-phosphate with ATP to give ribulose diphosphate, and ribulose diphosphate carboxylase which catalyses a reaction between ribulose diphosphate and

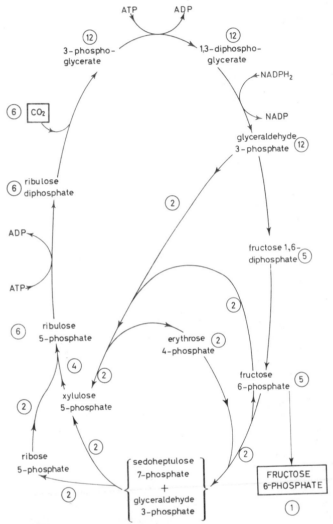

Figure 7.2 The Calvin cycle. The figures in circles indicate the number of molecules participating in one complete turn of the cycle. Although the cycle depicts the net synthesis of one molecule of fructose 6-phosphate from six molecules of carbon dioxide, several other intermediates in the cycle can be channelled into biosynthetic pathways

carbon dioxide to give two molecules of 3-phosphoglycerate.
The remaining Calvin-cycle enzymes are found in many if
not all micro-organisms in which they catalyse reactions on
the HMP pathway.

Although *Figure 7.2* depicts the Calvin cycle synthesising
one molecule of hexose from six molecules of carbon dioxide,
it is important to realise that several of the intermediates
in the cycle can be tapped off and used in the synthesis of
cell constituents. The main draw-off points are 3-phospho-
glycerate (which can lead to pyruvate synthesis by the EMP
pathway), erythrose 4-phosphate (leading to synthesis of
aromatic amino acids), ribose 5-phosphate (used in
nucleotide synthesis) and hexose phosphate. The amounts
of intermediates drained off must not exceed the amount
necessary to maintain the sequence of reactions in the
cycle.

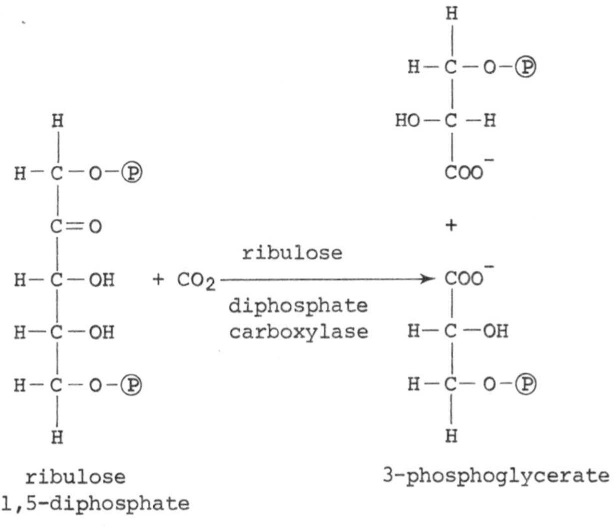

ribulose
1,5-diphosphate   3-phosphoglycerate

The discovery that in photosynthetic bacteria, such as
Chlorobium thiosulphatophilum and other anaerobic
bacteria, the first detectable product of carbon-dioxide
fixation are amino acids, suggested that the Calvin cycle
does not operate in these microbes. Further research led
to the discovery in these bacteria of enzymes which
catalyse two new carboxylation reactions, namely pyruvate
synthase, which catalyses the following reaction:

$$\begin{array}{cc}
\text{acetyl-CoA} + CO_2 & \text{pyruvate} + \text{CoA} \\
+ & + \\
\text{reduced ferredoxin} & \text{oxidised ferredoxin}
\end{array}$$

$$\begin{array}{cc}
\text{succinyl-CoA} + CO_2 & \alpha\text{-ketoglutarate} + \text{CoA} \\
+ & + \\
\text{reduced ferredoxin} & \text{oxidised ferredoxin}
\end{array}$$

These discoveries soon led to the formulation of a new cycle for net carbon-dioxide fixation, which is referred to as the *reductive carboxylic acid cycle* or more accurately as the *acetyl-CoA pathway* (*Figure 7.3*)

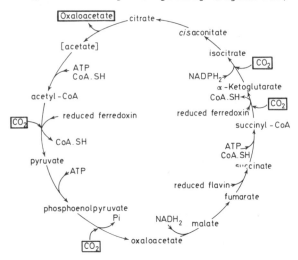

Figure 7.3   The acetyl-CoA pathway for carbon dioxide fixation

The cycle, which involves the reversal of several reactions of the TCA cycle (*see* page 198), results in net synthesis of one molecule of oxaloacetate from four molecules of carbon dioxide and a regeneration of the first carbon dioxide acceptor, acetyl-CoA.

Assimilation of reduced one-carbon compounds

Two quite different pathways operate when reduced C_1 compounds are assimilated by microbes. The charting of these pathways owes much to the research of Rodney Quayle and his colleagues at the University of Sheffield in

Britain. With one exception, namely *Pseudomonas oxalaticus* growing on formate, there is no evidence for operation of the Calvin cycle in microbes that are growing on reduced one-carbon compounds as sole source of carbon.

Many bacteria, growing on any one of a variety of reduced one-carbon compounds ranging in oxidation state between formate and methane, assimilate carbon atoms using a pathway a key reaction on which involves hydroxymethylation of glycine to produce serine, and is therefore known as the *serine pathway* (*Figure 7.4*). The

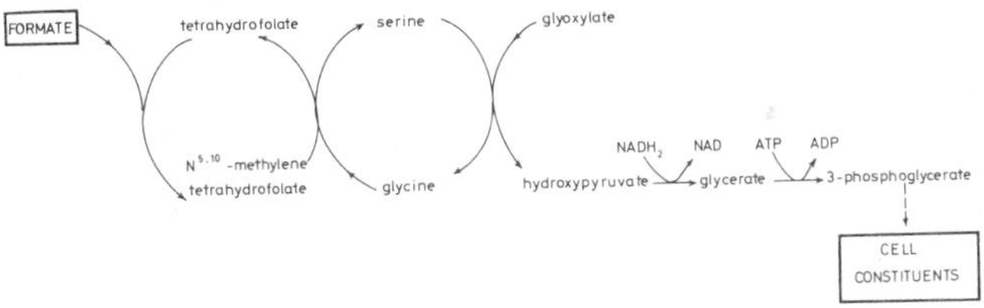

Figure 7.4   The serine pathway for converting formate and glycine into 3-phosphoglycerate, which is then incorporated into cell constituents

pathway accomplishes the conversion of a C_1 unit and a molecule of glycine to 3-phosphoglycerate. The C_1 compound is first oxidised to formate, which then reacts with tetrahydrofolate to give ultimately $N^{5,10}$-methylene tetrahydrofolate. This compound in turn reacts with glycine to produce serine in a reaction which is also used in glycine biosynthesis (*see* page 272). Glycine is then regenerated in a reaction which leads to formation of hydroxypyruvate and is catalysed by serine-glyoxylate amino transferase. Mystery surrounds the origin of the glyoxylate involved in this reaction but it must of course be synthesised from a C_1 compound. Subsequent reactions lead to the conversion of hydroxypyruvate to 3-phospho-glycerate which in turn is incorporated into cell constituents.

Examination of *Pseudomonas methanica* failed to reveal
activities of enzymes which catalyse reactions on the
serine pathway, and led later to the discovery of an
alternative pathway for assimilation of reduced C_1
compounds. This pathway is known formally as the *ribose
phosphate cycle of formaldehyde fixation* or colloquially
as the *allulose pathway,* since the fixation involves a
reaction between formaldehyde and ribose phosphate to
give allulose phosphate:

ribose 5-phosphate   allulose
                       6-phosphate

The fixation reaction is an acyloin condensation which
requires thiamine pyrophosphate. The cycle (*Figure 7.5*)
bears some resemblance to the Calvin cycle. The main
difference between it and the Calvin cycle is the by-
passing in formaldehyde fixation of the reductive step
(conversion of phosphoglycerate into glyceraldehyde
phosphate) which is necessary to reduce carbon dioxide to
the level of formaldehyde.
    Bacteria which oxidise methane do so using one, but not
both, of the pathways. Moreover, possession of one or
other of the pathways is of more than passing interest.
Methane-oxidising bacteria which have paired internal
membranes running through the cell or concentrated at
the periphery use the allulose pathway, whereas those
which have bundles of disc-shaped vesicles inside the
cell employ the serine pathway. This differentiation has
obvious evolutionary implications.

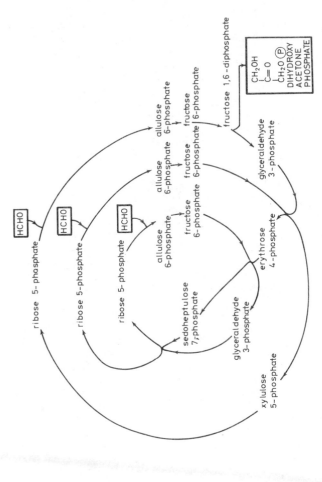

Figure 7.5 The allulose pathway for fixation of formaldehyde. The pathway shows fixation of three molecules of formaldehyde to give one molecule of dihydroxyacetone phosphate

7.3   NUCLEIC ACIDS AND PROTEINS

Proteins occupy a unique position in cell metabolism
since each of the metabolic reactions that takes place
in the cell is catalysed by a specific protein, i.e. an
enzyme. When considering the biosynthesis of proteins,
it is necessary therefore to explain not only how amino
acids are joined together to form polypeptides but how
the order in which they are joined is specified since
the sequence of amino acids is different in each protein
and determines the enzymic activity of the protein.
     It has been appreciated from the very early days of
biochemistry that the characteristics of a living organism
are determined by its ability to carry out certain
metabolic reactions and so, therefore, by its capacity to
synthesise specific enzymes. However, geneticists have
shown that these characteristics are also controlled by
the DNA in the cell. The essential identity of these two
points of view became clear when George Beadle formulated
his one gene-one enzyme hypothesis, which states that the
synthesis of any enzyme (or more correctly any polypeptide)
is controlled by one particular gene. Formulation of the
one gene-one enzyme hypothesis immediately established a
fundamental relationship between nucleic acids and
proteins, and it led inevitably to the question: is the
gene directly involved in synthesis of a specific protein?
Experiments with *Amoeba proteus* and the huge unicellular
alga *Acetabularia mediterranea,* in which it was shown
that protein synthesis continued even after the nuclei had
been removed experimentally, strongly suggested that DNA
is not directly concerned in protein synthesis.
Information about the nature of the intermediary between
DNA and protein came from experiments which showed a
direct correlation between the RNA content of a cell and
its growth rate, while the DNA content remained approximately constant.
     The DNA in a cell carries a set of coded instructions
for the 'running' of the cell. This information is passed
on to RNA which in turn translates the information into
protein. The DNA can be regarded therefore as forming the
legislative branch of cell metabolism, while RNA and
protein constitute the executive branch. The specificity
of a piece of nucleic acid is expressed solely by the
sequence of bases on that nucleic acid, and the sequence
is a simple code for the amino-acid sequence in a
particular polypeptide. This has been called the Sequence
Hypothesis.

Since the late 1940s, work on the mechanism of protein synthesis and the metabolic interrelationships between DNA, RNA and protein synthesis, has proceeded at a rapid and sometimes feverish pace. The nature of the principal processes involved, and also a great deal of information about many of these processes, are now firmly established. Nevertheless, as will become apparent from the following account, much remains to be discovered about the molecular mechanisms of many of the processes involved in protein synthesis, and particularly the ways in which the processes are regulated.

7.3.1 AMINO-ACID SYNTHESIS

All 20 of the amino acids required for protein synthesis in micro-organisms may be available in the medium; alternatively, the organisms may acquire a supply of these acids as a result of the action of extracellular or intracellular proteolytic enzymes. However, micro-organisms growing in media containing solely inorganic sources of nitrogen, or only a restricted number of amino acids, need to synthesise some or all of these acids from the available nitrogen-containing nutrients.

Inorganic nitrogen assimilation

The ultimate source of nitrogen for all forms of life is inorganic. The nitrogen atom occurs in natural compounds in several different states of oxidation, ranging from +5 (as in N_2O_5 or its hydrated form HNO_3), through +4 (NO_2), +3 (HNO_2), +2 (NO), +1 (hyponitrous acid, H_2N_2O), 0 (N_2), -1 (hydroxylamine, NH_2OH), -2 (hydrazine, $NH_2.NH_2$), to -3 (NH_3). It is incorporated into organic molecules mainly in the form of the ammonium ion. Micro-organisms growing in media containing the ammonium ion, or amino acids which are readily deaminated to yield ammonia, can incorporate the ion into organic compounds more or less directly. When however the available nitrogenous nutrients contain the nitrogen atom in one of the more oxidised states, the atom must be reduced before it can be incorporated into amino acids. The mechanisms used by micro-organisms for reducing some of the more commonly available forms of inorganic nitrogen are described briefly below.

Biological reduction of *nitrate* proceeds by a pathway involving only inorganic intermediates, two electrons being added in each enzymic step. The first step involves a reduction to nitrite in a process similar in many respects to that occurring during nitrate respiration (page 207). Nitrate reductase, the enzyme which catalyses this reduction, is a molybdenum-containing protein.

Assuming that the reduction of nitrate proceeds through further two-electron changes, it would pass through oxidation states of +1, -1 and thence to the ammonium ion (-3). The nature of the +1 intermediate formed, presumably, by the action of nitrite reductase   is unknown, although it has been suggested that it may be hyponitrous acid. The product of the reduction of this intermediate is likely to be hydroxylamine, and enzymes that catalyse hydroxylamine reduction have been obtained from several micro-organisms.

Nitrogen fixation

Nitrogen fixation, the process by which molecular nitrogen or dinitrogen is reduced to ammonia, is an extremely important process in the biosphere since it is the origin of most of the assimilable forms of the element nitrogen for living organisms. It has been calculated that between 10^9 and 10^{10} tons of dinitrogen are fixed globally every year, and that biological processes account for between 50 and 95% of this amount, the remainder being attributed to fixation during production of chemical fertilisers and formation of oxides of nitrogen in the atmosphere. We saw in Chapter 3 (page 102) that biological reduction of nitrogen is confined to prokaryotic microbes, either free living such as *Azobacter* species or in symbiotic association like species of *Rhizobium*.

Workers in a number of laboratories have isolated from bacteria purified proteins that are capable of reducing dinitrogen; these are known as *nitrogenases*. These preparations can be separated into two brown proteins both of which are necessary for nitrogen fixation and which have very similar properties. One component has a relatively high molecular weight (200 000-300 000 daltons) and contains iron, molybdenum, and labile sulphur roughly equivalent to the iron content. This protein is referred to as Fraction 1 or F_1. The other component is smaller (about 50 000 daltons) with less iron and labile sulphur, and no molybdenum; it is known as Fraction 2 or F_2. The smaller protein is irreversibly damaged by oxygen while the larger is relatively insensitive.

The mechanism of dinitrogen fixation has intrigued biochemists for many years. The fascinating aspect of the problem is the need to explain how dinitrogen, which is extremely inert chemically, can be reduced enzymically under relatively mild conditions. It has, however, proved one of the most intractable problems ever encountered in microbial physiology. The traditional methods used for studying metabolic pathways have proved surprisingly unfruitful and, in particular, have failed to reveal the formation of free intermediates in the reduction process. Recently, however, considerable progress has been made on the biochemistry of nitrogen fixation, and at last the process is beginning to reveal some of its molecular secrets.

Formation of the active enzyme requires one molecule of F_1 and two of F_2. A binding factor is not essential but is present in the *in vivo* enzyme. The participants in the reaction are dinitrogen, an electron donor, ATP, and Mg^{2+}. The original source of the electrons varies in different nitrogen-fixing bacteria. In anaerobic microbes, such as clostridia, they include formate, molecular hydrogen and pyruvate; in aerobic bacteria, such as *Azotobacter* species, the electrons originate in the TCA cycle. However, irrespective of the source of the reducing power, the electrons are transferred to a carrier of the ferredoxin or flavodoxin type which, in a reduced form, participates in the nitrogenase-catalysed reaction. The requirement for ATP is very puzzling. From thermodynamic considerations, ATP should not be required since the overall reaction for reduction of molecular nitrogen is an exergonic process. Moreover, the ATP requirement seems strangely high, values reported being between 4 and 20 moles ATP for each mole of dinitrogen reduced. Several suggestions have been made to explain this need for ATP. It has been proposed that ATP acts as a source of protons, as an electron activator, or as an inducer of conformational changes in the nitrogenase proteins. Which, if any, of these suggestions is correct is still unknown.

Nitrogenase appears to act in a stepwise fashion by adding two electrons at each step to the substrate. The evidence for this suggestion comes from the finding that other substrates, apart from dinitrogen, which are reduced by nitrogenase (such as acetylene which is reduced to

ethylene, and isocyanide which gives rise to a mixture
of ammonia and methane) are converted to products that
contain two more electrons. The substrate-binding site
on nitrogenase is thought to contain two metals (iron
and molybdenum) which only become accessible to the
substrate when the active complex is formed.

One of the most striking properties of nitrogenases is
their oxygen sensitivity. Anaerobic conditions are
mandatory when handling these enzymes, although F_1 is
only mildly sensitive to damage by oxygen. Exactly how
the enzyme is able to function in highly aerobic bacteria
such as *Azotobacter* species has intrigued microbial
physiologists for some time, and has elicited some
valuable data from John Postgate and his colleagues in
the University of Sussex, England. These workers have
shown that the enzyme can function in aerobic bacteria
because of the operation of two protective mechanisms.
Firstly, there is *conformational protection* by which the
enzyme becomes insensitive to oxygen (and unable to
catalyse dinitrogen fixation) as a result of undergoing
a conformational change. Secondly, the enzyme can be
protected, and maintained in a catalytically active form,
by the operation of an oxygen-scavenging process, such
as high respiratory activities. This is referred to as
respiratory protection.

Incorporation of ammonia into organic compounds

Four main mechanisms are known by which micro-organisms
can incorporate ammonia into organic compounds to give
amino acids. The first of these is catalysed by
glutamate dehydrogenase and is reversible. The reaction
is:

$$\alpha\text{-ketoglutarate} + NH_4^+ + NADPH_2 \rightleftharpoons L\text{-glutamate} + NADP$$
$$+ H_2O + H^+$$

This mechanism is used by some bacteria, and is
particularly prevalent in yeasts and fungi. The reaction
is thought to proceed in two stages, with the intermediate
formation of iminoglutarate. Glutamate dehydrogenases
linked to NAD are also found in micro-organisms, but these
appear to function catabolically in deamination of
glutamate. The second mechanism is a reaction catalysed
by alanine dehydrogenase:

$$pyruvate + NADH_2 + NH_4^+ \rightleftharpoons L\text{-alanine} + NAD + H_2O + H^+$$

Among *Bacillus* species and many actinomycetes, this
mechanism, rather than that involving glutamate
dehydrogenase, seems to be the principal mechanism for
incorporating ammonia into amino acids. The third
mechanism, which is of a different type, involves a
reaction catalysed by aspartate ammonia-lyase (usually
referred to as aspartase). This reaction can also be
used to generate fumarate:

$$\text{fumarate} + \text{NH}_4^+ \rightleftharpoons \text{L-aspartate} + \text{H}^+$$

The last mechanism was discovered quite recently in
Aerobacter aerogenes grown in a chemostat under conditions
of ammonia limitation. The mechanism *(Figure 7.6)*
involves initially synthesis of glutamine from glutamate

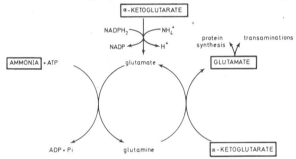

*Figure 7.6 The GOGAT pathway for assimilation of
ammonia by micro-organisms. See text for details*

in a reaction catalysed by glutamine synthetase and
involving ATP. This is the reaction used in synthesis of
glutamine required for protein synthesis. However, in
this ammonia-assimilation mechanism, the glutamine then
reacts with α-ketoglutarate in a transferase reaction
that leads to formation of two molecules of glutamate.
Glutamate can thus be considered to act catalytically in
this mechanism. The enzyme which catalyses this second
reaction is glutamine: α-ketoglutarate amino-transferase
which, when one uses the name oxoglutarate rather than
ketoglutarate, explains why the mechanism is often called
the *GOGAT pathway*. Expenditure of a molecule of ATP can
be considered the price that the microbe has to pay to
assimilate ammonia present in very low concentrations.
However, the mechanism is not confined to bacteria
growing under conditions of ammonia limitation. Indeed,
it is a constitutive pathway in the yeast *Schizosaccharomyces versatilis*.

There are also other reactions which lead to
incorporation of ammonia into organic compounds. Some
fungi incorporate ammonia into nitropropionic acid, but
this mechanism is not widespread among microbes.
Another compound which provides nitrogen atoms is
carbamoyl phosphate. This compound, which is synthesised
in the following reaction:

$$\text{glutamine} + \text{ATP} + \text{HCO}_3^- \longrightarrow \overset{-}{\text{NH}_2} - \overset{\overset{\text{O}}{\parallel}}{\text{C}} - \text{O} - \text{PO}_3\text{H}_2 + \text{ADP} + \text{glutamate}$$

is used to insert nitrogen atoms into pyrimidine rings.

Pathways of amino-acid synthesis

The pathways leading to synthesis of each of the 20
amino acids required for protein synthesis are well
charted. In the synthesis of some of these acids (such
as tyrosine and valine), the amino group, formed as a
result of incorporation of ammonia into alanine,
aspartate or glutamate, is introduced in the final
reaction of the pathway in reactions catalysed by
aminotransferases (otherwise known as transaminases).
Other amino acids arise by metabolic modification of
α-amino acids in pathways that do not require the
intervention of transaminases. In the pathways shown in
Figures 7.7 to *7.13*, it should be noted that no attempt
has been made to include all of the known intermediates.
   Both glutamate and aspartate act as starting materials
for synthesis of other amino acids. Glutamate gives
rise to arginine and proline. During arginine synthesis,
additional nitrogen atoms are introduced in reactions
with glutamate, carbamoyl phosphate and aspartate
(*Figure 7.7*).
   A pathway starting with aspartate (*Figure 7.8*) leads to
synthesis of lysine, methionine and threonine. Not all
micro-organisms use this pathway to produce lysine, for
certain moulds and yeasts synthesise this basic amino
acid using a pathway that involves formation of
α-aminoadipate (*Figure 7.9*).
   The starting materials on two other pathways are products
of carbohydrate catabolism, namely pyruvate and
3-phosphoglycerate. The first of these pathways leads to
synthesis of serine, glycine and cysteine (*Figure 7.10*).
The conversion of serine to glycine, a reaction catalysed
by serine hydroxymethyltransferase, requires tetrahydrofolate and pyridoxal phosphate. The enzyme which catalyses
the conversion of serine into cysteine - cysteine synthase -

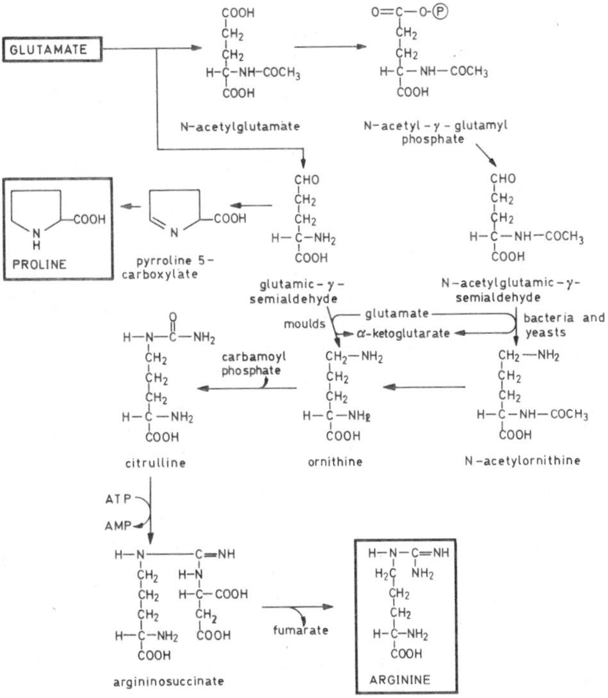

Figure 7.7 Pathways leading to synthesis of arginine and proline

also requires pyridoxal phosphate. The thiol group which participates in this reaction is formed by many microorganisms as a product of *assimilatory sulphate reduction*. This process differs from dissimilatory sulphate reduction (*see* page 207) in that adenosine 5-phosphosulphate (APS), which is formed in a reaction between SO_4^{2-} and ATP, is converted to phospho-adenosine 5-phosphosulphate (PAPS):

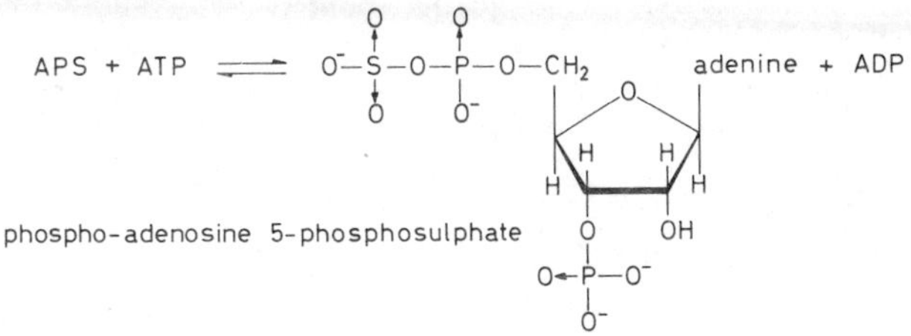

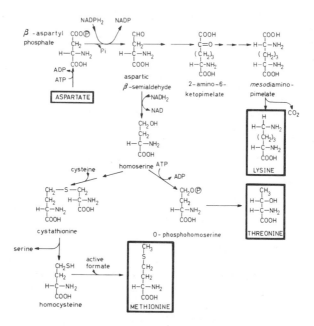

*Figure 7.8 Pathways leading to synthesis of lysine,
methionine and threonine*

The product is then reduced to sulphite which in turn is
further reduced to give a thiol group. Little is known
of the intermediates *en route* from sulphite to the thiol
group.

Pathways starting with pyruvate and threonine lead to
synthesis of leucine, isoleucine and valine (*Figure 7.11*).
An interesting feature of these pathways is that reactions
involving metabolism of analogues on the two pathways are
catalysed by the same enzymes. For example, the same
reducto-isomerase catalyses synthesis of both
α,β-dihydroxyisovalerate and α,β-dihydroxy-β-methylvalerate.

The three aromatic amino acids – phenylalanine,
tryptophan and tyrosine – as well as *p*-aminobenzoic acid
are synthesised by a fairly circuitous pathway (*Figure
7.12*). The starting materials are phosphoenolpyruvate and
erythrose 4-phosphate, and the key intermediate on the
pathway is chorismic acid. Benzene rings also occur in
other products of microbial metabolism (such as the mould

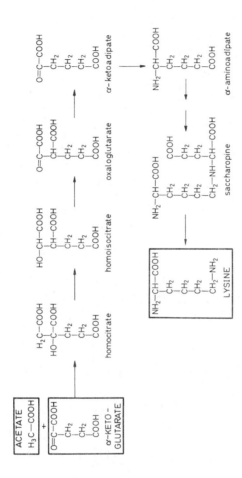

Figure 7.9 Pathway for lysine biosynthesis in moulds and yeasts

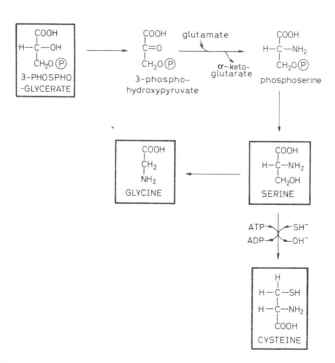

*Figure 7.10 Pathways leading to synthesis of serine,
glycine and cysteine*

product, alternariol). In some of these compounds, the
aromatic rings are formed not by the chorismic acid
pathway but by head-to-tail condensation of acetate
units (*see* page 322).

The pathway leading to synthesis of histidine (*Figure
7.13*) differs from other pathways leading to amino-acid
biosynthesis in that the carboxyl group in the amino
acid is formed during the last reaction on the pathway.

The pathways depicted in *Figures 7.7-7.13* were charted
as a result of studies on a relatively small number of
micro-organisms, notably *E. coli*. Many other micro-
organisms, particularly aerobic organisms, are also
thought to use these pathways. However, there is
evidence that certain anaerobic micro-organisms may use
alternative pathways for synthesis of at least some amino
acids. For example, when *Methanobacterium omelianskii* is
grown on ethanol and carbon dioxide as carbon sources,

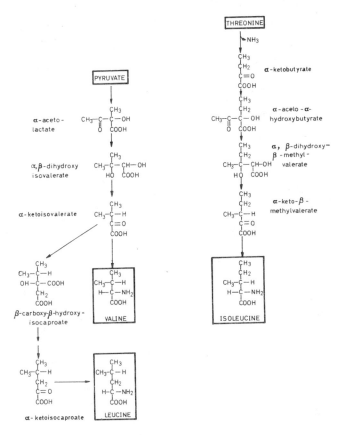

Figure 7.11 Pathways leading to synthesis of leucine,
isoleucine and valine

isoleucine appears to be synthesised by carboxylation of
a five-carbon precursor formed entirely from ethanol;
this pathway clearly differs from that depicted in *Figure
7.11*. Further research on the pathways of amino-acid
biosynthesis in obligate anaerobes could well lead to
the discovery of a whole new group of biosynthetic
pathways.

The pathways already described lead to the formation of
L-amino acids. But amino acids with the D-configuration

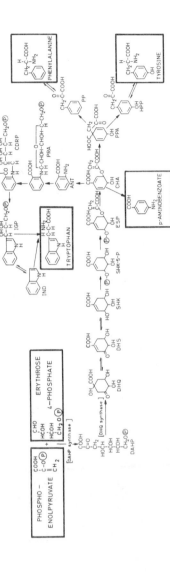

Figure 7.12 Pathways leading to synthesis of aromatic amino acids and p-hydroxybenzoate. The following abbreviations are used: ANT, anthranilate; CDRP, 1-(o-carboxyphenylamino)-1-deoxy-D-ribulose 5-phosphate; CHA, chorismate; DAHP, 3-deoxy-D-arabinoheptulosonic acid 7-phosphate; DHQ, 5-dehydroquinate; DHS, 5-dehydroshikimate; ESP, 3-enolpyruvylshikimic acid 5-phosphate; HPP, p-hydroxyphenylpyruvate; IGP, indoleglycerol 3-phosphate; IND, indole; PP, phenylpyruvate; PPA, prephenate; PRA, N-5'-phosphoribosyl anthranilate; SHK, shikimate

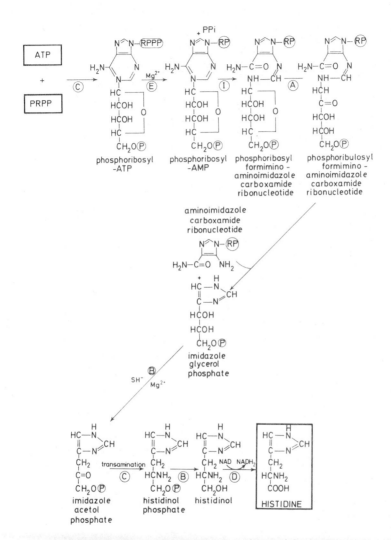

Figure 7.13 Pathway leading to synthesis of histidine. RP indicates a ribose phosphate residue; the letters in circles are the designations given to the genes that control synthesis of individual enzymes in bacteria

also occur in micro-organisms (as in the bacterial cell-
wall peptidoglycans and teichoic acids). D-Amino acids
can arise from the corresponding L-isomers by racemisation,
and glutamate- and alanine-racemases have been found in
extracts of several organisms. Transamination reactions
which involve D-amino acids are also known. Certain
strains of *Bacillus subtilis* and *B. anthracis* have
capsules consisting of poly-D-glutamate (page 22), and
these bacteria have been shown to possess a transaminase
that converts α-ketoglutarate to D-glutamate in the
presence of D-alanine or D-aspartate.

7.3.2 SYNTHESIS OF NUCLEOTIDES

The routes leading to synthesis of purine and pyrimidine
nucleotides are among the best charted of biosynthetic
pathways. The biosyntheses of purines and pyrimidines
proceed by completely independent routes but they have in
common that each pathway leads directly to the nucleotide.
The pathway of purine nucleotide synthesis is, not
surprisingly in view of the greater molecular complexity
of purines as compared with pyrimidines, the more lengthy
of the two (*Figure 7.14*). The product of this pathway is
inosinic acid (IMP), the ribonucleotide corresponding to
the purine hypoxanthine. The ribosyl phosphate part of
the molecule arises from phosphoribosyl pyrophosphate
(PRPP) which is formed from ribose 5-phosphate by the
following reaction:

$$\text{ribose 5-phosphate} + \text{ATP} \rightleftharpoons \text{PRPP} + \text{AMP}$$

The purine ribonucleotides adenylic acid (AMP) and
guanylic acid (GMP) which are required for RNA synthesis
arise from inosinic acid by the reactions shown in
Figure 7.15.

The pathway leading to synthesis of pyrimidine
ribonucleotides differs from that leading to purine
ribonucleotides not only in being shorter but also in
that PRPP, which again provides the ribosyl moiety
(*Figure 7.16*), enters the biosynthetic sequence at a
later stage when it combines with the pyrimidine orotic
acid to give orotidine 5-phosphate. Uridylic acid (UMP)
is formed directly from orotidine 5-phosphate by
decarboxylation. Extracts from several micro-organisms
have been shown to contain an enzyme that catalyses a
reaction between UTP, which is formed from UMP, and NH_4^+
to give CTP.

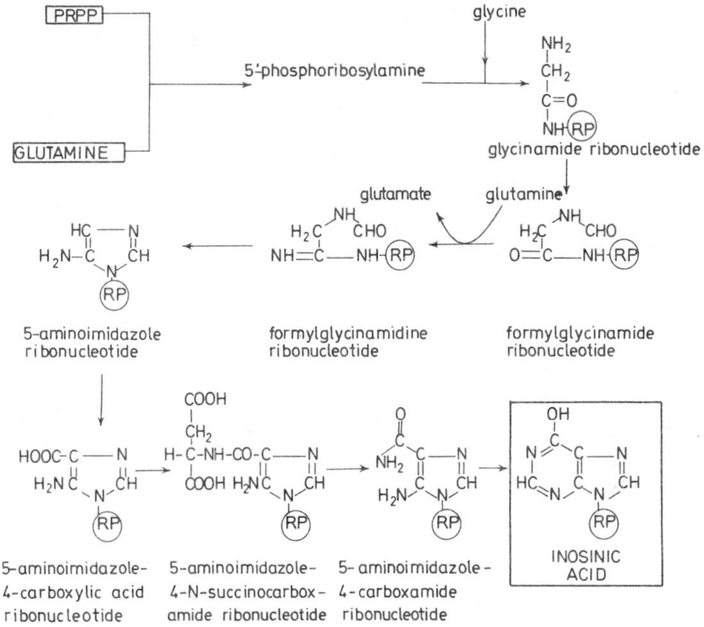

Figure 7.14 Pathway leading to synthesis of inosine monophosphate. RP indicates a ribose phosphate residue

The reactions leading to synthesis of deoxyribonucleotides are still poorly understood despite the pioneer work on this aspect of microbial metabolism by Peter Reichard and his colleagues in Stockholm, Sweden. It is clear, however, that the reactions vary in different micro-organisms. In *E. coli* the four naturally occurring ribonucleoside diphosphates are reduced to the corresponding 2'-deoxyribonucleoside diphosphates in reactions that are each catalysed by the same enzyme, a ribonucleoside diphosphate reductase. The reactions involve the reduced form of thioredoxin, which is regenerated by NADPH$_2$ in a reaction catalysed by the flavoprotein thioredoxin reductase. Action of the reductase also requires magnesium ions. Thioredoxin is a well studied protein with a molecular weight of about 12 000 daltons. *Lactobacillus leichmannii* is the only other microbe in which synthesis of deoxyribonucleotides has been studied in any detail. In this bacterium, the substrate in the reaction is the ribonucleoside triphosphate, and again the

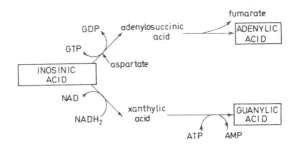

Figure 7.15 *Pathways leading to formation of adenylic and guanylic acids from inosinic acid*

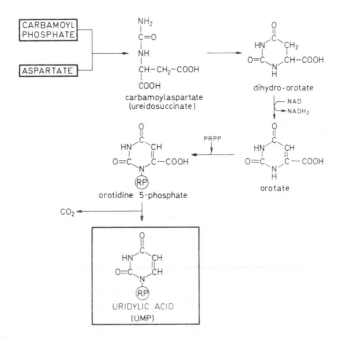

Figure 7.16 *Pathway leading to synthesis of uridylic acid (UMP). RP indicates a ribose phosphate group*

same enzyme catalyses reduction of all four naturally occurring ribonucleoside triphosphates. This enzyme has a molecular weight of around 100 000 daltons. However, the reaction differs yet again from that which operates in *E. coli* in that it involves coenzyme B_{12}. An interesting property of this reduction is that it takes place without rupture of the N-glycosidic bond in the substrate.

So far this account has been confined to the *de novo* synthesis of ribo- and deoxyribonucleotides starting from low molecular-weight compounds. But many micro-organisms are able to use purines or pyrimidines, or their nucleosides and nucleotides, as sources of RNA- and DNA-nucleotides. Some organisms are auxotrophic for these compounds because they are unable to synthesise one or more of the enzymes concerned on the *de novo* pathways. A variety of enzymes capable of catalysing interconversions among purines, pyrimidines and their nucleosides and nucleotides have been detected in extracts of micro-organisms. With some organisms, the free bases can react with PRPP to give a nucleotide, e.g.:

$$\text{adenine} + \text{PRPP} \rightleftharpoons \text{AMP} + \text{PPi}$$

Some rather interesting patterns of metabolism have been recorded among purine interconversions in micro-organisms. For example, *Candida utilis*, which does not require a purine for growth, incorporates exogenous adenine into both AMP and GMP, although it can use exogenous guanine only in the synthesis of GMP. An enzyme capable of converting cytosine directly to CMP has not yet been demonstrated in micro-organisms, and it would seem that those organisms that are capable of utilising cytosine as a source of pyrimidine nucleotides do so by first converting it to uracil (using cytosine deaminase) and then converting uracil to UMP.

Several antimicrobial compounds, particularly antibiotics, are known to interfere with reactions that lead to nucleotide synthesis. Two antibiotics produced by streptomycetes, namely azaserine and 6-diazo-5-keto-L-norleucine (known as DON), are analogues of glutamine. Both compounds prevent conversion of formylglycinamide ribonucleotide into formylglycinamidine ribonucleotide (*Figure 7.14*), a reaction that involves glutamine. The structures of these two antibiotics and glutamine are given on page 280. Another antibiotic, hadacidin, which is synthesised by species of *Penicillium,* inhibits

K

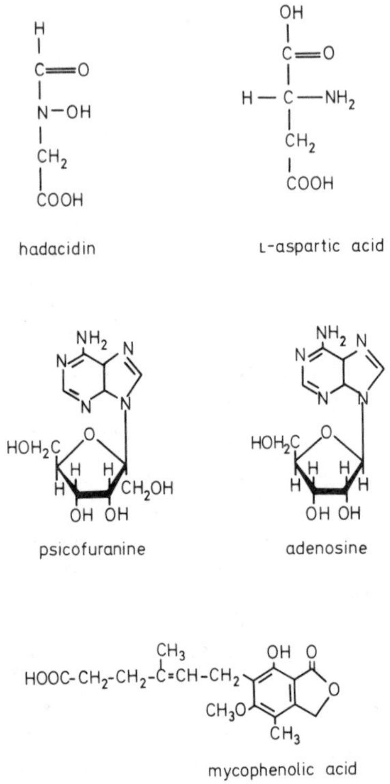

hadacidin

L-aspartic acid

psicofuranine

adenosine

mycophenolic acid

*Figure 7.17   Structural formulae of compounds that
inhibit nucleotide synthesis showing their structural
resemblance to intermediates on the biosynthetic
pathways*

conversion of IMP to adenylosuccinic acid (*Figure 7.15*).
It acts as an analogue of aspartic acid (*Figure 7.17*). •
Psicofuranine is an antibiotic synthesised by *Streptomyces
hygroscopicus* (*Figure 7.17*). Despite its structural
resemblance to adenosine, psicofuranine inhibits
nucleotide biosynthesis by blocking the final step in the
biosynthesis of guanylic acid (*Figure 7.15*). Another
reaction on this pathway, the conversion of inosinic acid
to xanthylic acid, catalysed by IMP dehydrogenase, is
inhibited by mycophenolic acid, an antibiotic produced by
Penicillium stoliniferum. It is difficult to understand

how mycophenolic acid acts since its structure does not
resemble either the substrate or the product of this
reaction.

7.3.3 POLYMERISATION OF NUCLEOTIDES

Most of the nucleotides synthesised by a microbe are
polymerised into DNA or RNA, but a small proportion
suffer a different fate and are used in the synthesis of
coenzymes and in the activation of low molecular-weight
compounds.

DNA synthesis

Replication of DNA is a unique event in the life of a
cell since it is the process by which the genetic
information in the cell is copied prior to the transfer
of a genome to a daughter cell. Since this genetic
information is encoded on the DNA molecule, DNA synthesis
must involve an *exact* replication of the parent DNA.
According to the Watson-Crick hypothesis (page 63), the
DNA molecule consists of two unbranched polynucleotide
chains wound round each other along one axis to form a
double-stranded helix. Since the model postulates a pair
of templates each complementary to the other, there is no
need for the production of a 'negative' copy of the DNA
during replication.

De novo synthesis of DNA using an enzyme from *E. coli*
was reported in 1956 by Arthur Kornberg and his colleagues
at Stanford University in California, U.S.A. The enzyme,
known as DNA nucleotidyltransferase or DNA polymerase,
catalyses a PPi-releasing reaction involving all four
deoxyribonucleoside triphosphates (corresponding to the
bases adenine, guanine, cytosine and thymine) and requires
the presence of template DNA in addition to Mg^{2+}:

$$\left.\begin{array}{c} nd\text{ATP} \\ + \\ nd\text{CTP} \\ + \\ nd\text{GTP} \\ + \\ nd\text{TTP} \end{array}\right\} + \text{DNA} \longrightarrow \left.\begin{array}{c} \text{DNA} \begin{array}{l} ---- \text{dAMP} \\ ---- \text{dCMP} \\ ---- \text{dGMP} \\ ---- \text{dTMP} \end{array} \end{array}\right\}_n + 4n\ PP_i$$

The unnatural nucleotide, dUTP, can also be polymerised by this enzyme, but this nucleotide is not found in DNA apparently because the kinase catalysing its formation from dUDP is not synthesised by micro-organisms. Methylation of bases in DNA (*see* page 65) is catalysed by DNA methylases. The methyl group donor in these reactions is S-adenosylmethionine, which is synthesised from methionine and ATP.

Although conceptually this is a most elegant and simple mechanism for replicating DNA, discovering the details of the process has to this day remained one of the liveliest topics in molecular biology. A key feature of the double helix is that the two chains of DNA run in opposite directions, and ever since the discovery of the Kornberg DNA polymerase it has been difficult to explain how the enzyme works forward on one strand of DNA and backwards on the other. However, the main advances in recent years have been concerned with the nature of the enzymes involved in DNA replication.

The enzyme discovered by Kornberg and his colleagues, and now referred to as DNA polymerase I, is capable of synthesising only short strands of DNA, and soon after its discovery doubts were cast on its role in DNA replication. Two other DNA polymerases, dubbed II and III, have since been discovered, DNA polymerase II interestingly by the son of Arthur Kornberg which explains why it has been called the Kornberg Jr enzyme. It has also been shown that DNA replication is preceded by, and dependent upon, a short burst of RNA synthesis. Protein synthesis is also known to be essential for the onset of replication. Currently DNA replication is thought to involve synthesis of a number of stretches of DNA which are then linked together. These stretches are known as 'Okazaki pieces' after their discovery in Japan by the late Reiji Okazaki. The process involves a plethora of enzymes: RNA polymerase to synthesise RNA primers; a DNA polymerase (thought to be polymerase III) to elongate the DNA primers; a ribonuclease to excise the RNA fragments; another DNA polymerase (possibly the I and II polymerases) to fill in the gaps with DNA; and finally a joining enzyme or ligase to stitch the whole chain together. Clearly, a great deal more work remains to be done on this most fundamental process in living cells. This account has been confined largely to the enzymology of DNA replication. In Chapter 9 (page 360) there is a discussion of the way in which the genome as an organelle is replicated in prokaryotes and eukaryotes.

Some antimicrobial compounds owe their activity to the ability to combine with DNA and to interfere with the

template function of this polymer. Many of these
antimicrobial compounds are planar molecules with fused
ring systems, and it has been suggested that they combine
with DNA by interposition of their fused ring systems
between adjacent stacked base pairs in the double helix,
as shown in *Figure 7.18*. The structures of some compounds
which are thought to react with DNA by intercalation are
shown in *Figure 7.19*. Mepacrine has found application
as a useful antimalarial drug, while ethidium bromide is
used to combat trypanosomes. Actinomycin D is widely
used to prevent DNA-dependent RNA synthesis in laboratory
experiments.

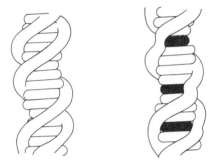

*Figure 7.18 Diagrams showing the secondary structure of
DNA (left) and of DNA containing intercalated molecules
(right). The base pairs (white) and the intercalated
molecules (black) appear in edgewise projection and the
deoxyribose phosphate backbone as a smooth coil*

RNA synthesis: transcription

The three types of RNA found in living organisms -
ribosomal or rRNA, messenger or mRNA and transfer or
tRNA - differ in size, composition and structure (*see*
page 83). Nevertheless, they are all copies of the
appropriate regions of one strand of DNA. The Sequence
Hypothesis (*see* page 262) states that the sequence of
bases in a piece of nucleic acid (DNA or RNA) codes for the
amino-acid sequence in a particular polypeptide. This
DNA-directed synthesis of polypeptides involves two
processes, termed *transcription* and *translation*. The
first of these processes, transcription, is one in which
DNA directs synthesis of RNA and in which genetic

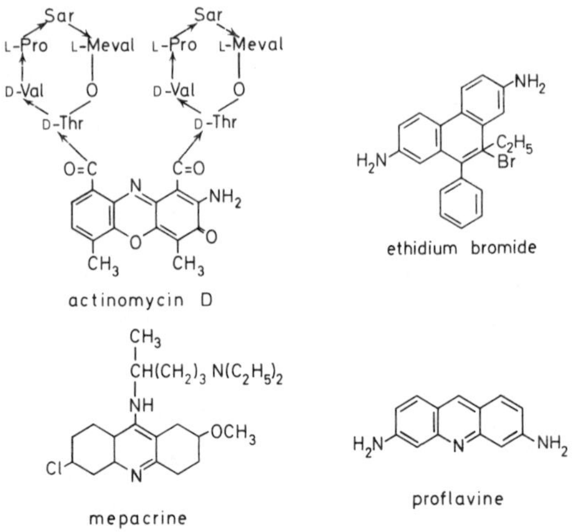

Figure 7.19   Structural formulae of some antimicrobial compounds that are thought to act by intercalating with DNA.   In the actinomycin D structure, Meval indicates a residue of N-methylvaline, and Sar one of sarcosine. Other abbreviations are explained on page (ix)

information that is encoded on DNA is transcribed onto RNA.   Transcription involves synthesis of rRNA, mRNA and tRNA, but it is in relation to mRNA synthesis that the term is most commonly used, simply because a large part of the genome codes for mRNA.   This fact was revealed by experiments in which RNA molecules and homologous DNA were annealed or hybridised.   Hybridisation revealed that only 0.2 and 0.02%, respectively, of the DNA in *E. coli* codes for rRNA and tRNA.

Although the various types of RNA have different functions and are coded for by different regions on the genome, their synthesis in bacteria at least is catalysed by the same enzyme, namely RNA nucleotidyltransferase or RNA polymerase.   The enzyme catalyses a PPi-releasing reaction in which all four ribonucleoside triphosphates take part, and which requires template DNA.

Information on the properties of RNA polymerase is most extensive for the enzyme from *E. coli*, although it is thought that the polymerases in other bacteria are similarly constituted.   The enzyme in *E. coli* is made up of five subunits.   There are two α chains, each with a

molecular weight of 41 000 daltons, one β chain of
155 000 daltons, one β' chain of 165 000 daltons and a
smaller (86 000 daltons) unit known as the σ component.
The complex constitutes the holoenzyme. However, the σ
component readily dissociates from the complex to give
the core polymerase which retains polymerising ability.
The functions of the individual subunits are far from
being understood, but it is believed that the β' subunit
contains the DNA-binding site and also possibly the
active site of the enzyme. The σ component, which is a
heat-labile acidic protein, locates the polymerase at
the correct starting point on the genome.

The activity of RNA polymerase is associated with
several other components known as *transcriptional factors*.
Termination of transcription is effected by the action of
the ρ factor, a protein with a molecular weight of
200 000 daltons. Another factor, the ψ factor, regulates
the amounts of the different types of RNA that are
synthesised. As much as 40% of the RNA made by *E. coli*
is rRNA, although we have seen only 0.2% of the genome
codes for this RNA. Clearly, there must be a very
efficient initiation of rRNA synthesis in the bacterium,
and it is believed that this is effected by the ψ factor.
The existence of several other transcriptional factors
has been postulated, but their functions are less clearly
understood.

A few antimicrobial compounds are known to act by
inhibiting action of bacterial RNA polymerase. The most
important of these are two groups of antibiotics, namely
the rifamycins, which are produced by *Streptomyces
mediterranei,* and the streptovaricins which are products
of other streptomycetes (*Figure 7.20*). Both types of
antibiotic combine with the β subunit, but are thought to
act differently in inhibiting action of RNA polymerase.

Very much less is known about the RNA polymerase from
eukaryotic micro-organisms. There is, however, reason
to believe that it differs from its prokaryotic counter-
part since α-amanitin, a toxic peptide produced by the
mushroom *Amanita phalloides* (*Figure 7.20*), inhibits the
eukaryotic but not the bacterial polymerase.

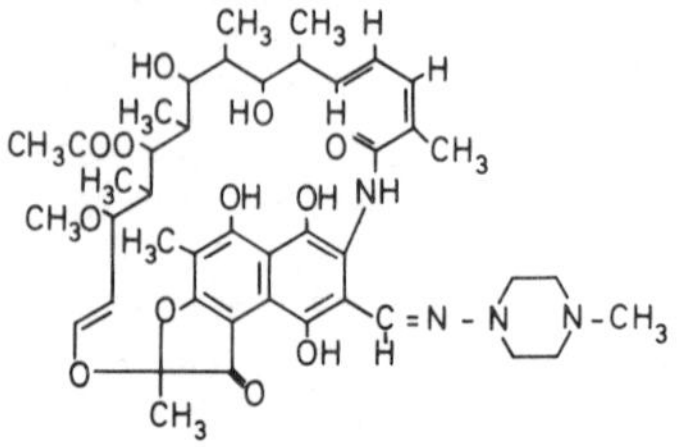

rifampicin

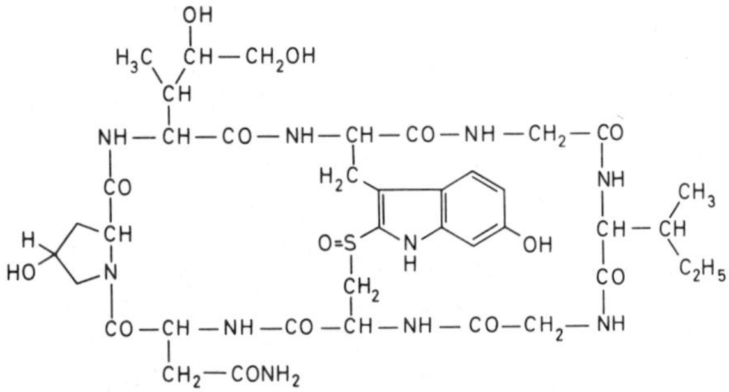

α - amanitin

Figure 7.20 Structural formulae of two antibiotics that inhibit action of RNA polymerase. Rifampicin is a semisynthetic derivative of the naturally occurring rifamycin B

7.3.4 PROTEIN SYNTHESIS: TRANSLATION

Truly spectacular progress has been made during the past couple of decades in elucidating the mechanism by which amino acids are polymerised into proteins. Once methods had been devised for preparing cell-free extracts capable of polypeptide synthesis, the way was paved for an analysis of the crude extracts to discover which of the components were involved in the polymerisation process. In recent years, this phase has given way to one in which details of the reaction mechanisms and the functioning of various factors are being studied. Animal and plant tissue, as well as micro-organisms, have been used in these studies, but there is every reason to believe that the mechanisms of protein synthesis are basically the same in all types of organism.

The most important fact concerning the translation process is that it is accomplished not by direct inter-action of amino acids with mRNA but through intermediate adapter molecules, namely tRNAs. Polymerisation of amino-acid residues attached to tRNA molecules takes place on polysomes, which consist of a group of 70S or 80S ribosomes, lying closely together and connected by a molecule of mRNA. Ribosomes do not have genetic specificity and they function quite non-specifically in the translation process. The mRNA molecule binds to the smaller 30S or 40S ribosomal subunits. Ribosomes begin translating mRNA while it is still being transcribed from DNA (*Figure 7.21*).

The necessary preamble to polymerisation of amino-acid residues on polysomes is the formation of amino-acyl-tRNA molecules. In every microbe which has been examined, there exist at least 20 different amino-acyl-tRNA synthetases which catalyse formation of an ester linkage between the carboxyl group of an amino acid and the 3'-hydroxyl group of the terminal adenosine residue of a specific tRNA molecule (*see* page 83). Formation of an amino-acyl-tRNA molecule is a two-stage process, involving first activation of the amino acid and then transfer to a tRNA molecule. The first reaction involves the amino acid and ATP, resulting in release of inorganic pyrophosphate:

amino acid + ATP + synthetase → synthetase-(amino-acyl-AMP) + PPi

In the amino-acyl-AMP complex, which remains attached to the synthetase, there is a mixed anhydride linkage between the carboxyl group of the amino acid and the phosphate of AMP. In a second reaction, the enzyme-bound amino-acyl-AMP reacts with the specific tRNA molecule to form the amino-acyl-tRNA:

synthetase-(amino-acyl-AMP) + tRNA → amino-acyl-tRNA + AMP + synthetase

The specificity of amino-acyl-tRNA synthetases, a prerequisite for faithful translation of the genetic message, must be very high. It is, in fact, exercised in both the activation and transfer reactions, the specificity in the latter reaction being higher than that in the former. For example, isoleucine-tRNA synthetase activates valine, while valine-tRNA synthetase activates threonine. However, these 'wrong' amino acids are not then transferred to tRNA molecules.

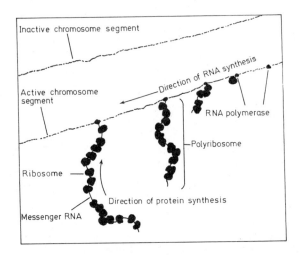

Figure 7.21 *The electron micrograph on the facing page
shows the chromosome of* Escherichia coli *in action. The
micrograph is interpreted in the drawing above. One sees
two strands of the bacterial chromosome, with only the
lower one being used for transcription. The micrograph
shows molecules of RNA polymerase, the one at the far
right being at the approximate initiation site;
successively transcribed mRNA molecules peel off towards
the left of the micrograph. Magnification is ×77 250.
(The micrograph was kindly provided by O.L. Miller)*

Polypeptide-chain synthesis involves three quite
distinct processes, namely initiation, elongation and
termination. Each of these processes will now be
considered separately, and as they take place in bacteria
information for which is the most extensive.

The first or N-terminal amino acid in almost all
polypeptide chains which are synthesised on polysomes is
N-formylmethionine (fMet), and *initiation* of polypeptide
synthesis involves binding of a molecule of N-formyl-
methionine-tRNA (fMet-tRNA) to the first or initiation
codon on the mRNA molecule, a process which also involves
GTP. There are two types of methionine-accepting tRNA
molecules in microbes, namely $tRNA_f^{Met}$ and $tRNA_m^{Met}$.

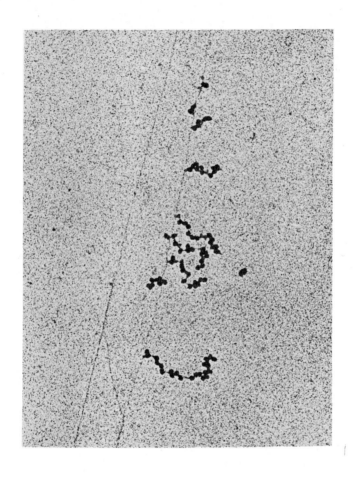

Figure 7.21   See caption on facing page for details

Methionine residues attached to tRNA$_f^{Met}$, but not those attached to tRNA$_m^{Met}$, can be formylated. Molecules of tRNA$_m^{Met}$ are involved in incorporation of methionine residues into the growing peptide chain. Formylation of tRNA$_f^{Met}$ is catalysed by a transformylase, with N^{10}-formyl-tetrahydrofolate as the formyl group donor

Once the fMet–tRNA molecule makes contact with the initiation codon, the larger ribosomal subunit becomes associated with the smaller subunit, in such a way that the fMet–tRNA molecule becomes attached to a specific site on the larger subunit, known as the *A* or *acceptor site*.

Formation of• the initiation complex involves not only mRNA, ribosomes, fMet-tRNA and GTP but also several *initiation factors,* or IF factors, which are themselves polypeptides. In bacteria, something is known of the roles played by certain of these IF factors. Factor IF-2, for example, directs binding of fMet-tRNA to the 50S ribosomal subunit, while IF-3 helps to bind the mRNA molecule to the 30S subunit. Eukaryotic micro-organisms have different initiation factors, but these are believed to have similar roles to those synthesised by bacteria. The final step in initiation of polypeptide-chain synthesis is the shifting of fMet-tRNA from site A on the 50S ribosomal subunit to another site on the same subunit usually referred to as *site P* (for polymerisation) or sometimes as the *donor site (Figure 7.22).*

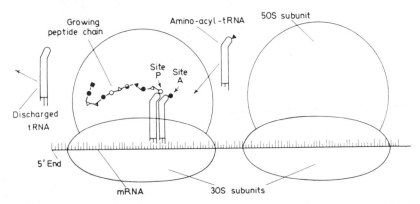

Figure 7.22 Diagram showing an early stage in polypeptide-chain extension on a bacterial polysome, with site A occupied by an amino-acyl-tRNA molecule, and site P by a polypeptidyl-tRNA molecule. See text for an explanation of the process

The first step in *chain elongation* involves attachment
at site A of an amino-acyl-tRNA specific for the codon
adjacent to the initiator codon. This attachment
involves hydrolysis of GTP to GDP and inorganic phosphate.
This and subsequent events in the chain-elongation
process also involve the operation of certain *elongation* or
transfer factors. Several such factors have been
identified in bacteria, and one of them catalyses
hydrolysis of GTP in the attachment process, while another
frees GDP from a complex with this first factor. After
the tRNA molecule from fMet-tRNA is discharged from the
ribosome at site P, the free carboxyl group so formed in
the methionine is linked by a peptide bond to the α-amino
group of the amino-acid residue in the amino-acyl-tRNA at
site A. The enzyme which catalyses this reaction,
peptidyl transferase, is a protein in the 50S subunit.
The peptidyl-tRNA is then moved or *translocated* from site
A to site P, and this is accompanied by movement of the
mRNA one codon relative to the ribosome in the 5' to 3'
direction. These movements require hydrolysis of another
molecule of GTP catalysed by yet another elongation
factor. This leaves site A empty, with site P occupied
by fMet-amino-acyl-tRNA. 'The stage is thus set for addition
of the peptidyl-tRNA to another amino-acyl residue at site
A by the same sequence of processes involving binding of
amino-acyl-tRNA molecules, peptide-bond synthesis and
translocation.

Termination of polypeptide-chain formation occurs when
translocation places one of the 'full stop' codons (*see*
page 331) near site A; then the ribosome does not bind
another amino-acyl-tRNA molecule. Instead, the 'full
stop' codon is recognised by a *termination* or *release
factor*. One of these isolated from bacteria and termed
R_1 is a single polypeptide chain of about 44 000 daltons
molecular weight, and it recognises the codons UAA or UAG.
A similar termination factor, dubbed R_2, recognises codons
UAA or UGA; it has a molecular weight of 47 000 daltons.
The activity of each of these recognition termination
factors is stimulated by yet another termination factor
known as the S protein. In the presence of these factors,
hydrolysis occurs and the polypeptide is released from
the ribosome.

Although almost all polypeptides synthesised on polysomes
in *Escherichia coli* are initiated by N-formylmethionine,
a large proportion of the polypeptides in this bacterium
have other amino-acid residues on the N terminus,

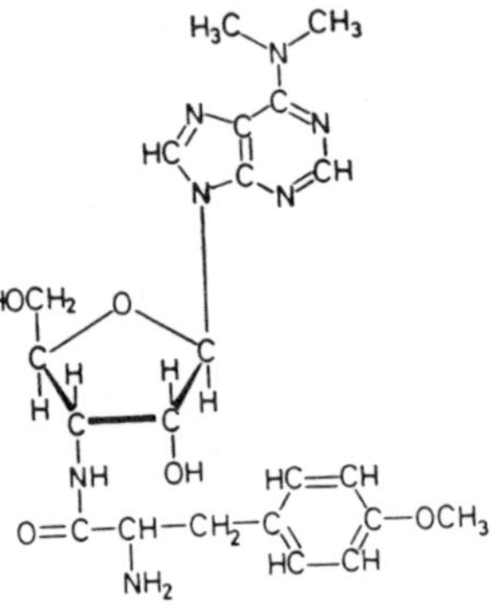

puromycin

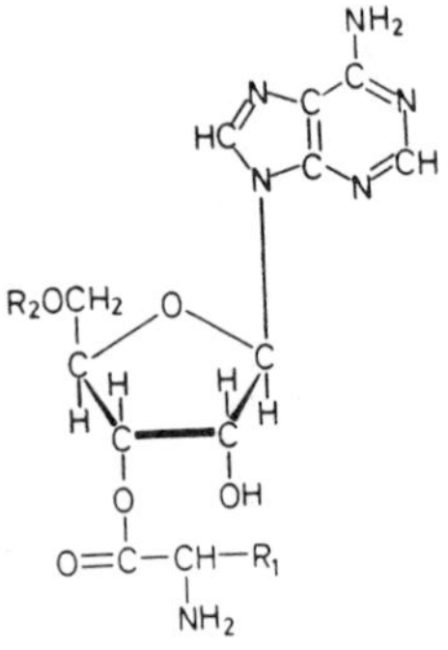

amino-acyl-tRNA
(R_1 = amino-acid residue;
R_2 = tRNA residue)

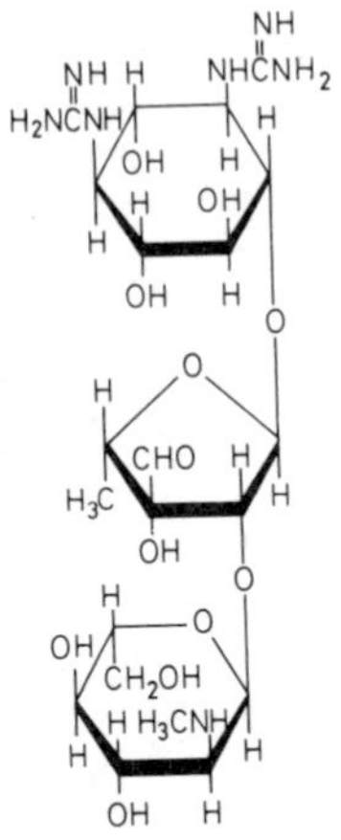

streptomycin

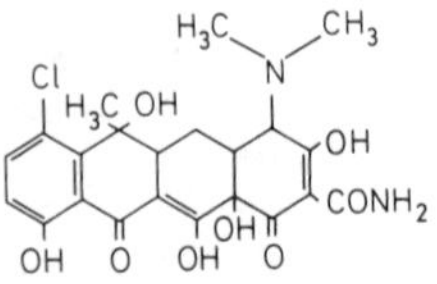

chlortetracycline

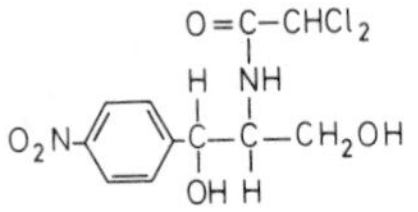

chloramphenicol

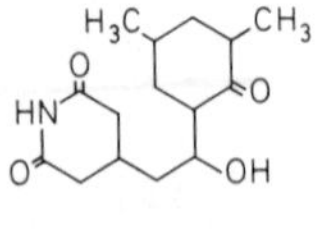

cycloheximide

Figure 7.23 Structural formulae of some antibiotics that inhibit various steps in the translation process. Also shown is the formula of amino-acyl-tRNA to illustrate its structural similarity to puromycin

particularly alanine, threonine or serine. Several
microbes have been shown to synthesise enzymes that
catalyse removal of the formyl residue, and often the
terminal methionine residue, from polypeptides.
A remarkably large number of antimicrobial compounds
have been shown to inhibit various steps in the
translation process. The antibiotic puromycin, a product
of *Streptomyces albo-niger*, is a unique inhibitor of
translation since the drug itself reacts to form a peptide
with the C-terminal end of the growing peptide chain on
the ribosome, thus prematurely terminating the chain.
Puromycin, of which the structural similarity to amino-
acyl-tRNA is shown in *Figure 7.23*, is equally active in
terminating protein synthesis on 70S and 80S ribosomes.
Several other antibiotics inhibit binding of amino-acyl-
tRNA to ribosomes. The tetracycline antibiotics –
chlortetracycline produced by *Streptomyces aureofaciens*
(*Figure 7.23*) and oxytetracycline by *Streptomyces
rimosus* – are equally effective in inhibiting binding of
amino-acyl-tRNA to the acceptor site on 70S and 80S
ribosomes. However, these antibiotics are more effective
against intact prokaryotic than eukaryotic cells.
Streptomycin, one of a group of aminoglycoside antibiotics
which includes gentamycin, kanamycin and neomycin, interacts
with a specific protein (called the P_{10} protein) on the
30S ribosomal subunit in bacteria, and leads to
conformational changes in the ribosome which cause
inhibition of translation. Streptomycin is produced by
Streptomyces griseus. Formation of peptide bonds on
prokaryotic ribosomes is inhibited when the antibiotic
chloramphenicol (*Figure 7.23*) – a product of *Streptomyces
venezuelae* which nowadays is obtained almost exclusively
by chemical synthesis – binds to the 50S ribosomal
subunit. An antibiotic which specifically inhibits
protein synthesis on 80S ribosomes is cycloheximide,
a product of *Streptomyces griseus* (*Figure 7.23*). There
is good evidence that cycloheximide interferes with
translocation of peptidyl-tRNA from the A site to the P
site on 60S ribosomal subunits.

7.3.5 EXTRACELLULAR PROTEINS

With one or two exceptions, little is known about the
synthesis of individual proteins or classes of proteins
in micro-organisms, or of the ways in which proteins, once
they have been synthesised, are incorporated into cell

organelles.   In this connection, one of the most
intriguing problems concerns the synthesis of extracellular proteins, which include exoenzymes and certain
toxins.   Here it is necessary to explain how large
molecules, which normally cannot penetrate the plasma
membrane in micro-organisms, are liberated from organisms
without causing an alteration in cell structure.

Studies on the physicochemical properties of microbial
extracellular proteins have shown that these are, as a
class, a rather unusual group of proteins.   Most of them
are of a low average molecular weight.   Microbial extracellular proteins, with one or two exceptions, contain
few or no cysteine residues.   Such residues confer
rigidity on a protein molecule by forming disulphide
bridges between residues in different parts of the folded
molecule.   An absence of cysteine residues in extracellular
proteins suggests that these proteins are more flexible
than the majority of intracellular proteins.   It has been
suggested that some measure of flexibility may be
conferred on extracellular proteins by Ca^{2+}, Mg^{2+} and Sr^{2+}
ions which are frequently required for stability and
activation of these proteins.

It is usually assumed that synthesis of extracellular
proteins takes place outside the permeability barrier of
the micro-organism, that is, either in or on the outer
surface of the plasma membrane.   If they are formed on the
outer surface of the membrane, then these proteins are
free to diffuse through the cell wall into the
environment, but if the proteins are synthesised within
the membrane, there must be some device by which they are
extruded through the lipid-protein membrane.   Several
ingenious explanations have been put forward to explain
this extrusion, including a reversal of pinocytosis and
the need to have a carbohydrate moiety attached to the
protein to assist passage through the cell wall.   As with
other aspects of membrane function, this will continue to
be a happy hunting ground for armchair microbial
physiologists until more satisfactory methods have been
developed for studying membrane function.

7.3.6   PEPTIDE SYNTHESIS

Despite exhaustive searches, no convincing evidence has
yet been reported to show that low molecular-weight
peptides are free intermediates in protein synthesis.
However, micro-organisms synthesise peptides ranging in

size from the simple tripeptide glutathione to much
larger molecules such as polypeptide antibiotics and the
polypeptide capsular material found in certain bacilli.
The mechanism of synthesis has not yet been worked out for
all of these microbial peptides, but from what is known it
appears that, unlike proteins, they are not synthesised by
pathways that involve mRNA or the formation of aminoacyl-
tRNA. Instead, these peptides are formed by a progressive
chain lengthening through reactions that do not involve
release of inorganic pyrophosphate.

The tripeptide glutathione, for example, is synthesised
by the following sequence of reactions:

glutamate + cysteine + ATP → γ-glutamylcysteine + ADP + Pi

γ-glutamylcysteine + glycine + ATP⟶ glutathione + ADP + Pi

It has also been shown that the polypeptide antibiotic,
gramicidin-S, is synthesised by enzymes in a cell-free
extract of *Bacillus brevis* in the presence of amino acids,
ATP, Mg^{2+} and a reducing agent. Since mRNA is not
involved in this synthesis, it must be assumed that the
enzymes concerned have sufficient structural specificity
to recognise an incomplete peptide chain as well as the
next amino acid to be incorporated.

7.4 OLIGOSACCHARIDES AND POLYSACCHARIDES

Micro-organisms, in common with higher organisms,
synthesise oligosaccharides and polysaccharides. The
amounts of these compounds in micro-organisms vary
considerably. As much as 60% of the dry weight of a
micro-organism can be accounted for by intracellular
polysaccharides, while an organism may produce extracellular
polysaccharides in amounts many times greater than its
dry weight. Microbial cell walls also contain large
amounts of polysaccharide. The range of different polysaccharides produced by micro-organisms is enormous and
only a few examples of their biosyntheses can be given in
this brief account.

7.4.1 SYNTHESIS OF SUGARS

If sugars are not available in the environment, they
must be synthesised by a micro-organism from the available
carbon sources. Autotrophic micro-organisms, growing on

carbon dioxide as the sole carbon source, employ the
reactions of the Calvin cycle to synthesise sugars (page
256). When heterotrophic micro-organisms are growing in
media containing C_2 and C_3 compounds as carbon sources,
hexose phosphates are synthesised from pyruvate by the
reactions of the EMP pathway (page 189). Certain of the
reactions in the EMP pathway are virtually irreversible.
During synthesis of sugar phosphates from pyruvate one
of these reactions, the conversion of pyruvate to
phosphoenolpyruvate, is catalysed by a different enzyme,
phosphoenolpyruvate synthase. Fructose 1,6-diphosphate
is hydrolysed by a specific phosphatase.

7.4.2   INTERCONVERSIONS AMONG SUGARS

Many different types of sugar residue occur in microbial
oligosaccharides and polysaccharides. These sugars are
formed from those immediately available in the environment,
and from sugar phosphates synthesised by the micro-
organism, in a variety of interconversion reactions. Most
of these interconversion reactions involve not the free
sugar but nucleoside diphosphate derivatives of sugars.
Glucose 1-phosphate, which is formed by isomerisation of
glucose 6-phosphate can, for example, react with UTP to
give UDP-glucose in a reaction involving the release of
inorganic pyrophosphate:

$$\text{glucose 1-phosphate} + \text{UTP} \rightleftharpoons \text{UDP-glucose} + \text{PPi}$$

UDP-glucose can then be metabolised to give a number of
other sugars. Galactose, for example, can be formed as a
result of a reaction catalysed by UDP-galactose
4-epimerase (*see* page 195):

$$\text{UDP-glucose} \rightleftharpoons \text{UDP-galactose}$$

The enzyme from *E. coli* K-12, which catalyses this
reaction, contains one mole NAD per molecular weight of
79 000 daltons. Other reactions of UDP-glucose involve
oxidation or reduction; e.g.:

$$\text{UDP-glucose} \xrightarrow[\text{NAD} \quad \text{NADH}_2]{} \text{UDP-glucuronic acid}$$

All of the common nucleosides form nucleoside diphosphate
sugars which are involved in interconversion reactions.
For instance, rhamnose (6-deoxy-L-mannose) is formed by
reduction of TDP-glucose:

TDP-glucose + NADPH$_2$ $\rightleftharpoons$ TDP-rhamnose + NADP + H$_2$O

Dideoxy sugars are formed by a similar type of reduction. In *Salmonella typhi*, CDP-glucose is the precursor of four of the five known dideoxy sugars (abequose, ascarylose, colitose and tyvelose); CDP-paratose is formed directly from CDP-tyvelose in a reaction which involves epimerisation at C-2 of the sugar residue.

A few interconversion reactions involve sugar phosphates. Mannose 6-phosphate is formed directly from glucose 6-phosphate. Glucosamine 6-phosphate is also formed from glucose 6-phosphate via fructose 6-phosphate; glutamine acts as the amino-group donor in this interconversion.

7.4.3 BIOSYNTHESIS OF OLIGOSACCHARIDES AND POLYSACCHARIDES

Oligosaccharides and polysaccharides are synthesised by micro-organisms by addition of sugar residues from nucleoside diphosphate sugars to acceptor molecules. A simple example is provided by the disaccharide trehalose, which contains two glucose residues joined by an α-1,1 linkage. In *Saccharomyces cerevisiae, Mycobacterium tuberculosis* and several other microbes, the disaccharide is formed in a reaction between UDP-glucose and glucose 6-phosphate but, in at least one streptomycete, GDP-glucose is used instead of UDP-glucose.

As regards polysaccharides, knowledge is most complete for synthesis of straight-chain polymers, although little is known of the mechanisms by which chain length is regulated. It is known, however, that during glycogen biosynthesis glucosyl units are transferred to the non-reducing terminus of the growing polymer. Almost nothing is known of the way in which branches are formed in polysaccharide chains. Extracts of several fungi have been shown to contain an enzyme, chitin synthase, which catalyses a reaction in which N-acetylglucosamine residues are transferred from UDP-N-acetylglucosamine to an acceptor molecule. Similarly, synthesis of yeast cell-wall mannan involves transfer of mannose residues from GDP-mannose to an acceptor. ADP-Glucose is involved in starch synthesis in *Chlorella* and in synthesis of glycogen in *Arthrobacter* spp.

Much less is known of the reactions involved in the biosynthesis of microbial heteropolysaccharides, especially highly branched molecules. Several groups of workers have,

however, achieved synthesis of certain type-specific
pneumococcal polysaccharides using cell-free extracts of
the appropriate strain of *Diplococcus pneumoniae*. For
example, Type III pneumococcal polysaccharide (*see* page
24) can be synthesised in a reaction mixture containing
UDP-glucose, UDP-glucuronic acid and a particulate fraction
from Type III *Diplococcus pneumoniae*. The sugar residues
from these nucleotides are incorporated in equal amounts
into a polysaccharide which can be precipitated by type-
specific antiserum. During the biosynthesis of this and
similar heteropolysaccharides, the sequence of sugar
residues is presumably determined by the specificities
of the transferases involved.

   Research on biosynthesis of a polysaccharide by
Aerobacter aerogenes has recently revealed a hitherto
unrecognised phenomenon in the biosynthesis of microbial
extracellular polysaccharides, namely the involvement of
a glycosyl carrier lipid with the structure

$$CH_3-\underset{\underset{CH_3}{|}}{C}=CH-CH_2-\left(CH_2-\underset{\underset{CH_3}{|}}{C}=CH-CH_2\right)_9-CH_2-\underset{\underset{CH_3}{|}}{C}=CH-CH_2-O-\underset{\underset{O}{\|}}{\overset{\overset{OH}{|}}{P}}-OH$$

The polysaccharide in question has the following tetrasaccharide repeating unit:

$$-\ (Gal\longrightarrow Man\longrightarrow Gal)\ -$$
$$\underset{GlcA}{\overset{\displaystyle\uparrow}{}}$$

It has been shown that in *A. aerogenes* there occurs a
sequential transfer of sugar residues from nucleotide
donors to a lipid. The lipid, which is undecaprenyl
phosphate, accepts two repeating units before these are
transferred to the growing polysaccharide chain. As we
shall see later, lipid intermediates are also involved in
synthesis of bacterial cell-wall polymers.

7.4.4   BIOSYNTHESIS OF BACTERIAL CELL-WALL COMPONENTS

Bacterial walls contain some unique and complex polymers,
notably peptidoglycans, teichoic acids and lipopolysaccharides (*see* page 32) and, during the past decade,
major advances have been made in our understanding of the
biosyntheses of these wall components. This research has
been prompted not only by the challenge which the molecular

complexity of these polymers offers the biochemist, but by the need to explain how synthesis of the polymers is controlled so as to give rise to an intricately structured organelle. It is known that the enzymes which catalyse synthesis of bacterial wall polymers are located in the plasma membrane. Detailed information on ways in which activities of these enzymes are regulated is lacking, but it has been established that synthesis of many wall polymers involves participation of a *glycosyl carrier lipid* which has been identified as undecaprenyl phosphate. Availability of this carrier in the membrane could well be a major regulatory process in biosynthesis of bacterial wall polymers. Finally, in studies on biosynthesis of wall polymers, the bacterial physiologist is spurred on by the knowledge that certain of these polymers are antigens of some importance to the microbiologist, and also that a number of clinically useful antibiotics interfere specifically with reactions on these biosynthetic pathways.

Peptidoglycans

Biosynthesis of cell-wall peptidoglycans has been studied mainly with strains of *Micrococcus lysodeikticus* and *Staphylococcus aureus* and, due largely to the work of Jack Strominger and his colleagues now at Harvard University, U.S.A., the pathways involved are reasonably well understood. The reactions involved in biosynthesis of the peptidoglycan in *Staph. aureus* are shown in *Figure 7.24*. The first stage involves synthesis of UDP-N-acetylglucosamine and UDP-N-acetylmuramyl acid pentapeptide. The assembly of the peptide portion onto UDP-muramic acid involves the stepwise addition of each amino acid by a process similar to that operating in the biosynthesis of glutathione and gramicidin-S. There is no evidence for the participation of aminoacyl-tRNAs in the synthesis of this pentapeptide. Moreover, ribonuclease does not affect incorporation of the amino acids. Presumably, the specificity for the amino-acid sequence resides in the enzymes that catalyse the incorporation. It is interesting to note that the peptide-synthesising system can be 'fooled' into incorporating 'wrong' amino acids, and this may eplain the deleterious effects which glycine and various D-amino acids have on cell-wall synthesis in bacteria.

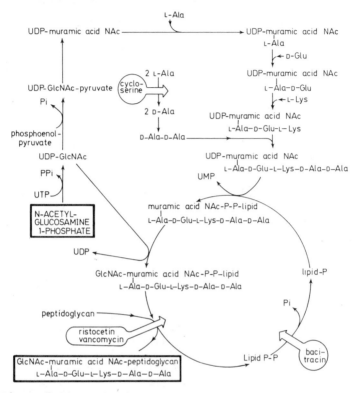

*Figure 7.24   Pathways leading to synthesis of cell-wall
peptidoglycan in* Staphylococcus aureus. *Reactions leading
to synthesis of the pentaglycine bridge are not shown.
The proposed loci action of certain antibiotics are
indicated.   For explanation of symbols see page (ix)*

After synthesis of the UDP derivatives of N-acetylglucosamine and N-acetylmuramyl pentapeptide, there
follows an orderly assembly of these residues into the
cell-wall peptidoglycan (*Figure 7.24*).   The process
involves the participation of a lipid carrier which has
been identified as again undecaprenyl phosphate.   After
a reaction in which N-acetylmuramyl pentapeptide is
linked to the lipid carrier by a pyrophosphate bridge,
an N-acetylglucosamine residue is donated by UDP-N-acetylglucosamine to give a disaccharide pentapeptide which
remains attached to the lipid carrier.   The disaccharide

pentapeptide unit is then inserted into the growing peptidoglycan chain, thereby liberating the pyrophosphate of the lipid carrier. This, in turn, gives rise to undecaprenyl phosphate which can then be used for another round of synthesis.

However, as we saw on page 33,, the peptidoglycans from many bacteria are cross-linked in the wall, and the disaccharide pentapeptide unit is inserted into the growing peptidoglycan chain only after the bridge has been attached to the peptide chain. This aspect of peptidoglycan biosynthesis has been studied most extensively in *Staph. aureus* which has a pentaglycine bridge (page 34). After the disaccharide pentapeptide has been synthesised on the lipid carrier, a pentaglycine chain is built onto the ε-amino group of the lysine residue on the lipid intermediate. Synthesis of the pentaglycine chain, unlike that of the peptide chain attached to the muramic acid residue, involves glycyl-tRNAs. These species of glycyl-tRNA differ from those that participate in polypeptide synthesis, and cannot in fact donate glycine residues in polypeptide synthesis. After insertion into the growing peptidoglycan chain, the pentaglycine group is cross-linked through the action of a transpeptidase that leads to elimination of the terminal D-alanine residue.

Elucidation of the reactions involved in peptidoglycan biosynthesis has enabled microbial physiologists to locate the reactions that are specifically inhibited by antibiotics, including phosphonomycin, oxamycin (also known as cycloserine), bacitracin, penicillin, ristocetin and vancomycin (*Figure 7.24*). Phosphonomycin, a product of *Streptomyces fradiae,* inhibits the reaction in which UDP-N-acetylglucosamine condenses with phosphoenol-pyruvate. D-Cycloserine is a competitive analogue of alanine, and inhibits action of the enzyme that catalyses formation of D-alanine from L-alanine (*Figure 7.25*). Bacitracin, a polypeptide antibiotic produced by a soil bacillus, inhibits release of inorganic phosphate from the pyrophosphate of the lipid intermediate (*Figure 7.24*). Ristocetin and vancomycin prevent transfer of the disaccharide peptide from the lipid carrier to the growing peptidoglycan chain by interacting with the acyl-D-Ala-D-Ala part of the peptide chain.

The mode of action of penicillin, in many ways the doyen of antibiotics and certainly still one of the most useful clinically, has interested microbial physiologists for well over 20 years. Evidence of an association between the

phosphonomycin   D-cycloserine

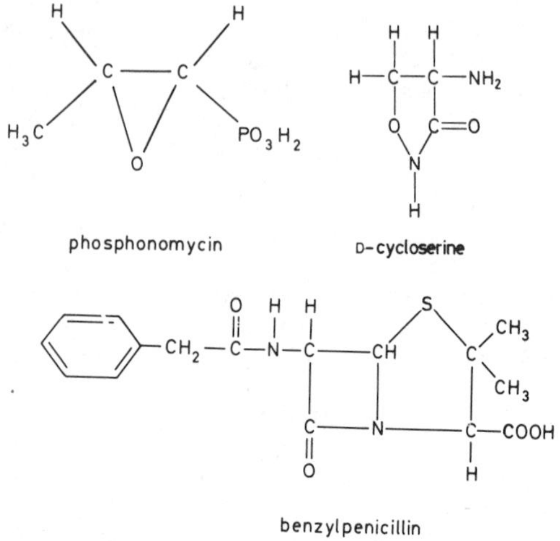

benzylpenicillin

Figure 7.25 Structural formulae of some antibiotics which inhibit synthesis of bacterial cell-wall peptidoglycans. The structure of D-alanine is also shown to indicate its similarity to that of cycloserine

action of penicillin and peptidoglycan biosynthesis came early in the 1950s from the work of James Park and his collaborators who found that staphylococci, grown in the presence of penicillin, accumulated UDP derivatives of muramic acid and muramic acid peptides, compounds which became known as *Park nucleotides*. It has since been established that, at least in *Staph. aureus*, penicillin prevents linking of the pentaglycine bridge in the terminal stages of the biosynthesis, specifically by inhibiting action of the transpeptidase.

Teichoic acids

The first step in the biosynthesis of those teichoic acids that have a poly-(glycerol phosphate) or poly-(ribitol phosphate) backbone is the formation of an unsubstituted backbone from CDP-glycerol or CDP-ribitol. Opinions differ with regard to synthesis of poly-(glycerol phosphate), but with synthesis of poly-(ribitol phosphate) it has been established that a lipid intermediate is not involved. Glycosylation of the backbone takes place in reactions which involve nucleoside diphosphate sugars and which are

catalysed by glycosyl transferases. Curiously, in
experiments with cell-free extracts, glycosylation of a
growing poly-(ribitol phosphate) backbone is more
efficient than glycosylation of an exogenously provided
poly-(ribitol phosphate). Until recently, little was
known of the manner in which alanine residues are attached
to the backbone. However, an enzyme has now been detected
in cell-free extracts of *Lactobacillus casei* which transfers
D-alanine residues to a membrane-bound acceptor, that may
well be a lipoteichoic acid.

In synthesis of teichoic acids which have sugar residues
in the backbone (*see* page 37), it has quite definitely
been established that lipid intermediates are involved in
synthesis of the backbone. With synthesis of the wall
teichoic acid in *Staphylococcus lactis* I3 (*Figure 2.13,*
page 37), the first intermediate is the lipid
pyrophosphate-N-acetylglucosamine. A glycerol phosphate
residue is then attached to the 4-hydroxyl group of the
N-acetylglucosamine residue to give a lipid intermediate
with a fully formed repeating unit (*Figure 7.26*). The
unit is then transferred to a growing backbone and lipid
phosphate is available for a further turn of the cycle.
It should be noted that in synthesis of these teichoic
acids, unlike that of peptidoglycans (*Figure 7.24*), free
lipid pyrophosphate is not formed, which explains why
teichoic acid biosynthesis is not affected by bacitracin.

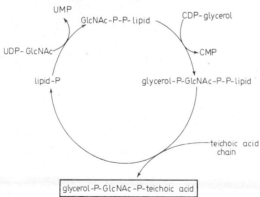

*Figure 7.26 Pathways leading to synthesis of the backbone
of the teichoic acid in the wall of* Staphylococcus
lactis *I3*

Lipopolysaccharides

The core polysaccharide and the O-specific repeating
units in bacterial lipopolysaccharides are synthesised by
different mechanisms. In fact, little is known of the
way in which the core polysaccharide is synthesised,
particularly the innermost portions. The enzymic system
for synthesis of L-glycero-D-mannoheptose (Hep; *see* page
41) has yet to be elucidated, although it has been
established that the D-mannoheptose residue arises from
sedoheptulose 7-phosphate. After residues of 2-keto-3-
deoxyoctonate and L-glycero-D-mannoheptose have been
incorporated into the inner core, the sugars that make
up the outer core are added successively in reactions
catalysed by specific glycosyl transferases.

By contrast, the O-specific repeating units, which
together make up the bulk of the lipopolysaccharide, are
synthesised as oligosaccharides on a lipid carrier before
they are incorporated into the lipopolysaccharide. The
main steps in the synthesis of the lipopolysaccharide in
strains of *Salmonella* are shown in *Figure 7.27*. The
lipid carrier involved in this synthesis is yet again
undecaprenyl phosphate (*see* page 298)

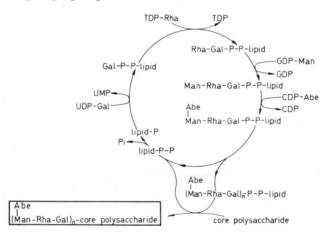

Figure 7.27 Probable pathways leading to synthesis of
Salmonella *lipopolysaccharide repeating units and their
attachment to the core polysaccharide. For explanation
of symbols see page (ix)*

There are reasons for believing that the lipopoly-saccharides are synthesised on the plasma membrane, anchored possibly by the hydrophobic lipid A, after which they are transferred to the outer membrane. Exactly how this process operates is far from clear, although electron micrographs show that there are bridges between the plasma and outer membranes, and these might be the transfer channels.

7.5 LIPIDS

Over the past two decades, research on biosynthesis of lipids has, after a sluggish start, proceeded at an astonishing pace with the result that the pathways leading to synthesis of the commoner lipids are now well charted. It is clear from the account of the lipid composition of microbes given in Chapter 2 (page 47) that there is a tremendous variety in the types of lipid synthesised by micro-organisms, and it is hardly surprising that the minutiae of the processes leading to synthesis of some of the less common lipids (such as plasmalogens and sphingolipids) remain somewhat obscure.

7.5.1 SYNTHESIS OF FATTY ACIDS

Synthesis of coenzyme-A esters of fatty acids is the first step in lipid biosynthesis. Long-chain even numbered fatty acids are formed by a series of reactions, often referred to as the *malonyl-CoA pathway*, in which C_2 units derived from malonyl-CoA are successively added onto a primer acetyl-CoA molecule. The overall reaction for synthesis of palmityl-CoA is:

$$CH_3.CO.S.CoA + 7COOH.CH_2.CO.S.CoA + 14NADPH_2$$
$$\downarrow$$
$$CH_3(CH_2)_{14}.CO.S.CoA + 7CO_2 + 7CoA.SH + 14NADP + 7H_2O$$

The first reaction on this pathway involves formation of malonyl-CoA. This intermediate can arise in several ways. One way is in a reaction catalysed by the biotin-containing enzyme, acetyl-CoA carboxylase:

$$H.CH_2.CO.S.CoA + CO_2 + ATP \rightleftharpoons HOOC.CH_2.CO.S.CoA + ADP + Pi$$

Malonyl-CoA can also be formed by a reaction between malonate and coenzyme A or succinyl-CoA; in *Clostridium kluyveri* it can arise by oxidation of malonyl semialdehyde-CoA.

Fatty-acid synthesis has been studied with extracts of several micro-organisms, particularly *E. coli* and *Sacch. cerevisiae*. The systems examined in bacteria are different in several respects from that in yeast. In bacteria, seven different proteins are involved in each turn of a cycle which gradually extends the fatty-acyl chain. Following synthesis of malonyl-CoA, the acyl group in this compound together with that from acetyl-CoA is transferred, by the action of transferases, to a thiol group on a protein known as *acyl carrier protein* (ACP). Acyl carrier proteins from bacteria are heat-stable and have a molecular weight of around 10 000 daltons. Moreover, they have 4'-phosphopantetheine as a prosthetic group, and bind acyl groups as thioesters. A third enzyme then catalyses condensation of acetyl.S.ACP and malonyl.S.ACP to yield acetoacetyl.S.ACP and carbon dioxide:

$$CH_3.CO.S.ACP + HOOC.CH_2.CO.S.ACP \rightarrow CH_3.CO.CH_2.CO.S.ACP$$
$$+ CO_2 + ACP.SH \quad + CO_2 + ACP.SH$$

There follows a reduction of acetoacetyl.S.ACP to give β-D-hydroxybutyryl.S.ACP:

$$NADPH_2 \quad NADP$$
$$CH_3.CO.CH_2.CO.S.ACP \longrightarrow CH_3.CHOH.CH_2.CO.S.ACP$$

β-D-Hydroxybutyryl.S.ACP in turn is dehydrated to crotonyl.S.ACP which, in a second reduction, is converted into butyryl.S.ACP:

$$H_2O$$

$$CH_3.CHOH.CH_2.CO.S.ACP \quad CH_3.CH=CH.CO.S.ACP$$

$$NADPH_2 \quad NADP$$
$$CH_3.CH_2.CH_2.CO.S.ACP$$

Butyryl.S.ACP then condenses with another molecule of malonyl.S.ACP and, by a comparable sequence of reactions, two more carbon atoms are added to the fatty-acyl chain. Further turns of the cycle take place until a fatty-acyl chain of the required length is formed.

The fatty acid-synthesising system in *Sacch. cerevisiae* differs from that found in bacteria in that it is isolated as a homogeneous multi-enzyme complex with a molecular weight of about 2.3 million daltons. To date, it has not been possible to fractionate this complex into individual enzymes, but there is evidence that the complex includes an acyl carrier protein similar to that found in bacterial fatty acid-synthesising systems. Largely as a result of studies by Feodor Lynen and his colleagues in Munich, West Germany, we now have a reasonably complete picture of the way in which the yeast fatty acid-synthesising system works. The system is a complex of seven individual enzymes, comprising 21 subunits, arranged around a central thiol group (*Figure 7.28*). The synthesis is initiated by transfer of an acetyl residue from acetyl-CoA to a peripheral thiol group, and transfer of a malonyl residue to the central thiol group. The acetoacetyl group formed by condensation remains attached to the central thiol group but, once a butyryl residue has been formed, it is transferred to a peripheral thiol group leaving the central group free to accept another malonyl residue.

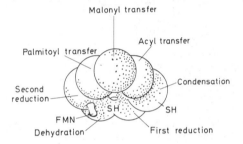

Figure 7.28   The proposed structure for the multi-enzyme fatty-acid synthetase in Saccharomyces cerevisiae

Although they are basically very similar, the yeast and bacterial systems also show other differences. The yeast system appears to have a strict requirement for acetyl-CoA as a primer molecule, whereas the bacterial system can use other coenzyme-A esters as primers. Some bacteria synthesise odd-numbered straight-chain fatty acids, and these arise from syntheses in which valeryl-CoA is the primer molecule. The ability of the bacterial system to use different primer molecules explains the formation by certain organisms of methyl-branched fatty acids (*see* page 50). For example, C_{15} and C_{17} *anteiso* methyl-branched acids in *Micrococcus lysodeikticus* are formed when 2-methylbutyryl-CoA is used as a primer molecule; 2-methylbutyryl-CoA is formed by catabolism of leucine. Isovaleryl-CoA acts similarly in a species of *Ruminococcus* to yield C_{15} and C_{17} isomethyl-branched C_{15} and C_{17} acids. Another branched fatty acid, tuberculostearic acid or

10-methylstearic acid, is formed in *Mycobacterium phlei*
by a different mechanism, namely the addition of a C_1 unit
from methionine to C-10 of oleic acid.

The products of fatty-acid synthesis in yeast also differ
from those in bacteria in that they are usually restricted
largely to acids 16 or 18 carbon atoms long. The bacterial
system often gives rise to acids with a much longer chain;
acids synthesised by mycobacteria may contain as many as
24-30 carbon atoms. Very little is known of factors which
determine chain length during fatty-acid synthesis, although
it is possible that they are based on a lack of affinity of
certain acids for the acyl carrier protein or an enzyme.
There is one other difference between the products of the
two systems which remains entirely unexplained. The product
of the yeast fatty acid-synthesising system is an acyl-CoA
ester, whereas the bacterial system gives rise to the free
acid.

The lipids in most micro-organisms contain a certain
proportion of fatty-acid residues with one, two or three
double bonds. The mechanisms by which double bonds are
introduced into fatty acids have been elucidated mainly
by Konrad Bloch and his associates working in Harvard
University, U.S.A. Two distinct fatty acid-desaturating
mechanisms operate in micro-organisms. The first of these
is used by eukaryotic organisms and by some groups of
bacteria, and requires molecular oxygen, $NADPH_2$ and the
coenzyme-A ester of the saturated fatty acid. For example:

$$CH_3(CH_2)_{14}.CO.S.CoA \text{ palmityl-CoA}$$

$$O_2 \downarrow NADPH_2$$

$$CH_3.(CH_2)_5CH=CH(CH_2)_7.CO.S.CoA \text{ palmitoleyl-CoA}$$

Additional double bonds can be introduced into the CoA
ester of the mono-enoic acid by a similar reaction which
may be catalysed by the same enzyme. A different
desaturation mechanism operates in many bacteria. It
differs from the aerobic mechanism in that desaturation
takes place at an earlier stage in synthesis of the fatty
acid, usually when the chain is 8-12 carbon atoms long.
The reaction involves dehydration of a β-hydroxyacyl
residue on the acyl carrier protein to produce a *cis*-β-
enoyl acyl residue. There follows a chain elongation of
the β-enoyl acyl residue to give mono-unsaturated fatty
acids of the appropriate chain length.

Fatty acids containing more than three double bonds are synthesised by some micro-organisms. Species of *Euglena* synthesise tetra-, penta- and hexa-enoic acids, as also do certain algae. The pathways leading to synthesis of these poly-unsaturated acids probably involve a combination of both of the desaturation mechanisms already described. Amoebae, for example, elongate linoleic acid to form a $11,14$-C_{20} di-enoic acid which is then progressively desaturated towards the carboxyl end of the chain to give a $5,6,11,14$-C_{20} tetra-enoic acid, known as arachidonic acid.

Cyclopropane fatty acid-residues, particularly C_{17} and C_{19} acids, are fairly common in bacterial lipids. These acids are formed from unsaturated acids by addition of a C_1 unit across the double bond; the C_1 donor is S-adenosylmethionine. In *Serratia marcescens* and *Clostridium butyricum*, formation of cyclopropane acids occurs when the unsaturated acid is present as a residue in phosphatidylethanolamine:

$$CH_2-O-CO-(CH_2)_9-\overset{\overset{\displaystyle H}{|}}{C}=\overset{\overset{\displaystyle H}{|}}{C}-(CH_2)_5-CH_3 \longrightarrow CH_2-O-(CH_2)_9-\overset{\overset{\displaystyle H}{|}}{C}-\overset{\overset{\displaystyle H}{|}}{C}-(CH_2)_5-CH_3$$

cis-vaccenic acid residue   lactobacillic acid residue

Some bacteria and yeasts can synthesise 3-hydroxy fatty acids, which are usually found in ester-bound or glycoside-bound extracellular lipids. For example, 3-hydroxy C_{16} and C_{18} acids are found in mannitol and pentitol esters produced by *Rhodotorula* species. These 3-hydroxy acids are probably formed by hydration of a double bond in a fatty acid.

7.5.2 SYNTHESIS OF LIPIDS

Glycerophospholipids

The principal reactions which lead to synthesis of
glycerophospholipids in micro-organisms are shown in
Figure 7.29. Phosphatidic acids play an important role
on this pathway. These compounds can be formed in one
of three main ways. The main way, shown in *Figure 7.29*,
involves a stepwise acylation of glycerol 3-phosphate with
coenzyme-A esters of fatty acids. Other ways in which
phosphatidic acids are formed are by phosphorylation of
a diglyceride with ATP in a reaction catalysed by a
diglyceride kinase, and phosphorylation of a monoglyceride
with ATP to give a lysophosphatidic acid which is then
acylated with the coenzyme-A ester of a fatty acid. The
last two mechanisms probably operate in microbes that
utilise reserves of triacylglycerols.

Nevertheless, the key participant in reactions leading
to synthesis of the major glycerophospholipids is
CDP-diglyceride which is produced by a reaction between
a phosphatidic acid and CTP. By reacting with glycerol
3-phosphate, serine or inositol, it can give rise,
respectively, to phosphatidylglycerol, phosphatidylserine
or phosphatidylinositol (*Figure 7.29*). A molecule of CMP
is released in each of these reactions. Cardiolipin, or
diphosphatidylglycerol, is formed in a reaction between
CDP-diglyceride and 3-phosphatidylglycerol 1'-phosphate.

Most microbes synthesise phosphatidylethanolamine by
decarboxylation of phosphatidylserine (*Figure 7.29*).
Phosphatidylcholine is then produced by a stepwise
N-methylation of phosphatidylethanolamine, S-adenosylmethionine being the molecule which donates the methyl
groups. This has been referred to as the *methylation
pathway* for synthesis of phosphatidylcholine. *Saccharomyces
cerevisiae,* and probably many other eukaryotic microbes
as well, can also use the *cytidine nucleotide pathway*
for synthesising phosphatidylethanolamine and phosphatidylcholine. In this latter pathway, ethanolamine (or choline)
is converted to the phosphoryl derivative by reacting with
ATP. The phosphorylated derivative is then converted into
the corresponding cytidine nucleotide derivative, to give
CDP-ethanolamine (or CDP-choline):

CTP + phosphorylethanolamine $\longrightarrow$ CDP-ethanolamine + PPi

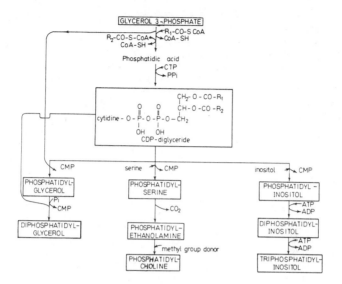

Figure 7.29 Pathways leading to synthesis of phospholipids

By reacting with a 1,2-diacylglycerol, CDP-ethanolamine or CDP-choline can give rise to phosphatidylethanolamine or phosphatidylcholine:

CDP-ethanolamine + 1,2-diacylglycerol

$\longrightarrow$ phosphatidylethanolamine + CMP

It seems likely that the cytidine nucleotide pathway evolved at a stage subsequent to the appearance of eukaryotic forms of life.

Triacylglycerols

Triglycerides or triacylglycerols, which are found in many eukaryotic micro-organisms, are synthesised from a phosphatidic acid which is first converted into a 1,2-diacylglycerol in a reaction catalysed by phosphatidic acid phosphatase. The coenzyme-A ester of a fatty acid then reacts with the 1,2-diacylglycerol to give a triacylglycerol. The second reaction is catalysed by a diacylglycerol acyltransferase. Although it has yet to be demonstrated experimentally in micro-organisms, it is

L

likely that in those organisms that synthesise triacylglycerols there is a turnover of these lipids, during
which a mono-acylglycerol is acylated to give a
diacylglycerol in a reaction catalysed by a mono-acylglycerol acyltransferase. This *mono-acylglycerol pathway*
leading to synthesis of triacylglycerols has been shown
to operate in mammalian intestinal mucosa, where the
preferred substrate is the 2-mono-acylglycerol rather
than 1-mono-acylglycerol.

Glycolipids

Studies on biosynthesis of glycolipids have been confined
largely to bacterial glycosyl diglycerides. These
compounds arise as a result of a reaction between a
1,2-diacylglycerol and a nucleoside diphosphate sugar.
For example, enzymes from *Micrococcus lysodeikticus*
catalyse formation of an α-D-mannosyl-(1,2)-diacylglycerol
and a dimannosyl-diacylglycerol when incubated with
GDP-mannose and a 1,2-diacylglycerol. In other bacteria,
different nucleoside diphosphate sugars are involved.
A particulate preparation from *Streptococcus faecalis,*
when incubated with UDP-glucose and a 1,2-diacylglycerol,
can catalyse synthesis of α-D-glucopyranosyl-diacylglycerol
and 2-O-α-D-glucopyranosyl-α-D-glucopyranosyl-diacylglycerol.

7.5.3 SYNTHESIS OF POLY-β-HYDROXYBUTYRATE

Although it is not a lipid, poly-β-hydroxybutyrate (PHB)
is the component of the so-called lipid granules found in
certain bacteria (*see* page 81). The reactions leading
to synthesis of PHB from acetyl-CoA are shown in *Figure
7.30*. Acetoacetyl-CoA can be synthesised from acetyl-CoA
either by a condensation catalysed by a β-ketothiolase or
by condensation with a molecule of malonyl-CoA catalysed
by a β-keto-acyl-CoA synthetase. The former route is
thought to be favoured by most bacteria that synthesise
PHB. Action of acetoacetyl-CoA reductases is linked,
preferentially, either to $NADH_2$ as in *Hydrogenomonas
eutropha* or to $NADPH_2$ as with *Azotobacter beijerinckia*.
The final reaction on the pathway leading to PHB synthesis
is catalysed by an enzyme bound to the PHB granule.
   Synthesis of PHB can be seen to involve reactions that
are related to those that take place during fatty-acid
biosynthesis. However, they are in fact quite different

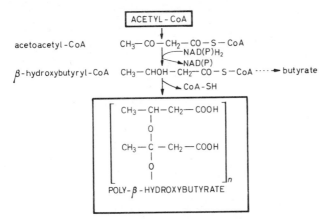

Figure 7.30. Reactions leading to synthesis of poly-β-hydroxybutyrate

reactions, witnessed by the finding that acetoacetate, or D-(-)-3-hydroxybutyrate, when linked to an acyl carrier protein, cannot participate in reactions that lead to PHB synthesis.

7.6 TETRAPYRROLES AND RELATED COMPOUNDS

Micro-organisms contain compounds which have as their basic structural unit the pyrrole ring. In many of these compounds, four pyrrole rings are joined together to give a tetrapyrrole structure which, when the rings are joined by carbon bridges, is known as the *porphyrin nucleus*. In most biologically active porphyrins, the tetrapyrrole nucleus is co-ordinated with a metal such as iron (which occurs in haemoproteins including cytochromes and catalase) or magnesium (as in the chlorophylls; page 77). The corrin ring in the cobamide (vitamin B_{12}) molecule is co-ordinated with cobalt, but this is not a true porphyrin nucleus since two of the pyrrole rings (A and D) are joined directly. The phycobilins, which are prosthetic groups in the photosynthetically active biliproteins found in algae, are linear tetrapyrroles (*see* page 77). Prodigiosin, the blood-red pigment which occurs in *Serratia marcescens* and in certain actinomycetes, is a linear tripyrrole (*see Figure 2.34,* page 86) which is synthesised by the coupling of monopyrrole and bipyrrole precursors.

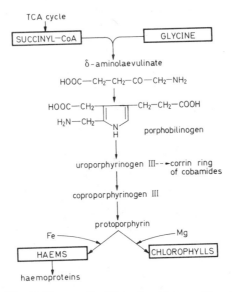

*Figure 7.31 Pathways leading to the synthesis of
tetrapyrroles and related compounds*

The major steps leading to synthesis of porphyrins and
related compounds in micro-organisms are shown in *Figure
7.31.* The key intermediate on this pathway is δ-aminolae-
vulinate, two molecules of which condense to form the
substituted monopyrrole porphobilinogen. Four molecules
of porphobilinogen then combine to form a tetrapyrrole
nucleus into which iron or magnesium is ultimately
co-ordinated to give **haems** or **chlorophylls.** The steps
leading to synthesis of the corrin ring in cobamides are
not fully understood, but are thought to involve formation
of uroporphyrinogen III as an intermediate. Phycobilins
are probably synthesised by an opening of the tetrapyrrole
ring.

7.7   TERPENES AND RELATED COMPOUNDS

Micro-organisms synthesise representatives of a group of
compounds which, although they apparently have no common
structural features, have been shown by biosynthetic
studies to be related. These are the *terpenes* which are
defined as compounds that have a distinct architectural

and chemical relationship to the branched five-carbon isoprene or 2-methyl-1,3-butadiene molecule:

$$CH_3 \diagdown C-CH=CH_2$$
$$CH_2 \diagup$$

Terpenes are divided into groups depending upon the number of isoprene units in the molecule; the groups include monoterpenes $(C_{10}H_{16})$, sesquiterpenes $(C_{15}H_{24})$, diterpenes $(C_{20}H_{32})$, triterpenes $(C_{30}H_{48})$ and polyterpenes $(C_5H_8)_n$. The number of terpenes and terpene-like compounds synthesised by micro-organisms is extremely large. They include carotenoids (polyterpenes), sterols (triterpenes) and members of the coenzyme-Q family (e.g. ubiquinone).

The principles involved in terpene biosynthesis are embodied in the 'isoprene rule'. The biosynthesis involves four main stages (*Figure 7.32*): (a) Synthesis of the key intermediate mevalonate from either acetyl-CoA or leucine. (b) Dehydration and decarboxylation of mevalonic 5-pyrophosphate to give the five-carbon 'active isoprene', isopentenyl pyrophosphate:

mevalonic 5-pyrophosphate   isopentenyl pyrophosphate

This reaction is followed by the condensation of isoprene units to form acyclic terpenes of various chain lengths. (c) Cyclisation of acyclic structures, e.g. of squalene in sterol biosynthesis. (d) Further modifications of the cyclised structure.

Although the principles of terpene biosynthesis have been formulated, with the exceptions of carotenoid and sterol biosyntheses, little is known of the nature of the intermediates involved. Even with biosynthesis of carotenoids and sterols, knowledge of the biosynthetic pathways which lead to individual compounds is far from complete. The important reaction in sterol biosynthesis

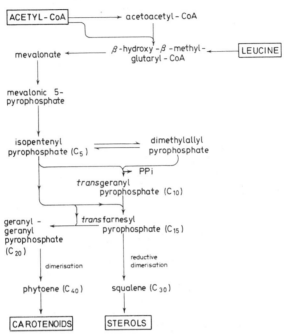

*Figure 7.32 Pathways leading to the biosynthesis of
carotenoids and sterols*

is that in which squalene is cyclised to give lanosterol.
Modification of lanosterol to yield sterols which are
found in microbial membranes probably involves more than
one pathway.

7.8   B-GROUP VITAMINS

The only property which B-group vitamins have in common
is that they each function metabolically as coenzymes or
as part of coenzymes. The chemical complexity of many
B-group vitamins, together with the fact that they are
synthesised by micro-organisms only in very small amounts,
has made a study of their biosynthesis difficult, and many
details of the pathways leading to synthesis of B-group
vitamins in micro-organisms have still to be discovered.
   The starting materials for the biosynthesis of most
B-group vitamins are the low molecular-weight compounds
that are used in the synthesis of other cell constituents.

An example is provided by the biosynthesis of pantothenate, details of which have been worked out in some detail (*Figure 7.33*). The vitamin is synthesised from pantoate and β-alanine. Pantoate is formed from α-ketoisovalerate, an intermediate in valine biosynthesis (*see* page 273). Two independent pathways have been shown to lead to synthesis of β-alanine. Some micro-organisms (e.g. *Rhizobium* spp.) decarboxylate aspartate to form β-alanine, but other micro-organisms use a pathway that is also found in mammalian cells and involves a transamination reaction between an amino acid and malonic semialdehyde (formed from propionyl-CoA).

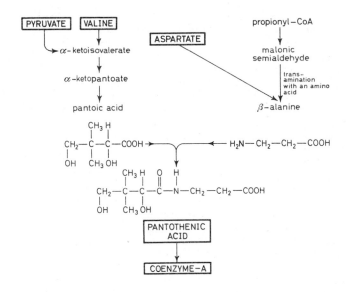

Figure 7.33 Pathways leading to the biosynthesis of pantothenate and coenzyme-A

7.9   SECONDARY METABOLITES

In this brief account of the biosynthetic activities of
micro-organisms, I have so far been able to deal only
with the major chemical components of microbial cells.
But, in many ways, the biosyntheses so far described
provide poor testimony to the tremendous synthetic
capabilities possessed by many microbes. The diversity
of the chemical structures that are produced by micro-
organisms is probably best illustrated by considering a
group of compounds that microbes produce seemingly
gratuitously.

Many fungi and some bacteria, as soon as they enter the
stationary phase of growth in batch culture, begin to
synthesise and often to excrete compounds which, although
they have no obvious function in cell growth and
multiplication, are often produced in such quantity as
to account for an appreciable proportion of the ATP
produced by the microbe. These compounds have been dubbed
by John Bu'Lock of the University of Manchester, England,
as *secondary metabolites* to distinguish them from primary
metabolites which are synthesised during exponential
growth of microbes. A distinction between primary and
secondary metabolites was made many years ago by plant
physiologists who recognised compounds that are
synthesised by all plants (primary metabolites, such as
chlorophylls) and those synthesised only by particular
species (for example camphor).

The range of chemical structures found among microbial
secondary metabolites is truly tremendous. Their names
are often bizarre. They adorn indexes of chemical
compounds with names that range from abietic acid and
ajmaline to zierone and zingiberine. In spite of this
diversity in chemical structure, secondary metabolites
have in common the property that they are
synthesised from low molecular-weight compounds that,
during exponential growth, are used in synthesis of cell
constituents. Five main classes of microbial secondary
metabolite are recognised as indicated in *Figure 7.34.*

Metabolites included in the first class (sugar
derivatives) are compounds synthesised from pentoses and
hexoses which, during exponential growth, would have been
incorporated into polysaccharides and nucleic acids. The
peptides and alkaloids which go to make up the second
class are derived from compounds such as TCA-cycle
intermediates that are precursors in synthesis of amino

acids (page 268). The third class is really a special
category of the second, since the compounds included in
the class are synthesised from precursors of aromatic
amino acids or from these amino acids themselves (page
274). Polyketides, which form the fourth class, are
compounds which are grouped together because they are all
derived biosynthetically from a β-polyketide chain. The
polyketide chain may consist of C_2 units, formally related
to acetate, $(CH_2.CO)_n$, or of propionate units $(CH_2.CH_2.CO)_n$.
Formation of two-carbon units by an exponentially growing
micro-organism is often a prelude to synthesis of fatty
acids (page 305). But formation of these units in the
stationary phase of growth can give rise to a vast array
of secondary metabolites, two examples of which are given
in *Figure 7.34*. Finally, secondary metabolism of
precursors in the biosynthesis of terpene (page 314)
gives rise to compounds which make up the fifth class of
microbial secondary metabolite. Not surprisingly, some
microbial secondary metabolites do not fall neatly into
just one of these classes. Novobiocin, for example
(*Figure 7.34*), contains a residue of noviose, a sugar
derived from glucose, in addition to a residue of
3-aminocoumarin (derived from shikimate) and an
isopentenyl group which originates from isoprenoid
precursors.

Microbial secondary metabolites have for some years
provided an almost never-ending challenge to the natural-
product chemist. More recently, biochemists have started
to study the biosynthesis of these compounds with, as yet,
only moderate success no doubt because of the molecular
complexity of these compounds. Data from experiments
using radioactively labelled precursors have not only
revealed the nature of precursors but, in some instances,
ways in which the precursors may give rise to the
metabolite. For example, the oxytetracycline molecule is
thought to be synthesised as a result of head-to-tail
linkage of acetate units (*Figure 7.35*).

It is not only the microbial biochemist who is interested
in secondary metabolites formed by microbes. As the
examples given in *Figure 7.34* indicate, many of the
compounds formed by micro-organisms are commercially
valuable as drugs (penicillin, streptomycin, ergometrine)
or in agriculture and industry (gibberellic acid).

One of the most intriguing aspects to production of
secondary metabolites by micro-organisms are the reasons
for their synthesis. It is of course conceivable that

Class	Examples	Producing Organism
I. Sugar derivatives	Streptomycin	*Streptomyces griseus*

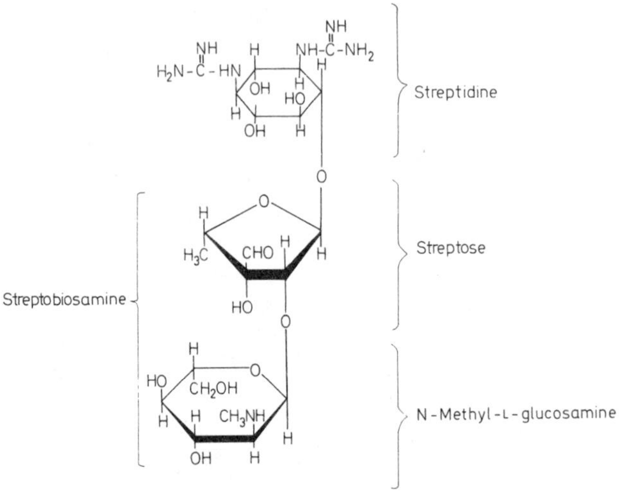

Streptidine

Streptose

Streptobiosamine

N-Methyl-L-glucosamine

| | Novobiocin | *Streptomyces niveus* |

| II. Polyketides and alkaloids | Benzylpenicillin | *Penicillium chrysogenum* |

| | Ergometrine | *Claviceps purpurea* |

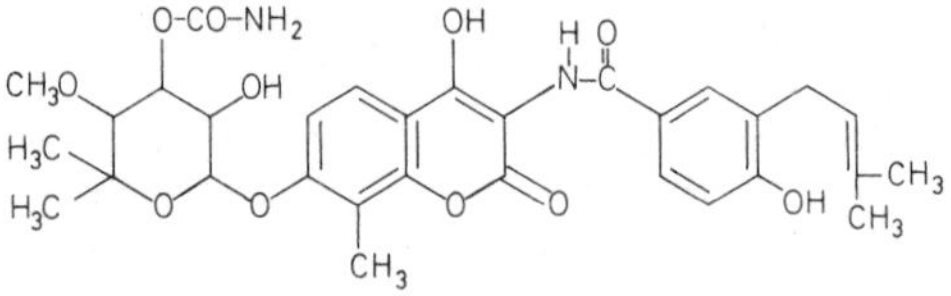

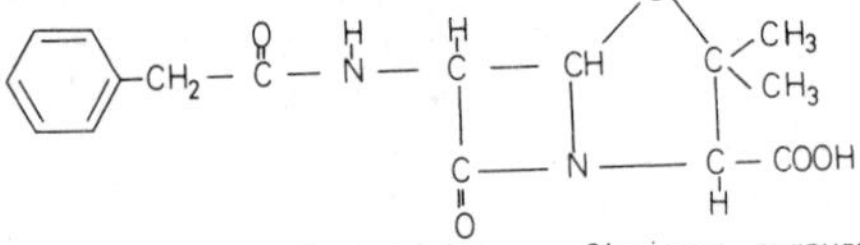

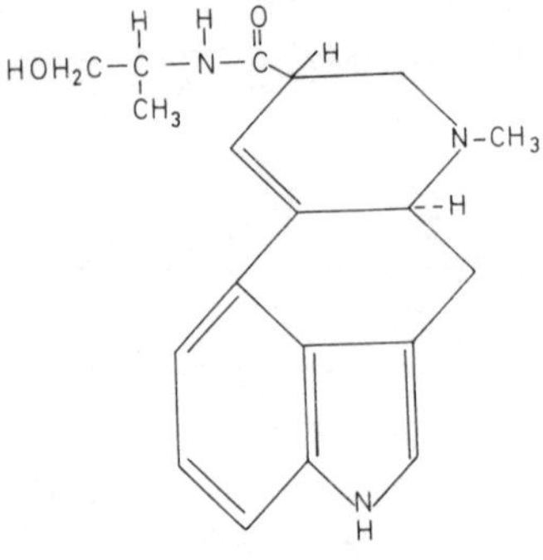

III. Shikimic
metabolites

Gliotoxin   *Gliocladium fimbricatum*

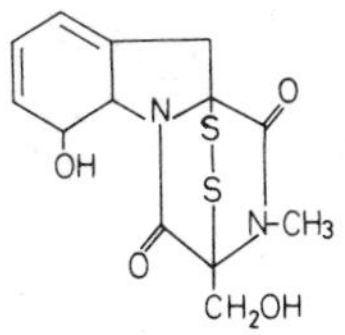

IV. Polyketides

Rutilantinone   *Streptomyces* spp.

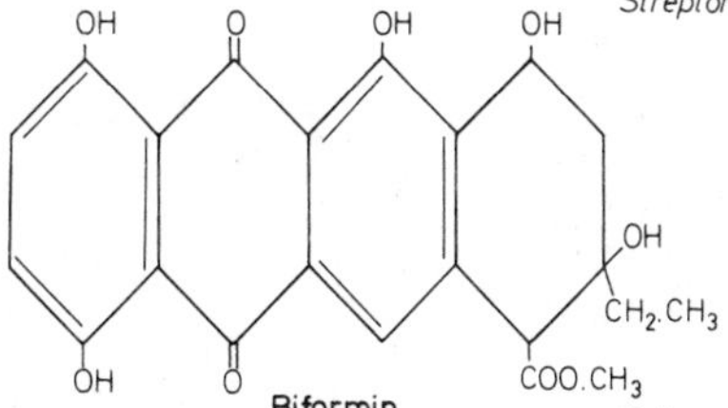

Biformin   *Polyporus biformis*

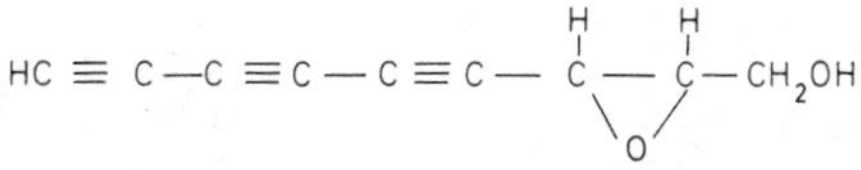

$$HC \equiv C - C \equiv C - C \equiv C - \overset{H}{\underset{O}{C}} - \overset{H}{C} - CH_2OH$$

V. Isoprenoids

Gibberellic acid   *Gibberella fujikuroi*

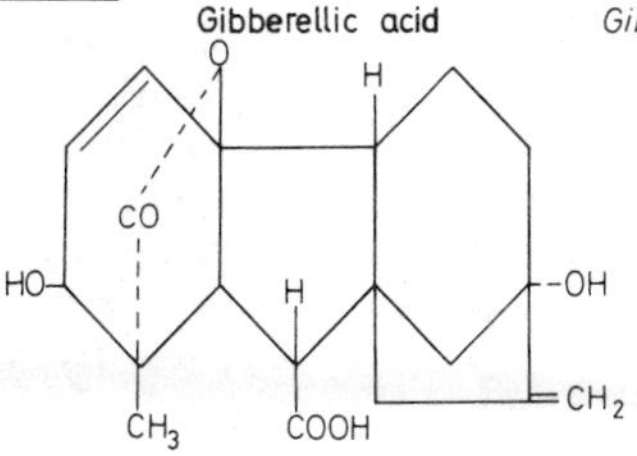

Figure 7.34   Classification, and some examples and producing organisms, of microbial secondary metabolites

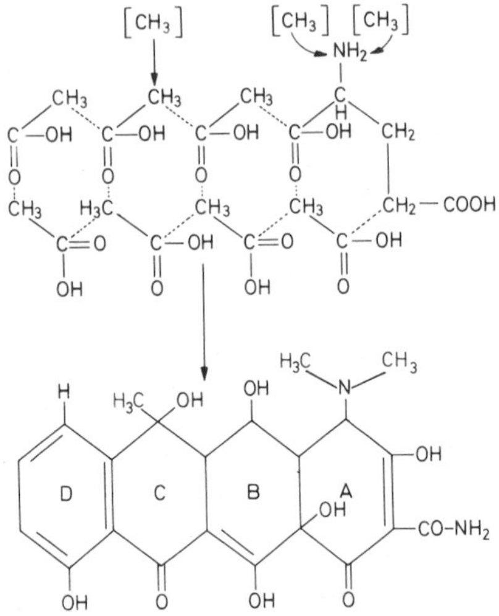

*Figure 7.35 Diagram showing how oxytetracycline could
arise by head-to-tail linkage of acetate units. Certain
of the atoms in ring A arise from glutamate. The N-methyl
and C-methyl groups arise from the methyl group of
methionine*

these compounds represent the results of 'errors' in
metabolism and therefore serve no useful function.
However, many microbial physiologists hold that the
process of secondary metabolism is more important to the
microbe than the specific products. Despite the operation
of efficient metabolic control mechanisms (see Chapter 8,
page 327), microbes in cultures which enter the stationary
phase of growth might be faced with death because of the
accumulation in cells of very high concentrations of low
molecular-weight compounds. Secondary metabolism can be
viewed as a process which leads to relief of such a crisis
by converting these intermediates into inocuous products.
It has also been suggested that certain secondary
metabolites may have a metabolic role, such as mediating
passage of certain molecules across the plasma membrane
(valinomycin, alamethicin), or in combating other microbes
in natural micro-environments. However, there is little
experimental evidence to support these proposals.

FURTHER READING

THE ATP BALANCE SHEET

FORREST, W.W. and WALKER, D.J. (1971). The generation
and utilization of energy during growth. *Advances in
Microbial Physiology,* 5, 213-274
LEHNINGER, A.L. (1971). *Bioenergetics,* 2nd edn, 270 pp.
Menlo Park, California; W.A. Benjamin Inc.
PAYNE, W.J. (1970). Energy yields and growth of
heterotrophs. *Annual Review of Microbiology,* 24, 17-52
PINE, M.J. (1972). Turnover of intracellular proteins.
Annual Review of Microbiology, 26, 103-126

RAW MATERIALS FOR BIOSYNTHESES

KORNBERG, H.L. (1966). Anaplerotic sequences and their
role in metabolism. *Essays in Biochemistry,* 2, 1-31
QUAYLE, J.R. (1972). The metabolism of one-carbon
compounds by micro-organisms. *Advances in Microbial
Physiology,* 7, 119-203
RIBBONS, D.W., HARRISON, J.E. and WADZINSKI, A.M. (1970).
Metabolism of single carbon compounds. *Annual Review
of Microbiology,* 24, 135-158

NUCLEIC ACIDS AND PROTEINS

Amino-acid synthesis

BROWN, C.M., MACDONALD-BROWN, D.S. and MEERS, J.L. (1974).
Physiological aspects of microbial inorganic nitrogen
metabolism. *Advances in Microbial Physiology,* 11, 1-52
EADY, R.R. and POSTGATE, J.R. (1974). Nitrogenase.
Nature (London), 249, 805-810
POSTGATE, J.R. (1972). *Biological Nitrogen Fixation,*
61pp. Watford; Merrow Publishing Co. Ltd.
STREICHER, S.L. and VALENTINE, R.C. (1973). Comparative
biochemistry of nitrogen fixation. *Annual Review of
Biochemistry,* 42, 279-302

Synthesis of nucleotides

O'DONOVAN, G.A. and NEUHARD, J. (1970). Pyrimidine
metabolism in micro-organisms. *Bacteriological
Reviews,* 34, 278-343

Polymerisation of nucleotides

BURGESS, R.R. (1971). RNA polymerase. *Annual Review of
Biochemistry,* 40, 711-740
KLEIN, A. and BONHOEFFER, F. (1972). DNA replication.
Annual Review of Biochemistry, 41, 301-332
PATO, M.L. (1972). Regulation of chromosome replication
and the bacterial cell cycle. *Annual Review of
Microbiology,* 26, 347-368

Protein synthesis: translation

HASELKORN, R. and ROTHMAN-DENES, L.B. (1973). Protein
synthesis. *Annual Review of Biochemistry,* 42, 397-438
LENGYEL, P. and SÖLL, D. (1969). Mechanism of protein
biosynthesis. *Bacteriological Reviews,* 33, 264-301
LUCAS-LENARD, J. and LIPMANN, F. (1971). Protein
biosynthesis. *Annual Review of Biochemistry,* 40,
409-448

Peptide synthesis

BODANSZKY, M. and PERLMAN, D. (1969). Peptide antibiotics.
Science (New York), 163, 352-358
PERLMAN, D. and BODANSZKY, M. (1971). Biosynthesis of
peptide antibiotics. *Annual Review of Biochemistry,*
40, 449-464

OLIGOSACCHARIDES AND POLYSACCHARIDES

ARCHIBALD, A.R. (1974). The structure, biosynthesis and
function of teichoic acids. *Advances in Microbial
Physiology,* 11, 53-95
BLUMBERG, P.M. and STROMINGER, J.L. (1974). Interaction
of penicillin with the bacterial cell: penicillin-
binding proteins and penicillin-sensitive enzymes.
Bacteriological Reviews, 38, 291-335
GLASER, L. (1973). Bacterial cell surface polysaccharides.
Annual Review of Biochemistry, 42, 91-112

LENNARZ, W.J. and SCHER, M.G. (1972). Metabolism and
function of polyisoprenol-sugar intermediates in membrane-
associated reactions. *Biochimica et Biophysica Acta,*
265, 417-441
ROTHFIELD, L. and ROMEO, D. (1971). Role of lipids in the
biosynthesis of the bacterial cell envelope.
Bacteriological Reviews, 35, 14-38
SUTHERLAND, I.W. (1972). Bacterial exopolysaccharides.
Advances in Microbial Physiology, 8, 143-213

LIPIDS

DAWES, E.A. and SENIOR, P.J. (1973). The role and
regulation of energy reserve polymers in micro-organisms.
Advances in Microbial Physiology, 10, 135-266
GOLDFINE, H. (1972). Comparative aspects of bacterial
lipids. *Advances in Microbial Physiology,* 8, 1-58
HUNTER, K. and ROSE, A.H. (1971). Yeast lipids and
membranes. In: *The Yeasts,* Eds. A.H. Rose and
J.S. Harrison, Vol.2, pp.211-270. London; Academic
Press
MCMURRAY, W.C. and MAGEE, W.L. (1972). Phospholipid
metabolism. *Annual Review of Biochemistry,* 41, 129-160

B-GROUP VITAMINS

GOODWIN, T.W. (1963). *Biosynthesis of Vitamins and
Related Compounds,* 366 pp. London; Academic Press

SECONDARY METABOLITES

BU'LOCK, J.D. (1965). *The Biosynthesis of Natural
Products. An Introduction to Secondary Metabolism,*
149 pp. New York; McGraw-Hill
TURNER, W.B. (1971). *Fungal Metabolites,* 446 pp.
London; Academic Press
WEINBERG, E.D. (1970). Biosynthesis of secondary
metabolites; roles of trace metals. *Advances in
Microbial Physiology,* 4, 1-44

8

REGULATION OF METABOLISM

It is self-evident that the metabolism of micro-organisms, and indeed of all living organisms, is under strict *in vivo* control. Micro-organisms show a general reproducibility in structure and composition, one that is on the whole maintained even when they are grown in media of widely different chemical composition. Moreover, rapidly growing micro-organisms usually accumulate intracellularly, or excrete into the environment, only very small quantities of compounds that could be used in synthesis of new cell material, which indicates that there is extremely little wastage during cell metabolism. There are many other examples that could be quoted to show the remarkable degree of control and economy that operates in the metabolism of micro-organisms.

It is reasonable to infer from these observations that not all of the metabolic pathways described in the two previous chapters operate in all micro-organisms, and that each pathway may be used to a different extent in those organisms in which it operates. The regulation of metabolism is a subject which is still in its infancy, but already certain basic control mechanisms have been discovered and these are discussed in this chapter. However, it is as well to stress at the outset that the microbial physiologist is still a long way from being able to explain the molecular basis of many of the regulatory mechanisms that apparently operate in micro-organisms.

8.1   GENETIC CONTROL

The genetic material in a micro-organism, known
collectively as the *genome*, is made up of DNA which has
encoded on it the instructions for the 'running' of the
cell. Synthesis of each enzyme, or more accurately each
polypeptide chain, is controlled by a separate gene
(a structural gene) in the genome. Clearly, therefore,
the genome determines qualitatively the enzymic
composition of a micro-organism. If the gene for a
particular enzyme is not present in the genome, then that
enzyme cannot be synthesised by the micro-organism.

   In the previous chapter, we saw how the coded
instructions on DNA are transcribed and translated into
protein molecules. The following account of the genetic
control of microbial metabolism will be confined to a
discussion of the nature of the genetic code and of the
ways in which the DNA in micro-organisms is subject to
variation.

8.1.1   THE GENETIC CODE

The Sequence Hypothesis (page 262) states that the amino-
acid sequence in a polypeptide is determined by the
sequence of nucleotides in a piece of DNA known as a
gene. Since there are only four main bases in microbial
DNA, namely adenine, guanine, cytosine and thymine, the
genetic code must have only four letters. The problem
then is to discover the sequence of bases - the *codon* -
that codes for each of the twenty amino acids that are
incorporated into proteins.

   The genetic code was cracked in the mid 1960s, but long
before this, many general properties of the code were
known mainly from the results of genetic experiments with
bacteria. It was fairly well established that each codon
consists of three consecutive bases on DNA or mRNA, in
other words that it is a *triplet code*. The maximum number
of codons is therefore 64 (i.e. 4^3). A doublet code would
have provided too few codons (16, i.e. 4^2) and a quadruplet
code the embarrassingly large number of 256 codons (i.e.
4^4). With 64 different codons in a triplet code, it
seemed likely - and it has since been confirmed - that the
code is *degenerate* with more than one codon for several of
the twenty amino acids. Genetic and mutation experiments
also revealed that the code is *non-overlapping* in that

none of the bases in a codon is shared with either of the adjacent codons. There were also grounds for believing that the code is *universal* in living organisms. The most compelling evidence in favour of universality came from so-called 'chimaera' experiments in which, in a cell-free protein-synthesising system, reaction mixtures were used in which the activating enzyme preparation was from one living organism and the tRNA from another organism.

It was long suspected that the most likely way of transferring information from gene to protein is by sequential read-out, in other words, that the nucleic acid and the polypeptide it specifies should be colinear. Proof of this *colinearity* was established first by Charles Yanofsky and his colleagues at the California Institute of Technology, U.S.A. and later by Sydney Brenner and his associates in Cambridge, England. Yanofsky and his group determined the amino-acid sequence in a stretch of one of the subunits (the A protein) in tryptophan synthase from *Escherichia coli*. The stretch was about 75 residues long, and they located on it separate amino-acid alterations produced by nine different mutations. They also found, by purely genetic methods, the order of these mutations on the genetic map. The results showed that the two orders were the same; moreover, sites close together on the polypeptide chain were also close together on the genetic map. Brenner and his colleagues obtained a similar finding using chain-terminating mutants of the gene for the head protein of bacteriophage T4.

The first information on the key to the genetic code came in 1961, and the breakthrough was made by Heinrich Matthaei and Marshall Nirenberg at Bethesda, Maryland, U.S.A., who were studying synthesis of protein by cell-free extracts of *E. coli*. The reaction mixture contained all of the necessary fractions and cofactors including all 20 amino acids, one or more of which were labelled with ^{14}C. The experiment was started by adding mRNA, and the mixture was incubated at 37°C for a brief period before the protein was precipitated and its radioactivity counted. The amount of protein formed was always small because of the instability of the mRNA. Matthaei and Nirenberg therefore decided to use synthetic polyribo-nucleotides as messengers instead of mRNA. When they used polyuridylic acid (poly-U) in the reaction mixture there was an enormous increase in the incorporation of phenylalanine but not of any other amino acid. This led them to suggest that the codon for phenylalanine is

(UUU) on mRNA and therefore (AAA) on DNA. These
experiments of Matthaei and Nirenberg completely
revolutionised the biochemical approach to the coding
problem since, in theory, it provided a way of discovering
the codon for each of the 20 amino acids.

The experiments were quickly extended by Nirenberg and
his colleagues, and by Severo Ochoa and his group in New
York, U.S.A. Polyadenylic acid was shown to yield a
polymer of lysine, and polycytidylic acid to direct
polymerisation of proline. More information came later
from the use of simple heteropolyribonucleotide messengers.
Knowing the frequency of all possible triplets in a
polymer, it was possible to predict the composition of
the polypeptide formed and hence to discover the nature
of previously unknown codons.

Subsequent research on the genetic code received a
tremendous impetus from the discovery that aminoacyl-tRNA
molecules will recognise not only an appropriate
nucleotide region in a mRNA molecule attached to ribosomes,
but also a trinucleotide attached to ribosomes. Thus,
ribosomes together with triuridylic acid (UUU) form a
complex with phenylalanine-tRNA but not with any other
aminoacyl-tRNA nor with any free amino acid. Nirenberg
and his colleagues, and Gohbind Khorana and his group at
the University of Wisconsin, U.S.A., very quickly
exploited this technique, and the data which they
reported made the final contribution to the cracking of
the genetic code.

The key to the genetic code is given in *Table 8.1*. One
of the most striking features of the code is that in many
codons the third base in the triplet is not rigidly
defined. The significance of this particular form of
degeneracy intrigued Francis Crick who formulated what
has come to be known as the 'wobble' hypothesis. This
hypothesis stipulates that contact of codons with
anticodons on tRNA molecules is governed by base pairing
with just two of the bases in the codon. The presence of
odd bases in tRNA molecules (*see* page 85) allows for
apparent distortions in the normal base pairing in the
third position. The wobble hypothesis also explains how
more than one tRNA molecule can code for the same amino
acid.

Another important feature of the genetic code is the
existence of the three codons (UAA, UAG and UGA) which
terminate polypeptide chain extension. The first chain-
terminating codon to be discovered, UAA, was dubbed by
the colourful epithet *amber*. A respect for neatness and

Table 8.1 The key to the genetic code

2nd →	U	C	A	G	3rd
1st					
U	Phe	Ser	Tyr	Cys	U
	Phe	Ser	Tyr	Cys	C
	Leu	Ser	ochre	umber	A
	Leu	Ser	amber	Tryp	G
C	Leu	Pro	His	Arg	U
	Leu	Pro	His	Arg	C
	Leu	Pro	GluN	Arg	A
	Leu	Pro	GluN	Arg	G
A	Ileu	Thr	AspN	Ser	U
	Ileu	Thr	AspN	Ser	C
	Ileu	Thr	Lys	Arg	A
	Met	Thr	Lys	Arg	G
G	Val	Ala	Asp	Gly	U
	Val	Ala	Asp	Gly	C
	Val	Ala	Glu	Gly	A
	Val	Ala	Glu	Gly	G

The four RNA bases adenine, guanine, cytosine and uracil
are represented by the letters A, G, C and U respectively.
The 20 amino acids are represented by their standard
abbreviations. The first base in any triplet is indicated
on the left, the second base at the top, and the third at
the right of the table. For explanation of the terms
'amber', 'ochre' and 'umber', see the text

uniformity prompted the discoverers of the other two
chain-terminating codons to adopt equally distinctive
appellations, namely *ochre* and *umber*. Chain-termination
codons have also been called 'nonsense' codons since they
do not code for an amino acid.

Not all codons in the genetic code are used to the same
extent. For reasons which are not very clear, certain of
the synonymous codons for some amino acids tend to be
used in preference to others. In *E. coli*, for example,
there appears to be a preference for using AAA and GAA
codons for, respectively, lysine and glutamic acid over
their AAG and GAG synonyms. But on the whole, the

frequency with which amino acids appear in polypeptide
molecules correlates roughly with the number of
available codons.   The one main exception is arginine.
With six codons for arginine in the code, the expected
percentage for arginine residues in microbial proteins
is 10%, whereas the analytical value is 3-4%.   The reason
for this sparing use of arginine codons is not known,
although it has been suggested that arginine may have
been a late-comer into the genetic code.   Before arginine
arrived, it is conceivable that biosynthetic precursors
of arginine (such as ornithine; *see Figure 7.7*) or lysine,
which occurs in proteins rather more frequently than is
to be expected from the number of lysine codons, were
preferentially incorporated into microbial proteins.

8.1.2   INDIVIDUALITY OF MICROBIAL DNA

The sequence of bases in DNA differs to some extent in
each species and strain of micro-organism.   Although
methods are available for extracting DNA from micro-
organisms in a comparatively undenatured state, it is
not yet possible to determine the base sequence in this
nucleic acid using chemical methods.   Meanwhile, studies
have been made on the overall base composition of DNAs
from different micro-organisms, and it has been shown
that there are some interesting relationships between
the base compositions of DNAs from taxonomically related
micro-organisms.

   Although the molar proportions of guanine:cytosine and
adenine:thymine in DNA always equal unity, the proportions
of (G + C) residues in the DNA, expressed as a percentage
of the total bases in the nucleic acid, vary widely in
different micro-organisms.   *Micrococcus lysodeikticus,*
for example, has 72% (G + C) while *Tetrahymena pyriformis*
has only about 25%.   Several different methods have been
devised for determining the proportion of (G + C)
residues in extracted DNA.   One way is to hydrolyse the
DNA to give the free purine and pyrimidine bases, and then
separate these bases chromatographically and determine the
relative proportions present.   It has also been shown that
there exists a proportionality between the molar percentage
of (G + C) in a DNA and the buoyant density of the nucleic
acid in a gradient of caesium chloride, and this has been
used as the basis of another method.   The most adaptable
method involves measuring the thermal denaturation
temperature of the DNA.   When native DNA is heated in

solution, there is a sharp increase in the extinction of
the solution around the temperature at which the
transition takes place from the native double-stranded
structure to the denatured state. The temperature
corresponding to the mid-point of the rise in extinction
is known as the T_m value or colloquially as the *melting
point of the DNA*. The values are related linearly to the
average DNA base composition, the higher the proportion
of (G + C) the greater the thermal stability. Values for
T_m can be measured using quartz spectrophotometers fitted
with constant-temperature housings for the cuvettes. The
method is particularly useful in that it requires only
10-50 μg DNA.

The DNAs from micro-organisms belonging to many different
taxonomic groups have been examined for base composition
with some quite interesting results. It was found, for
example, that in the genus *Proteus* all species except
P. morganii had around 39% (G + C) whereas *P. morganii* had
50%. From the results of other taxonomic tests,
P. morganii had previously been thought to be only remotely
related to the other species in this genus, and a new genus,
Morganella, had been proposed for it. A study of the base
compositions of DNAs from *Bacillus* spp. revealed that the
molar proportions of (G + C) vary from 37 to 50%. This
range is unusually wide for the organisms from one genus,
and it lends support to the view that Bacillus should not
be considered as a single genus but rather as a collection
of genera or biotypes.

8.1.3 EVENTS WHICH LEAD TO CHANGES IN DNA COMPOSITION

The process of evolution in a population of living
organisms depends upon the formation and subsequent
selection or rejection of members of the population that
acquire new or different genetic material in their
genomes. There follows a discussion of the biochemical
aspects of processes that lead to changes in the
composition of DNA in micro-organisms.

Mutation

Mutation can be defined as an abrupt change in the
composition of a genome in a living cell caused by
chemical or physical factors in the environment. The
vast majority of mutations that arise or are caused are

point mutations, that is mutations which affect only a
very small part, often one nucleotide, of the genome.
There are two main classes of mutation, which are
referred to as frameshift and base-pair substitutions.
Frameshift mutations involve either deletions or
additions of small numbers of base pairs to the genome.
Base-pair substitutions can be further divided into
transitions and transversions. A *transition* is the
replacement of a purine residue by a different purine
residue, or a pyrimidine residue by a different
pyrimidine residue. A *transversion* is the replacement
of a pyrimidine residue by a purine residue, or a purine
residue by a pyrimidine residue.

The number of chemical compounds which give rise to
mutations in microbial genomes is vast. Mutations can
also be induced by various types of radiation, and these
effects are discussed in Chapter 3, page 126. Among the
most widely used mutagenic reagents with micro-organisms
are the *alkylating agents* ethylmethane sulphonate and
N-methyl-N'-nitro-N-nitrosoguanidine. The main chemical
reaction between these alkylating agents and the
microbial genome is an alkylation at the N-7 position of
guanine residues on the DNA, and to a lesser extent at
the N-3 position of adenine residues. The product of
alkylation at N-7 on guanine residues ionises differently
from unsubstituted residues, and a pairing error is
introduced in which the alkylated guanine pairs with
thymine instead of cytosine, thereby introducing a
transitional type mutation. Other alkylating agents
bring about transversions rather than translations.
A somewhat more specific chemical mutagen is *hydroxylamine*
which reacts mainly with cytosine and guanine residues,
and brings about transitional type mutations and
mispairing. Another commonly used mutagen is *nitrous acid*
which deaminates adenine, guanine and cytosine. Conversion
of adenine to hypoxanthine, and of cytosine to uracil,
causes transitions, but conversion of residues of guanine
to xanthine residues probably has little effect as both
guanine and xanthine preferentially pair with cytosine.
Base analogues, particularly bromodeoxyuridine, are
mutagenic. Bromodeoxyuridine replaces all of the thymine
in DNA when free thymidine is lacking, as with auxotrophic
mutants, and the incorporated analogue forms base pairs
with adenine, as of course does thymine. But the presence
of the analogue in place of thymine increases the rate of
pairing mistakes, since the analogue occasionally pairs
with guanine. The reaction of *acridines* with DNA has been

referred to in Chapter 7 (page 283). Intercalation of
acridines with the DNA double helix causes a lengthening
of the DNA molecule, and the production of frameshift
mutations.

Transfer of genetic material

Another way in which the composition of the DNA in
microbes can be changed is by transfer of a portion of
DNA into the micro-organism, and the incorporation or
recombination of this DNA with that in the recipient
microbe. There are four methods by which genetic
material can be transferred to micro-organisms, and they
differ in the manner in which the donor DNA is made
available and in the amount of it which combines with the
receptor genome.

Conjugation. Conjugation is the most widely distributed
mechanism especially among eukaryotic micro-organisms for
the transfer of genetic material. This process, which
involves the fusion of gametes during sexual reproduction,
is of extreme importance since it results in the formation
of a diploid zygote which, with the subsequent process of
meiosis, permits the recombination of genetic material
donated by the gametes. As a result, there is a
randomisation of maternal and paternal genes, a reshuffling
of the pack of chromosomes. However, little if anything
is known of the molecular mechanism of this process.
Organisms which reproduce sexually have a cycle in which
the haploid and diploid states alternate. They vary in
the predominance of one state over the other but most
algae, fungi and protozoa are haploid.
    Details of the conjugation process vary in different
micro-organisms. In algae, fungi and protozoa, the entire
genomes of the two gametes are involved; also the gametes
are often differentiated morphologically. In bacteria
only a fraction of the genetic material in the donor
organism is transferred to the recipient cell which
contributes not only its entire genome but also its
cytoplasm. The recipient bacterium is therefore a zygote
(strictly speaking a *merozygote*) which is incompletely
diploid with the result that recombination is restricted
to the diploid part of the bacterial chromosome. Several
reports have been published on the physiology of

conjugation in bacteria. The DNA that is transferred
has been shown to be newly replicated in the male. There
is evidence, too, that conjugation is accompanied by DNA
replication in the female. Both male and female bacteria
consume metabolic energy during conjugation, although the
way in which this energy is expended is far from clear.

In a few micro-organisms, including some aspergilli,
there occurs a cycle of events, known as the *parasexual
cycle,* which leads to the production of recombinants by
a process that is quite distinct from, and much rarer
than, the sexual cycle.

Transformation. Transformation occurs when DNA isolated
from one strain of a micro-organism enters the cell of
another strain of the same organism and directs synthesis
of certain enzymes in the recipient cell. The process
also operates with DNA extracted from a phage for a
bacterial strain, but it is then referred to as
transfection. The discovery of transformation dates from
the work of the British microbiologist Fred Griffith, who
in 1928 found that rough strains of pneumococci could be
converted into smooth strains by substances isolated from
dead smooth organisms. The transforming substance was
subsequently identified as DNA by Oswald Avery and his
colleagues in the U.S.A. Transformation has since been
observed in a number of bacteria, including members of the
genera *Bacillus, Diplococcus, Haemophilus, Neisseria,
Rhizobium* and *Streptococcus*. It appears that the process
is confined to bacteria. As a means of transferring
genetic material from one micro-organism to another it is
therefore more restricted than conjugation.

Transformation involves two quite separate processes.
The first is uptake of DNA by the recipient microbe, and
the second integration of some or all of this DNA into
the genome of the organism.

To date, much more information has been reported on the
first of these processes. The ability of a bacterium to
be transformed depends on the physiological state of the
organism, and one which is capable of taking up DNA is
said to be *competent*. Research with strains of *Diplococcus*
and *Haemophilus*, and more recently of *Bacillus subtilis,*
has high-lighted several properties which differentiate
competent from non-competent bacteria. The more important
of these are a diminished capacity to synthesise DNA and
RNA, a lower density, and the possession of chromosomes with
replication points arrested near the replicative origin,

and of mesosomes which extend from the plasma membrane
to the region of the genome. Moreover, acquisition of
competence is associated with autolytic enzymes, action
of which presumably leads to the appearance on the surface
of the bacterium of molecules which act as receptors for
the DNA molecules. During most experiments on
transformation, only between 1 and 10% of the bacteria
in the population are transformed. About ten seconds is
usually required for DNA to enter the recipient bacterium.
The size of the DNA required for transformation varies,
but the mean molecular weight of transforming pneumococcal
DNA is about 5×10^6 daltons which corresponds to about
one two-hundredth of the total genetic material in the
donor bacterium. In *Bacillus subtilis,* it has been shown
that the transforming DNA becomes associated with a
mesosome in the recipient cell, and is then transported
to the genome region through the mesosomal structure.
But the mechanisms of transport across the plasma membrane,
and of progression along the mesosomal structure, are
clothed in mystery.

The ability of DNA to bring about transformation depends
upon the successful operation of the second process,
namely integration of the donor DNA with the genome of the
recipient bacterium. The mechanism of this process is
still incompletely understood. Using transforming DNA
labelled with ^{32}P it has been shown that half of the DNA
entering a pneumococcus is converted to a single-stranded
form while the other half is degraded to dialysable
oligonucleotides. The single strand of DNA is subsequently
incorporated into the recipient genome.

Transduction. During the lytic cycle of a bacteriophage,
fragments of the genetic material of the host bacterium
are occasionally incorporated into the progeny particles
which can therefore act as vectors for transferring this
host DNA into those bacteria which they subsequently
infect. In other words, the phage takes over the role of
a gamete. This process, which has been termed *transduction,*
has been reported to occur with a number of Gram-negative
bacteria. It resembles transformation in that it is a
fairly restricted mechanism for bringing about transfer
of genetic material, but the fragment of DNA transferred
during transduction is usually larger than that involved
in transformation and represents about one-hundredth of
the donor genome.

8.2   REGULATION OF ENZYME ACTION

The genome controls qualitatively the enzymic composition
of a micro-organism in that an organism can synthesise
only those enzymes for which it has appropriate genes.
However, if a micro-organism is to respond rapidly to
changes in the chemical and physical properties of the
environment, it is clearly essential for it to possess
mechanisms by which the activities of enzymes can be
regulated so that the reactions which they catalyse
produce just sufficient of the various metabolic end-
products. Several different mechanisms have been
discovered by which enzyme action, in *in vitro* experiments,
can be regulated, and these are described in the following
pages. Still, much remains to be discovered about
regulation of enzyme action in microbes, not least the
extent to which the mechanisms which have been shown to
operate in experiments using extracts of micro-organisms
act and interact in intact cells.

8.2.1   ACCESS OF SUBSTRATES TO ENZYMES

A fairly obvious, but often overlooked, way in which
enzyme action is regulated in living cells is by control
of the access of substrate to the appropriate enzyme.
This is effected in a number of different ways, ranging
from regulation of the rate at which a substrate is
transported into a microbe to the formation, in eukaryotic
microbes, of organelles such as mitochondria that are
bounded by membranes. In strains of *Saccharomyces
cerevisiae*, the vacuoles in the cell contain a wide range
of enzymic activities, principally hydrolases, and it is
suggested that these organelles function as a type of
lysosome (*see* page 80). Clearly, therefore, the
permeability of the vacuolar membrane in *Sacch. cerevisiae*
is a very important factor in the control of enzyme action.

8.2.2   FEEDBACK INHIBITION OF ENZYME ACTION

The most extensively studied of the mechanisms for
regulating enzyme action in microbes is *feedback
inhibition,* which is a process that micro-organisms as
well as other cells use to prevent overproduction of low
molecular-weight intermediates such as amino acids and
purine and pyrimidine nucleotides. If a metabolic product,

E, is synthesised through a series of reactions in which
A is converted into *B*, *B* to *C*, and so on to *E*, then, by
a process of feedback inhibition, the end-product *E* can
inhibit the action of enzymes that catalyse one or more
of the earlier reactions in the sequence and thereby
prevent overproduction of *E*. The low molecular-weight
end-products which interact with enzymes are referred to
as *effectors*.

The way in which effector molecules interact with
enzymes has interested many molecular enzymologists. The
substrate of the enzyme which is inhibited is usually
quite different structurally from the effector, and this
suggests that the effector does not combine with the enzyme
at the catalytic or active site but at a separate
regulatory site. This site on the enzyme at which the
effector combines has been termed the *allosteric site*.
A combination between the enzyme and the effector at the
allosteric site causes the enzyme to lose its catalytic
activity as a result of a conformational change in the
structure of the protein. Enzymes which combine with
effectors in this manner are often referred to as
allosteric proteins. With many allosteric proteins, it
appears that the amino-acid residues at the allosteric
site differ in nature from those at the active site. For
example, crystalline aspartate carbamoyltransferase, an
enzyme involved in synthesis of pyrimidine nucleotides
(*see* page 278), loses its sensitivity to feedback
inhibition by CTP after treatment with the thiol
inhibitor *p*-chloromercuribenzoate, although the catalytic
activity of the protein is unaffected by this treatment.

One of the first examples of feedback inhibition was
discovered during work on the pathway used by *E. coli* for
synthesising isoleucine. Initially it was shown that
threonine was an intermediate on the pathway to isoleucine
since a mutant strain of the bacterium that was auxotrophic
for threonine was found to utilise exogenous threonine to
synthesise isoleucine (*see Figure 7.11*). When this mutant
strain was grown in a medium containing isoleucine,
threonine was less efficiently utilised than in a medium
lacking isoleucine. A possible explanation of this
sparing effect was that isoleucine was inhibiting the
action of the enzyme that catalyses the first reaction
in the conversion of threonine to isoleucine, namely
L-threonine hydro-lyase. When cell-free extracts of the
mutant bacterium were examined it was indeed found that
the action of L-threonine hydro-lyase was inhibited by
L-isoleucine.

As soon as feedback inhibition had been shown to
operate on biosynthetic pathways, it was clear that a
modified type of mechanism must operate on branched
pathways which lead to the production of two (or more)
end-products (*F* and *H*) but in which certain reactions are
shared:

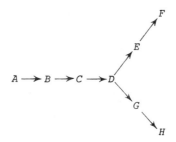

Clearly, overproduction of one of these end-products
(e.g. *F*) could, by feedback inhibition of the enzyme
catalysing the conversion of *A* to *B*, also restrict
synthesis of *H* although the cell may not have
synthesised sufficient of *H* to satisfy its immediate
metabolic requirements. An example is provided by the
pathway leading to synthesis of the amino acids lysine,
methionine and threonine (*Figure 8.1*). One of the main
ways in which enzyme action is regulated on branched
pathways is by feedback inhibition of enzymes that
operate at branch-points on the pathway. For example,
threonine inhibits action of the enzyme that converts
homoserine into O-phosphohomoserine (*Figure 8.1*). But
there are also other ways in which control is effected
in these situations. Data from experiments using cell-
free extracts of *E. coli* indicated that this bacterium
synthesises no less than three different types of
aspartate kinase, which is the enzyme that catalyses the
first step on the pathway, namely formation of β-aspartyl
phosphate from aspartate. One of these enzymes is
selectively inhibited by lysine, and one by threonine.
A partial purification of the first two of these
aspartate kinases has shown that they can be
further differentiated on the basis of their stability to
heat. Enzymes of this type which, although they are
produced by the same organism and catalyse the same enzyme
reaction, are nevertheless different proteins and are
termed *isoenzymes*. Several other examples of isoenzymes

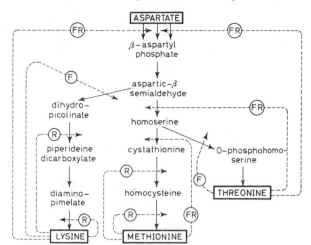

Figure 8.1 Diagram showing the sites of regulation of lysine, methionine and threonine biosynthesis in Escherichia coli *by feedback inhibition (F) and repression (R) of enzyme synthesis*

have been reported in micro-organisms and in plant and animal cells.

However, full control of this branched pathway in *E. coli* obviously requires the operation of other regulatory mechanisms and several of these have been discovered (*Figure 8.1*). They have been shown to involve not only feedback inhibition of enzyme action but also repression of enzyme synthesis (*see* page 347). Consider, for example, the situation in which the bacterium is provided with an excess of lysine. The immediate effect is a diminution in the synthesis of β-aspartyl phosphate as a result of feedback inhibition of the lysine-sensitive aspartate kinase. The diminished amount of aspartyl phosphate produced by the action of the other two aspartate kinases is subsequently prevented from being converted to lysine by feedback inhibition of the enzyme that catalyses conversion of aspartic semialdehyde to dihydropicolinate, and a decreased synthesis of the enzymes which catalyse, respectively, conversion of dihydropicolinate into piperideine dicarboxylate and of diaminopimelate into lysine.

The number of variations which have been reported on the basic phenomenon of feedback inhibition is quite large. One has been shown to operate on the pathway

leading to synthesis of lysine, methionine and threonine.
Bacillus polymyxa and *Rhodopseudomonas capsulatus,* unlike
E. coli, synthesise just one aspartate kinase which is
inhibited little or not at all by individual amino-acid
end-products, but strongly by lysine and threonine
together. This phenomenon has been termed *concerted
feedback inhibition.*

The action of enzymes which catalyse reactions involved
in several different pathways must obviously be subject
to very sophisticated regulatory mechanisms. This is
well illustrated by considering the enzyme glutamine
synthetase which catalyses the reaction:

$$\text{glutamate} + NH_4^+ + ATP \rightleftharpoons \text{glutamine} + ADP + Pi + H^+ + H_2O$$

Glutamine is a nitrogen donor in the biosynthesis of a
wide range of compounds including ATP, carbamoyl phosphate,
CTP, glucosamine 6-phosphate, histidine, NAD, NADP and
tryptophan; alanine and glycine may also arise from
glutamine-dependent transaminations. Experiments with
extracts from a number of prokaryotic and eukaryotic
microbes have shown that the total residual activity of
glutamine synthase in reaction mixtures containing
several different feedback inhibitors is equal to the
product of the fractional activities manifest when each
is present separately in the reaction mixture. This
effect has been dubbed *cumulative feedback inhibition,*
and is possible because of the existence on the enzyme of
several separate non-interacting binding sites.

Feedback inhibition of enzyme action also operates on
catabolic pathways. The main products of these pathways
are ATP and reduced nicotinamide nucleotides. Almost
nothing is known of the mechanisms that operate in
regulating synthesis of $NADH_2$ and $NADPH_2$, and there is
an urgent need for more data in this area. More is
known about the effects of ATP on the rate of enzyme
action in microbes. We saw in Chapter 6 (page 185) that
the energy content of a microbe is conveniently expressed
in the form of the *adenylate charge* which describes the
extent to which the ATP-ADP-AMP system is 'filled' with
high-energy phosphate bonds. *In vitro* experiments have
shown that the activity of many enzymes associated with
ATP generation (for example the isocitrate dehydrogenase
of *Sacch. cerevisiae*) is inhibited when the energy charge
in the reaction mixture (which can be adjusted by
including appropriate concentrations of each of the
nucleoside phosphates) has a value of 0.8 or above.
Conversely, the rate of action of enzymes which catalyse

reactions on anabolic pathways and which consume ATP
(such as aspartokinase) is increased when the adenylate
charge of the reaction mixture is high. The physiological
significance of control of enzyme activity by adenylate
charge is well illustrated by the regulatory mechanisms
which operate at the important branch-point in carbohydrate
metabolism at which phosphoenolpyruvate is either
dephosphorylated to give pyruvate (*see Figure 6.2*) or
carboxylated to give oxaloacetate (*see Figure 7.1*).
A high adenylate charge inhibits action of the former
reaction, and so leads to a decreased synthesis of ATP;
under the same conditions, activity of the carboxylase
is unaffected and indeed the reaction may even increase
in extent because of the greater amount of
phosphoenolpyruvate available.

A process very similar to feedback inhibition may
operate on transport proteins. The apparent K_m value for
uptake of glucose by *Sacch. cerevisiae* is greater for
organisms grown aerobically rather than anaerobically.
The decreased affinity of the permease system for glucose
in yeast grown aerobically appears to explain part of the
Pasteur effect. In 1861, Louis Pasteur found that, when
yeast (*Sacch. cerevisiae*) was transferred from an
anaerobic to an aerobic environment, growth was
accelerated while uptake of sugar was diminished. The
effect on growth is perhaps easy to explain for, under
aerobic conditions, the yeast is able to produce far more
ATP from each molecule of glucose than when growing under
anaerobic conditions (*see* Chapter 6), and could therefore
synthesise new cell material at a faster rate. The second
part of the Pasteur effect - the diminution in the rate
of uptake of sugar under aerobic conditions - can now be
explained by these observations. However, the nature of
the effector - whether it is molecular oxygen, glucose or
some product of glucose catabolism - and the nature of the
protein on which it acts in the glucose-uptake system, are
far from clear.

8.2.3 MODIFICATION TO ENZYME STRUCTURE

Recent research, originally with mammalian tissues but
since extended to microbes, has shown that the activity
of several enzymes can be regulated as the result of a
modification to the enzyme protein. The process can
entail *chemical modification* when specific groups are
attached covalently to the enzyme molecule causing a change

M

in the catalytic properties of the protein, or *physical modification* when effectors bind non-covalently to the enzyme and cause a conformational change and an alteration to the catalytic activity.

Chemical modification

This process was discovered during research on glycogen phosphorylase from mammalian tissue, when it was shown that the enzyme can exist in two chemically distinct forms, formation of which is controlled by an enzyme-catalysed phosphorylation-dephosphorylation cycle. The cycle has also been shown to operate in *Neurospora crassa* and probably in other microbes. Glycogen phosphorylase exists in a phosphorylated form (called the *a* form), which can be fully active in the absence of 5'-AMP, and a dephosphorylated or *b* form which has an obligatory requirement for 5'-AMP for activity. Both forms exist in *N. crassa,* and they are interconverted through the action of a specific kinase and phosphatase. Interconversion of two different forms of glycogen synthetase by enzymic phosphorylation and dephosphorylation has also been demonstrated with extracts of *Sacch. cerevisiae* and *N. crassa.* The two forms differ in their dependence on glucose 6-phosphate for activity.

Another form of chemical modification involves adenylylation, and this operates in several bacteria on the enzyme glutamine synthetase. The reaction involves the enzyme and ATP, and leads to formation of a 5'-adenylyl-O-tyrosyl residue in the protein. The catalytic properties of glutamine synthetase are radically altered by complete adenylylation. The fully adenylylated form has an absolute requirement for Mn^{2+}, unlike the non-adenylylated enzyme, as well as a different pH optimum and altered affinities for various end-products of glutamine metabolism. Bacteria which use this form of chemical modification also synthesise a deadenylylating enzyme, activity of which is influenced by several different effectors, including α-ketoglutarate.

Physical modification

This type of control is probably of common occurrence in intact microbes, but relatively few examples from *in vitro* studies have been fully documented. An example is provided by the ADP-glucose synthetase from *E. coli* which shows marked differences in kinetic behaviour depending on whether Mg^{2+} or Mn^{2+} is the cofactor. But one of the most interesting examples of the effect of cations on regulation of enzyme action comes from work on the glutamine synthetase of *E. coli*. When isolated from the bacterium, this enzyme is a protein with a molecular weight of around 592 000 daltons and consisting of 12 apparently identical subunits arranged in two hexagonal layers. When isolated in the presence of Mg^{2+} to protect the enzyme from inactivation, the preparation also contains Mn^{2+}. In this state, the enzyme has catalytic activity, and is said to be in the *taut* form. When Mn^{2+} is removed from the enzyme, the *relaxed* and catalytically inactive form is produced. The relaxed form has the same molecular weight as the taut form, but is a less compact structure. As yet, however, it has to be established whether this switch-on and switch-off mechanism, controlled by Mn^{2+} ions, has a regulatory function in the intact bacterium.

8.2.4 DEGRADATION OF INDIVIDUAL ENZYMES

Arguably the most sophisticated mechanism yet discovered for regulating action of enzymes in microbes is that in which highly specific proteolytic enzymes degrade enzymes which are no longer required for metabolism of a micro-organism. Two of the best documented examples of this type of regulatory process have come from work on *Sacch. cerevisiae*.

One enzyme in *Sacch. cerevisiae* which can be inactivated in this manner is tryptophan synthase. The inactivation occurs when cultures of the yeast enter the stationary phase of growth in batch culture, that is when the environmental conditions are such that synthesis of tryptophan is no longer necessary. However, the inactivating enzyme can also be detected in exponentially growing yeast, but in these cells action of the degrading enzyme is specifically inactivated by a heat-stable protein. It has yet to be discovered how this regulatory protein ceases to be associated with the proteolytic enzyme when the rate at which the cells are growing declines.

A closely similar mechanism operates in regulation of fructose 1,6-diphosphatase in *Sacch. cerevisiae*. Yeast grown in glucose-free medium has a high activity of this enzyme, but addition of glucose to the culture causes an abrupt loss of activity in the cells. This loss is attributable to the action of a highly specific proteolytic enzyme. Again, the yeast synthesises a heat-stable inhibitor of the proteolytic enzyme, which presumably is effective only when the cells are in a glucose-free medium.

8.3 REGULATION OF ENZYME SYNTHESIS

Feedback inhibition of enzyme action provides a rapid and sensitive method for preventing overproduction of low molecular-weight end-products by micro-organisms. But it is fundamentally an inefficient mechanism, in that enzymes are synthesised only to be prevented from functioning. However, it is known that microbes employ other more efficient regulatory mechanisms which act by controlling enzyme synthesis and thereby conserve cell protein by ensuring that organisms stop synthesising those enzymes that are not required. Some idea of the physiological importance of regulation of enzyme synthesis is seen by the fact that *E. coli* can synthesise up to 10% of its dry weight in the form of the enzyme β-galactosidase when grown under conditions that prevent these regulatory mechanisms from operating. Two main types of regulation of enzyme synthesis have been discovered, and these are known respectively as *induction* and *repression of enzyme synthesis*.

8.3.1 INDUCTION OF ENZYME SYNTHESIS

Some enzymes are synthesised by micro-organisms irrespective of the chemical composition of the environment. These are known as *constitutive enzymes* to distinguish them from *induced enzymes* which are synthesised by a micro-organism only in response to the presence in the environment of an inducer, which is usually the substrate for the enzyme or some structurally related compound. When an inducer is present in the environment, not one but a number of inducible enzymes may be formed by a micro-organism. This occurs when the pathway for metabolism of the compound is mediated by a sequence of inducible enzymes. The micro-organism responds to the inducer by synthesising the appropriate enzyme, and the intermediate formed as a product of the reaction induces synthesis of a second enzyme; the intermediate formed in the reaction catalysed by the second enzyme then induces synthesis of a third enzyme, and so on. This phenomenon has been termed *sequential induction*.

Induced synthesis of enzymes by micro-organisms was reported during the 1930s but it was not until a decade later that serious thought was given to the mechanism of the process. Most of the work was concerned with induced synthesis of β-galactosidase in *E. coli*. When this bacterium is grown in a lactose-free medium it is unable to break down lactose, but when it is grown in a lactose-containing medium it synthesises an enzyme, β-galactosidase, which catalyses hydrolysis of lactose into glucose and galactose. Two important facts were established regarding induction of β-galactosidase synthesis in *E. coli*. First, growth of the bacterium in a medium containing lactose as the sole carbon source was shown not to be the result of the selection of mutant strains which all along had the capacity to synthesise the enzyme. Rather, it was shown that each of the bacteria present synthesised β-galactosidase. Secondly, the enzyme was shown to be synthesised by the bacteria *de novo* from amino acids and not by some modification of preformed protein.

8.3.2 REPRESSION OF ENZYME SYNTHESIS

One of the first examples of repression of enzyme synthesis was discovered during studies on the pathway of methionine synthesis in *E. coli* which, as shown in *Figure 8.1*, proceeds from homoserine via cystathionine and homocysteine. When the bacterium was grown in a medium containing only glucose and inorganic salts, extracts of the bacteria contained an enzyme that catalyses conversion of homocysteine into methionine. But when methionine was included in the growth medium, extracts of the bacteria were devoid of this enzyme. Thus, synthesis of the enzyme had been prevented by growing the bacterium in the presence of methionine, the bacterium preferring to use the methionine supplied rather than synthesise its own. Numerous other examples of end-product repression of enzyme synthesis have since been reported.

End-product repression of enzyme synthesis on a pathway differs from feedback inhibition of enzyme activity in that all of the enzymes involved on the pathway are usually affected. On some pathways, such as that leading to synthesis of histidine in *Salmonella typhimurium,* synthesis of each of the enzymes is repressed by the end-product to the same extent; this has been termed *coordinate repression.* Other examples have been reported in which synthesis of each of the enzymes on a pathway is repressed to a different extent. One such pathway is that leading to arginine synthesis in *E. coli.* With several of the pathways which show co-ordinate repression, the genes that direct synthesis of the enzymes on the pathway are thought to lie clustered together on the chromosome. End-product repression of enzyme synthesis also differs from feedback inhibition of enzyme action in that, with pathways leading to synthesis of certain amino acids including histidine and valine, the effector molecules that activate the repression mechanism are tRNA derivatives of the amino acids rather than the free amino acids.

An interesting example of enzyme repression has been found to operate on pathways leading to synthesis of the branched-chain amino acids leucine, valine and isoleucine and the B-vitamin pantothenate (*see Figure 7.11*). These two separate pathways involve a homologous series of reactions, and enzymes catalyse reactions on both pathways. Studies on enzyme repression on these pathways in *E. coli* and *S. typhimurium* have shown that there is no repression of synthesis of the common enzymes until each of the end-products valine, leucine, pantothenate and isoleucine is simultaneously present in excessive concentrations. This type of concerted action has been termed *multivalent repression.*

Repression of enzyme synthesis is not confined to processes in which the effector is the end-product of a pathway. Synthesis of many enzymes is known to be repressed by low molecular-weight compounds which are not directly associated with those pathways on which reactions are catalysed by the enzymes concerned. The best studied of these processes is *catabolite repression.* It has been known for well over 30 years that synthesis of certain enzymes by micro-organisms can be prevented by including glucose in the growth medium, which explains why the effect was originally referred to as the *glucose effect.* For example, if *E. coli* is grown in a mineral salts medium containing glycerol as the carbon source for growth,

addition of the compound isopropylthiogalactoside, which
like lactose induces synthesis of the enzyme
β-galactosidase, induces a high rate of synthesis of this
enzyme. If glucose is subsequently added to the culture,
there is a rapid and severe repression of synthesis of
β-galactosidase. After several generations of growth,
there may be a partial recovery to a somewhat higher rate
of enzyme synthesis but this is always much lower than
the originally observed induced rate. These two types of
repression are referred to as, respectively, *transient
catabolite repression* and *permanent catabolite repression*.

Catabolite repression of enzyme synthesis operates on
many metabolic pathways. It operates simultaneously
with induction of enzyme synthesis and with repression
caused by end-products of pathways. Operation of these
various mechanisms for regulating synthesis of enzymes in
micro-organisms is very well illustrated by the pathways
which pseudomonads (*Pseudomonas aeruginosa* and *Ps. putida*
have been the favoured species in these studies) use for
breaking down aromatic compounds (*see* page 223). Synthesis
of the enzymes on these pathways is regulated by a process
of sequential induction of a group of enzymes by
intermediates on the pathways as well as by catabolite
repression. Starting with the aromatic compound mandelic
acid, synthesis of each of the five enzymes which catalyse
steps in the conversion of mandelate to benzoate is
induced by mandelate, while at the same time subject to
catabolite repression by benzoate, catechol and succinate
(*Figure 8.2*). Benzoate itself can be used as a growth
substrate, and it induces synthesis of an enzyme which
catalyses its conversion to catechol. Synthesis of this
enzyme is subject to catabolite repression by catechol or
succinate. Further steps in the breakdown of catechol
involve conversion to *cis, cis*-muconate and β-ketoadipate (*see* page 350), before this last intermediate is
cleaved to give acetate and succinate. Compounds which
induce or repress synthesis of enzymes which catalyse
these reactions are indicated on *Figure 8.2*.

The advantage of this simultaneously operating system
of sequential induction and repression of enzyme synthesis
is that intermediates on the pathway for degradation of
mandelate can induce synthesis, in sufficient amounts, of
enzymes required for their own metabolism without
necessitating synthesis of enzymes that catalyse earlier
steps in the breakdown. Also, different starting
substrates can induce synthesis of their own specific
enzymes, again with the operation of catabolite repression,
in amounts that are needed.

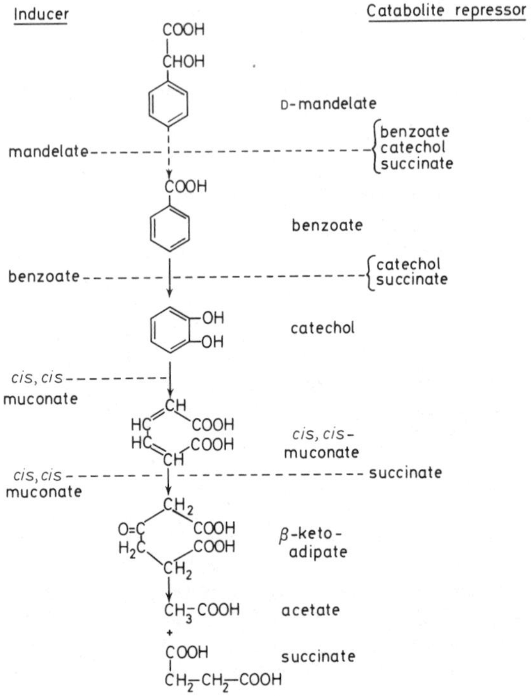

Figure 8.2   Regulation of synthesis of enzymes on the pathway used by Pseudomonas putida *for degradation of mandelate*

Regulation of enzyme synthesis can also take place in microbes by mechanisms which, although related to catabolite repression, differ in that the effector is not an intermediate on a catabolic pathway. In species of *Hydrogenomonas,* using molecular hydrogen as the energy source, synthesis of several enzymes is subject to catabolite repression with molecular hydrogen being the effector. Moreover, in many fungi, the ammonium ion regulates synthesis of some enzymes and transport proteins, a phenomenon which has been termed *ammonia metabolite repression.*

8.3.3 MOLECULAR MECHANISMS IN REGULATION OF ENZYME SYNTHESIS

Induction and repression of enzyme synthesis have in common that each involves regulation of synthesis of large molecular-weight proteins by relatively low molecular-weight effector molecules. It was obvious from the start that the effectors must act at the transcriptional or translational level of protein synthesis (*see* page 281) alongside the regulatory mechanisms (mainly involving protein factors; *see* Chapter 7, page 285) that have already been described. Some brilliant genetic experimentation by François Jacob and Jacques Monod, at the Institut Pasteur in Paris, led these workers to formulate a hypothesis in 1961 for the molecular mechanisms involved in induction and repression of enzyme synthesis in bacteria. The Jacob-Monod hypothesis has, in the intervening years, been subjected to the most rigorous experimental examination, and is now firmly accepted as the only established explanation for induction and repression of enzyme synthesis, certainly in prokaryotes. Its discovery must surely represent the most important biological achievement of the Twentieth Century.

Jacob and Monod proposed that synthesis of enzymic proteins (or more accurately polypeptides) is controlled by *structural genes,* related groups of which lie closely together on the genome. They suggested, further, that there exist separate *regulator genes* responsible for producing compounds (which we now know are proteins) which act by controlling the expression of a group of structural genes. These are the *repressor proteins*. The well established fact that synthesis of enzymes with quite different catalytic activities is quite frequently co-induced or co-repressed to the same extent by a low molecular-weight compound suggested that repressor proteins act not on individual structural genes but on a separate region of the genome where groups of genes can be switched on or switched off. This site at which repressor proteins act on the genome is known as the *operator gene,* and each of these genes controls transcription of a group of structural genes. Another genetic element, the existence of which was not proposed by Jacob and Monod but which was later discovered from the results of genetic experiments, is the *promotor region* which lies alongside the operator gene and is the site on the genome at which RNA polymerase becomes attached.

A group of structural genes, together with the associated regulator and operator genes and promotor region, is known as an *operon*. Operons on bacterial genomes differ tremendously in size and the degree to which the different genes in the operon are juxtaposed. *Figure 8.3* shows an idealised structure for a small operon.

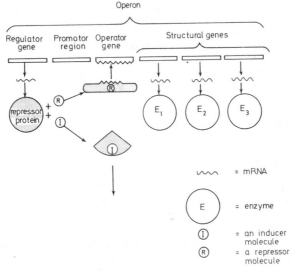

Figure 8.3 Diagrammatic representation of the Jacob-Monod scheme for the molecular basis of induction and repression of enzyme synthesis. See text for a detailed explanation

An operon very similar in structure to the idealised arrangement shown in *Figure 8.3* is the lactose (or *lac*) operon in *E. coli* which formed the subject of the pioneer research by Jacob and Monod. Since their early studies, the *lac* operon in *E. coli* has been researched very intensively, and is now by far the best understood part of any cellular genome. The elements in the *lac* operon in *E. coli*, reading from left to right in *Figure 8.3*, are the regulator (or *i* gene), the promotor region (*p*), and the operator gene (*o*), followed by three structural genes, namely β-galactosidase (the *z* gene), the lactose transport protein gene (*y*), and a gene which specifies thiogalactoside transacetylase (*a*) an enzyme which, although it is regulated in concert with the first two structural genes, has no known metabolic role in *E. coli*.

The fine structure of the *lac* operon in *E. coli* is
rapidly being revealed, largely as a result of research
by Walter Gilbert and his colleagues at Harvard
University in Boston, U.S.A. The operator gene in this
operon is only 27 base pairs long. Interestingly, the
arrangement of these base pairs in the gene shows a
remarkable symmetry. Gilbert and his colleague Benno
Müller-Hill have also isolated the repressor protein for
this operon. This is a tetramer protein with a molecular
weight of about 150 000 daltons and 347 amino-acid
residues.

Other bacterial operons are very different in structure
from the *lac* operon in *E. coli*. In the same species of
bacterium and in *S. typhimurium*, the histidine (*his*)
operon is truly a giant with 10 structural genes, all
close together, and one operator gene; this operon is
sometimes referred to as the *histidon*. At the opposite
extreme, the arginine (*arg*) operon in *E. coli* has
structural genes scattered all over the genome, but with
only one operator gene. In the operon that specifies
enzymes for galactose utilisation in *E. coli* (the *gal*
operon) there are two operator genes.

The next question, of course, is exactly how are genes
switched on or off? Work on this has proceeded at a
feverish pace over the past 15 years. As a result it has
been possible to recognise two separate types of control
mechanism, namely *negative* and *positive control of protein
synthesis*.

Negative control of protein synthesis

In their original scheme, Jacob and Monod proposed that
induction of protein synthesis is possible only when the
repressor protein of the operon is not in contact with
the operator gene. When an inducer is not available,
either because it is not supplied in the medium or because
it cannot be synthesised by the microbe, the repressor
protein remains in contact with the operator gene, thereby
preventing expression of, that is mRNA synthesis on,
associated structural genes. The repressor protein is
thought to have this effect because it occludes a
part of the promotor region next to the operator gene
and so prevents RNA polymerase from making contact with
the promotor region. When an inducer is available to a
micro-organism, either by being provided in the medium or
because it has been synthesised in the cell, it combines

with the repressor protein which, since it is an
allosteric protein, changes conformation and loses its
affinity for the operator gene (*Figure 8.3*).

Repression of enzyme synthesis was explained by Jacob
and Monod by assuming that, in the absence of the
effector molecule, the repressor protein has no affinity
for the operator gene, but that it acquires an affinity
when it undergoes a conformational change following
interaction with the effector (*Figure 8.3*). The common
feature to these suggested mechanisms is that gene
expression is possible only when the operator gene is
free. For this reason, these control mechanisms are said
to be negative in nature. Negative control of protein
synthesis is well established at the molecular level in
many bacterial operons, notably induction of synthesis of
proteins of the *lac* operon in *E. coli* grown in the presence
of lactose and absence of glucose. Induction of enzyme
synthesis can therefore be viewed as a derepression
process.

Positive control of protein synthesis

Positive control of protein synthesis is said to occur
when an association between a protein and a part of the
regulatory region of an operon, which may not necessarily
be the operator gene, is essential for expression of
related structural genes in the operon. Examples of
positive control are not so well documented as with
negative control. One of the better understood is in
relation to expression of the genes concerned with
utilisation of arabinose by *E. coli* (the ara operon),
mainly because of work on this operon by Ellis Englesberg
and his associates of the University of California at
Santa Barbara, U.S.A. The first three genes on this
operon specify synthesis of, respectively, an epimerase,
an isomerase and a kinase; the respective genes are *ara*
D, *ara* A and *ara* B. Another linked gene, *ara* C, does not
determine the structure of an enzyme but behaves as a
regulator gene. It is suggested that the product of the
ara C gene is a protein which combines with the inducer
arabinose to form an activator molecule which acts at the
initiator site on the operon. This activator molecule is,
therefore, responsible for positive control of expression
of the *ara* structural genes. When the protein is unable
to combine with arabinose, it acts as a repressor protein,
like the product of the *i* gene in the *lac* operon, and

combines with the operator gene in the *ara* operon to prevent expression of structural genes.

Positive control of protein synthesis also operates during catabolite repression. For a long time, the molecular basis of the effect of glucose on synthesis of enzymes in microbes puzzled microbial physiologists. It was suspected that the effector, which presumably acts on the genome during catabolite repression, might be a compound, or a group of compounds, synthesis of which is regulated by the concentration of glucose in the growth medium. It is now known that this assumption was correct, but identification of the elusive effector (or group of effectors) was helped in large measure by the results of research on the effect of mammalian hormones on cells. These researches pointed to a compound, cyclic adenosine monophosphate (cAMP), as a possible candidate for the role of effector in microbial catabolite repression because of the similar role which this compound plays in the control of enzyme synthesis in mammalian cells. The first evidence that cAMP is important in regulation of enzyme synthesis in bacteria was the finding that addition of this compound to growth media can overcome transient catabolite repression of β-galactosidase synthesis in *E. coli*.

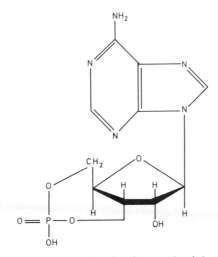

cyclic adenosine monophosphate

It has since been established that cAMP combines with
a protein, called the *catabolite gene activator* (CGA)
protein, to give a complex that facilitates binding of
RNA polymerase to the promotor region on the operon.
The CGA protein from the *lac* operon in *E. coli* has been
isolated and shown to be a protein dimer, with a strong
affinity for DNA which is enhanced considerably when the
protein is combined with cAMP. The situation is further
complicated in that cells synthesise a second catabolite
repression effector, cyclic guanosine monophosphate (cGMP).
This compound acts in exactly the opposite way to cAMP,
and a combination of cGMP with CGA protein lowers the
affinity of the protein for the promotor region on the
operon.

Cyclic AMP and cGMP are made from the corresponding
nucleoside triphosphates (ATP, GTP) in reactions catalysed,
respectively, by adenyl cyclase and guanyl cyclase.
Moreover, both cAMP and cGMP can be broken down in
reactions catalysed by esterases. Clearly, therefore, the
activities of these enzymes concerned with synthesis and
breakdown of cAMP and cGMP are extremely important in the
regulation of enzyme synthesis.

FURTHER READING

GENETIC CONTROL

The genetic code

WATSON, J.D. (1970). *Molecular Biology of the Gene,* 2nd
    edn., 662 pp. New York; W.A. Benjamin Inc.
WOESE, C.R. (1967). *The Genetic Code: the Molecular
    Basis of Genetic Expression,* 200 pp. New York; Harper
    and Row
WOESE, C.R. (1970). The genetic code in prokaryotes and
    eukaryotes. *Symposium of the Society for General
    Microbiology,* 20, 39-54

Individuality of microbial DNA

HILL, L.R. (1966). An index to deoxyribonucleic acid
    compositions of bacterial species. *Journal of General
    Microbiology,* 44, 419-437

Events which lead to changes in DNA composition

ARCHER, J.L. (Ed.) (1973). *Bacterial Transformation*,
414 pp. London; Academic Press
CURTISS, R. (1969). Bacterial conjugation. *Annual
Review of Microbiology*, 23, 69-136
DRAKE, J.W. (1970). *Molecular Basis of Mutation*, 273 pp.
San Francisco; Holden-Day
LEGATOR, M.S. and FLAMM, W.G. (1973). Environmental
mutagenesis and repair. *Annual Review of Biochemistry*,
42, 683-708

REGULATION OF ENZYME ACTION

Feedback inhibition of enzyme action

ATKINSON, D.E. (1969). Regulation of enzyme function.
Annual Review of Microbiology, 23, 47-68
HARRISON, D.E.F. (1976). Regulation of respiration rate
in growing bacteria. *Advances in Microbial Physiology*,
14, in press
SANWAL, B.D. (1970). Allosteric controls of amphibolic
pathways in bacteria. *Bacteriological Reviews*, 34,
20-39

Modification to enzyme structure

HOLZER, H. and DUNTZE, W. (1971). Metabolic regulation
by chemical modification of enzymes. *Annual Review of
Biochemistry*, 40, 345-374

Degradation of individual enzymes

SHAPIRO, B.M. and STADTMAN, E.R. (1970). The regulation
of glutamine synthesis in microorganisms. *Annual
Review of Microbiology*, 24, 501-524

REGULATION OF ENZYME SYNTHESIS

CALVO, J.M. and FINK, G.R. (1971). Regulation of
biosynthetic pathways in bacteria and fungi. *Annual
Review of Biochemistry*, 40, 943-968

CLARKE, P.H. (1972). Positive and negative control of
  bacterial gene expression. *Science Progress (Oxford)*,
  6Q, 245-258
LEWIN, B. (1974). Gene expression. 1. Bacterial
  genomes. 642 pp. London; John Wiley and Sons
RICKENBERG, H.V. (1974). Cyclic AMP in prokaryotes.
  Annual Review of Microbiology, 28, 353-369
SCHLEGEL, H.G. and EBERHARDT, U. (1972). Regulatory
  phenomena in knallgasbacteria. *Advances in Microbial
  Physiology*, 7, 205-242
STANIER, R.Y. and ORNSTON, L.N. (1973). The β-ketoadipate
  pathway. *Advances in Microbial Physiology*, 9, 89-151

9

GROWTH AND SURVIVAL

The co-ordinated synthesis of cell constituents by a
micro-organism leads to growth and reproduction and
ultimately to the formation of a culture or colony of
micro-organisms. This chapter describes the biochemical
events that take place during growth and reproduction of
micro-organisms, and at the same time deals with factors
that affect survival of microbes.

9.1 GROWTH

Growth can be defined as an orderly increase in the
numbers of all of the components of living cells. As
such, it involves replication of all cell structures,
organelles and components, and occasionally also
biogenesis of certain organelles, such as mitochondria
and chloroplasts, in micro-organisms that are suddenly
exposed to conditions (presence of oxygen or light) where
these organelles are essential. At present, our understanding of the molecular events involved in replication
and biogenesis of many cell structures and organelles is
limited. However, it is an area of microbial physiology
that is currently being studied intensively in many
laboratories, and there is every reason to believe that
our understanding of these replication and biogenesis
processes will be considerably extended during the coming
decade.

9.1.1   REPLICATION AND BIOGENESIS OF ORGANELLES

In studying replication and biogenesis of structures and
organelles, the microbiologist is primarily concerned
with discovering how macromolecules and other constituents
of these cell components are transferred from the site of
synthesis and assembled into new structures and organelles.
The extent to which macromolecules are transferred within
a micro-organism seems to vary with each structure and
each organism. Many micro-organisms contain what are to
some extent autonomous organelles. These organelles,
which include mitochondria and chloroplasts, contain their
own DNA- and RNA-synthesising enzymes, and are therefore
able to synthesise *in situ* at least some of their own
enzymes and organelle constituents. At the other extreme,
one has the situation in certain protozoa in which the
Golgi apparatus seems to be the site of synthesis of
certain organelles which, once they have been formed,
move to their appropriate location in the cell. The
surface scales of some flagellates, such as *Chrysochromulina*
spp., which have a characteristic shape and ornamentation,
can be seen to accumulate at the Golgi body before they
migrate to the cell surface.

Replication of genomes

Of all the processes involved in growth of cells, none
is as important as replication of the genome for, in the
absence of chromosome replication, the information for
biosynthesis of other organelles and structures cannot be
transferred to future generations. It should be noted,
however, that inhibiting biosynthesis of individual
organelles does not necessarily block genome replication.

Replication in prokaryotes. In view of the large amount
of research done on the biosynthesis of nucleic acids and
proteins in bacteria, it is not altogether surprising
that the molecular events that take place during
replication of the genome in prokaryotes are better
understood than those involved in replication of
eukaryotic genomes.

The genome in bacteria consists of a single circular
chromosome which is attached to the plasma membrane (*see*
page 65). The first insight into the mechanism by which
this organelle replicates itself came from the autoradio-

graphs prepared by John Cairns (*see* page 66) which
revealed the presence of *forks* in the replicating genome.
Initiation of a round of replication in a prokaryotic
cell occurs when the cell has grown to a certain critical
size, and when there is a fixed ratio of cell mass to DNA
content. Initiation is also dependent on protein
synthesis. Except during very rapid growth, there is a
single replicating fork in the chromosome which moves
around the chromosome starting at an origin. The
replicating fork is thought to be the point at which the
genome is attached to the plasma membrane. In rapidly
growing bacteria, two further forks have been seen to
start at the origin before the first one has completed a
round of replication. In many bacteria, the origin for
replication lies in the lower left quadrant of the
conventional chromosome map, and replication proceeds
clockwise. Replication begins at the 3'-hydroxyl end of
the DNA strand and terminates at the 5'-phosphoryl end.
Although the evidence available is not nearly so
extensive, it is thought that replication of plasmids
proceeds in a similar manner to that of the main genome.

Following the discovery by Cairns of a replicating fork
in the bacterial genome, a number of different models
have been put forward to explain the mechanism of
replication at the growing point in the fork. The Cairns
model (*Figure 9.1*) shows a single replicating fork with a
growing point at which both strands of DNA are replicated.
A modification of this model, put forward by the Japanese

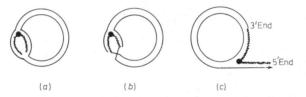

(a)   (b)   (c)

*Figure 9.1   Models for DNA replication in prokaryotes.
(a) Shows the Cairns-type model; (b) the Yoshikawa model;
and (c) the rolling circle model. The wavy line
represents newly synthesised DNA, and the filled-in
circles the growing point of the chromosome*

worker Hiroshi Yoshikawa, proposes that the newly
synthesised strands of DNA are covalently linked to the
termini of the parental chromosome (*Figure 9.1*). A third
model, called the *rolling circle model,* proposes that one

of the two strands of parental DNA is cut by action of
an endonuclease to generate free 3'-hydroxyl and
5'-phosphoryl ends. The 5' end is then displaced from
the circle and the 3' end is extended by the action of,
presumably, DNA polymerase III. The strand ending in the
5' end may also be copied. Which if any of these models
represents truthfully the events which actually occur
when a bacterial genome replicates is still not known.
Intriguingly, there is very recent evidence that genome
replication in prokaryotes may be not unidirectional but
bidirectional.

It has quite firmly been established, however, that
genome replication in bacteria is semi-conservative,
that is the two sister duplexes produced each contain one
old and one newly synthesised DNA strand. This was first
demonstrated in what is now acknowledged as a classical
experiment by Matthew Meselson and Fred Stahl performed
at the Massachusetts Institute of Technology, U.S.A., in
1958. *Escherichia coli* was grown for several generations
in a 'heavy' medium, that is one containing $^{15}NH_4^+$ as the
sole source of nitrogen. The bacteria were then
transferred to a 'light' medium containing $^{14}NH_4^+$, and
portions of the culture were removed at intervals and the
DNA extracted and analysed for specific density by
centrifugation through a gradient of caesium chloride.
During the first generation of growth in the 'light'
medium, heavy (^{15}N-^{15}N) DNA was gradually replaced by
half-heavy (^{15}N-^{14}N) DNA. However, the quantity of half-
heavy DNA in the cells at the end of the first generation
in the light medium persisted in subsequent generations
during which 'light' (^{14}N-^{14}N) DNA accumulated. Density-
gradient centrifugation of heat-treated DNA showed that
the half-heavy species always contained one heavy and one
light strand.

Replication in eukaryotes. Techniques similar to that
used by Cairns with the bacterial genome have been extended
to nuclei from a number of higher eukaryotic organisms,
but so far only to *Saccharomyces cerevisiae* among
eukaryotic microbes. It seems that the process of
chromosome replication is very similar in various types of
eukaryote. Replicating forks or 'bubbles' have been
detected in a small proportion of labelled chromosomes in
randomly growing and synchronous cultures of *Sacch.
cerevisiae*. Most of these forks are located internally,
and **they** appear to extend in both directions, as shown
in *Figure 9.2*. Beyond this, little is known about genome
replication in eukaryotes. It has however, been established
that, as with genome replication in bacteria, protein

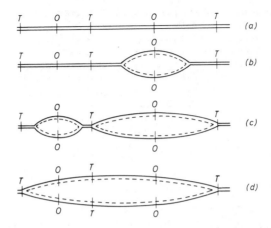

Figure 9.2 Bidirectional model for DNA replication in eukaryotes. Each pair of horizontal lines represents a segment of double helical DNA. Continuous lines represent parental DNA; dashed lines newly synthesised DNA; O indicates an origin; and T a terminus. The diagrams represent stages from the beginning (a) to the end (d) of replication

synthesis is required. But how the process is initiated, and whether or not the signal comes from the cytoplasm, has yet to be discovered.

9.1.2 BIOGENESIS OF RIBOSOMES

The enormous strides made recently in the elucidation of the molecular architecture of the bacterial ribosome (*see* page 69) have led to extensive research on the manner in which ribosomal particles are synthesised. All three types of ribosomal RNA are synthesised on the genome, but not as such; the transcription products for each type are larger. For example, in *E. coli* the precursor of 16S rRNA is larger by about 50 000 daltons, and that of 23S rRNA by 50 000–100 000 daltons. After synthesis, each precursor product undergoes cleavage and methylation (to give rise to methylated bases), a process referred to as *maturation*. The sequence of events is not too clear, but cleavage is thought to precede methylation. The need for co-ordinated synthesis of all three types of rRNA is

obvious, and it is possible that the respective genes
have an identical promotor region.

Ribosomal proteins are deposited onto rRNA molecules to
give the 30S and 50S subunits.   The assembly process is
not a random one but involves a stepwise addition of the
proteins in a definite sequence.   Many of the details of
this process have come from work in the laboratory of
Masayasu Nomura in the University of Wisconsin, U.S.A.
It has proved difficult to catch or identify precursor
particles although, as a result largely of research on
mutants of *E. coli* that are defective in ribosome assembly,
the process is thought to proceed as follows:

$$\text{nascent 16S RNA} \longrightarrow 21\text{S} \longrightarrow 26\text{S} \longrightarrow 30\text{S subunit}$$
$$\text{nascent 23S RNA} \longrightarrow 32\text{S} \longrightarrow 43\text{S} \longrightarrow 50\text{S subunit}$$

The need for a sequential addition of proteins is taken
to indicate that addition of each protein induces a
conformational change in the growing particle that is
necessary for addition of the next protein in the sequence.

In most bacteria, RNA synthesis takes place only when
there is available in the cell an adequate supply of all
amino acids.   When strains that are auxotrophic for an
amino acid are starved of this essential nutrient, net
protein synthesis ceases and RNA synthesis is decreased
to about 10% of that in non-starved organisms.   Mutant
strains of bacteria which continue to synthesise RNA
when starved of an essential amino acid have been isolated.
These are described as *relaxed* strains to distinguish them
from *stringent* strains which require amino acids for RNA
synthesis.   Although they contain 16S and 23S RNA, the
particles formed by relaxed strains grown under conditions
of amino-acid starvation differ from mature 30S and 50S
ribosomal subunits since they contain a smaller proportion
of protein.   Similar RNA-containing particles are formed
by bacteria grown in the presence of drugs, such as
chloramphenicol and puromycin, that inhibit protein
synthesis.

Much less is known about ribosome biogenesis in
eukaryotic microbes, particularly as regards addition of
ribosomal proteins to rRNA molecules.   But it has been
established that all three types of rRNA arise by cleavage
of a single molecule of 45S RNA.

9.1.3   REPLICATION AND BIOGENESIS OF MITOCHONDRIA

Mitochondria contain DNA (mDNA) and it is clear, therefore, that these organelles potentially have a certain genetic autonomy. Moreover, mitochondria possess their own protein-synthesising machinery, and in their sedimentation coefficients mitochondrial ribosomes resemble bacterial ribosomes rather than those present in the cytoplasm of the eukaryotic microbe. In light of these observations, some intrepid individuals have suggested that mitochondria originated following a symbiotic association between primitive eukaryotic organisms and prokaryotic ancestors.

During the cell cycle of aerobically growing micro-organisms, the mitochondria divide to produce a population of organelles that enters the daughter organisms. Virtually nothing is known about the physiology and biochemistry of *mitochondrial replication*. Probably the most suitable organism in which to study this process is the alga *Micromonospora*, which contains just one mitochondrion.

Much more is known about the molecular events which occur during *biogenesis of mitochondria*, a process which has been studied most intensively in the yeasts *Sacch. cerevisiae* and *Sacch. carlsbergensis*. When grown anaerobically, these organisms cease to produce mitochondria of the type seen in electron micrographs through thin sections of aerobically grown yeast cells (*see* page 61).   Yeast cells grown under anaerobic conditions do, however, produce particles which contain mitochondrial DNA. These structures, which are very fragile and can only be visualised in a morphologically intact form in cells that have been fixed with glutaraldehyde, have been termed *promitochondria*. They lack many of the characteristics of mitochondria, including a double membrane structure, DNA-dependent RNA polymerase, mitochondrial rRNA and a protein-synthesising machinery.

When a suspension of anaerobically grown *Sacch. cerevisiae* in growth medium is aerated, mature mitochondria can be detected in the cells after 6-8 h incubation. The experiment is usually done using a growth medium that contains a fairly low concentration of glucose, because the process of mitochondrial biogenesis is subject to catabolite repression. Several workers have followed the changes in cytology and enzyme activity that take place during biogenesis of mitochondria. After 0.5-1.0 h, vesicles appear in the yeast, presumably from

promitochondria. Initially, these vesicles are electron-
dense but they gradually become less dense and, after 2-3 h,
are enclosed by a double unit-membrane; mature
mitochondria appear some 3-4 h later. During the first
hour of aeration there is an appreciable synthesis of
cytochromes, but thereafter cytochrome synthesis occurs
less rapidly. It would seem from these findings that
cytochrome synthesis and development of mitochondria are
closely linked.

We now know that a most subtle and sophisticated control
is exerted over synthesis of mitochondrial proteins in
micro-organisms, in that some subunits of the proteins are
synthesised in the organelle and others in the cytoplasm
coded by the nuclear DNA. For example, the mitochondrial
ATPase in *Sacch. cerevisiae* contains 10 subunits, six of
which are coded by nuclear DNA and four by mDNA. Likewise,
the mitochondrial cytochrome oxidase in this yeast has
seven subunits, four of which are coded by the nucleus
and three by the mDNA. This situation raises many
problems, not the least of which is the need to explain
how polypeptides that are coded by the nuclear DNA pass
across the mitochondrial membranes to take up their
location in the organelle.

9.1.4   REPLICATION AND BIOGENESIS OF THE PHOTOSYNTHETIC
        APPARATUS

The position regarding the photosynthetic apparatus in
microbes is similar to that with mitochondria in that
almost nothing is known of the way in which they are
replicated in micro-organisms. Much more is known about
the process in which these organelles are produced *de novo*
when microbes, which have been grown under conditions
when they do not synthesise the photosynthetic apparatus
(in the dark), are transferred to an environment (exposure
to visible radiation) where these organelles are required.

Among eukaryotes, the most systematic attempt to study
chloroplast biogenesis has been with mutants (*y* mutants)
of *Chlamydomonas reinhardii* which, unlike the wild type,
do not synthesise chlorophylls or thylakoid membranes in
the dark. On being illuminated they rapidly become green
in colour and acquire a full complement of thylakoids and
photosynthetic activity in six to eight hours. It has
been established that there is a stepwise assembly of the
chloroplast structure, that is there is first a development
of the thylakoid membranes containing a certain minimum

number of components after which other components are
added when the construction has reached the appropriate
stage. This means that full photosynthetic activity is
not achieved until near the end of the eight-hour period
of biogenesis.

Experiments with photosynthetic bacteria have been
confined mainly to members of the Athiorhodaceae (species
of *Rhodopseudomonas* and *Rhodospirillum*) which can grow
in the dark or in the light. The bacteria are grown
aerobically in the dark, and biogenesis of the photosynthetic apparatus is studied after the cultures have
been transferred to anaerobic-light conditions. The
lamellae of the structures are formed by invagination of
the plasma membrane; the invaginations then grow parallel
to the membrane to produce flat vesicles. Biogenesis of
the photosynthetic apparatus in bacteria is linked to
protein synthesis because of the need for synthesis of
additional enzymes and membrane proteins. In
Rhodopseudomonas spheroides, haemoproteins and
ubiquinone are synthesised concomitantly with chlorophylls,
although synthesis of ribulose diphosphate carboxylase
is, for some unknown reason, delayed.

9.1.5 GROWTH OF PLASMA MEMBRANES AND CELL WALLS

Because a microbe needs continually to maintain an intact
plasma membrane, and to protect this membrane with a cell
wall, it is clear that growth of plasma membranes and of
cell walls must be very intimately connected processes.
Unfortunately, very little indeed is known about the way
in which a plasma membrane grows. Growth of cell walls,
on the other hand, is a process that is more amenable to
investigation at the cellular level. Nevertheless, until
just over a decade ago, information on this process was
meagre and largely confined to observations on cell-wall
staining and autoradiography of walls. Then, some 15 years
ago, Roger Cole in Bethesda, Maryland, U.S.A. and John May
in the University of Western Australia, working
independently, applied the fluorescent antibody technique
to the problem. As a result of these studies, considerable
progress has been made in our understanding of cell-wall
biogenesis.

The fluorescent-antibody technique involves preparing
an antiserum against a cell-surface component of a micro-
organism. This antigen is then labelled with an
appropriate fluorochrome, such as fluorescein; experiments

have also been made using antibody labelled with electron-
dense ferritin. There are several ways of using the
fluorescence-labelled antibody. One of the commonest is
to grow a micro-organism in a medium containing labelled
antibody and then to transfer the organisms to fresh
medium containing unlabelled antibody. Alternatively,
the organisms may be grown initially in medium containing
unlabelled antibody and then transferred to medium
containing labelled antibody. Smears are then made of the
organisms, and the wall that has been formed in the
presence of the labelled antibody will fluoresce because
of the retention of labelled antibody. Use of the technique
is based upon a number of assumptions, the more important
of which are that the presence of the labelled antibody
on the organism does not alter the pattern of cell-wall
growth, and that all cell-wall structures are replicated
in concert with the cell-surface antigen that is located
with the labelled antibody. To date, the technique has
been applied to only a small number of bacteria and
yeasts.

Not all micro-organisms replicate their cell walls in
precisely the same manner. In *Streptococcus pyogenes*,
new cell-wall material is laid down centripetally, so
that the new hemispheres of the cocci are initiated back-
to-back (*Figure 9.3*). This pattern of cell-wall
replication was observed in *Strep. pyogenes*, regardless
of whether the labelled antibody was specific for the
type-specific microcapsular M protein or for the group-
specific C polysaccharide which is thought to lie more
deeply in the cell wall. These findings suggest that
microcapsular materials, at least in this bacterium, are
replicated in concert with other cell-wall components.
In *Salmonella* new wall material is inserted continuously
into the old wall along the length of the cell by a process
of diffuse intercalation. These data indicate that the
enzymes that catalyse reactions concerned in synthesis of
bacterial cell-wall polymers can be either widely
distributed in the plasma membrane or confined to certain
loci.

Experiments have also been carried out with the fission
yeast, *Schizosaccharomyces pombe*, which lays down new
cell wall by a process of apical extension predominantly
at one end not unlike that operating in *Strep. pyogenes*.
With budding yeasts, such as *Sacch. cerevisiae*, application
of the immunofluorescence technique has shown that the
wall of the bud is almost entirely newly synthesised and
contains very little old cell-wall material. In yeasts and

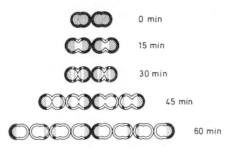

Figure 9.3 A diagrammatic representation of the
immunofluorescent appearance of the cell wall of
Streptococcus pyogenes during growth. The bacteria
were grown initially in a medium containing a fluorescent
antibody against the type-specific microcapsular M
protein, and were then incubated in a medium containing
the same antibody lacking a fluorescent label. Dark
areas indicate fluorescence; light areas an absence of
fluorescence. The times of incubation in the medium
containing the non-fluorescent antibody are indicated on
the diagram

fungi, it is believed that at least some enzymes concerned
in synthesis of new wall material, which include lytic
enzymes such as β-glucanases as well as biosynthetic
enzymes, are synthesised inside the cell and transported
to the plasma membrane in vesicles.

9.1.6 BIOGENESIS AND REPLICATION OF CAPSULES

Very few reports have appeared on the biogenesis of
capsules on micro-organisms. However, one of these has
shown that Bacillus anthracis becomes capsulated only
towards the end of the exponential phase of growth in
batch culture. Fully capsulated bacteria inoculated into
broth form chains of cells which are non-capsulated except
at their tips and at occasional junctions between
neighbouring cells. These observations suggest that
capsules are partitioned between cells following cell
division.

9.2   GROWTH OF INDIVIDUAL MICRO-ORGANISMS

The co-ordinated replication of individual structures, organelles and components of a micro-organism constitutes growth of the organism. Numerous studies on the kinetics of growth of individual organisms have been made, many of which have aimed at describing parameters such as the rate of increase in cell size or cell volume, and ascertaining whether these processes occur continuously or discontinuously, linearly or non-linearly. They have also set out to describe the kinetics of synthesis of major cell macromolecules during the *interdivision time* or *cell cycle* which is the period between the formation of a cell by division of its mother cell and the time when the cell itself divides to form two daughters. Although studies of this type were first reported over 70 years ago, the data available today are still fragmentary mainly because of the difficulties encountered in making these observations.

9.2.1   OBSERVATIONS ON INDIVIDUAL CELLS

The method most commonly used for studying the kinetics of growth of individual microbes has been direct microscopic measurement of cell length or cell volume, as a function of time. Ward, as long ago as 1898, carried out such a study and reported steady exponential increase in the length of *Bacillus ramosus* between divisions. Later studies had the advantage of more sophisticated microscopy, but on the whole they did not greatly extend our understanding of the kinetics of growth of individual organisms. They have shown that the size of individual micro-organisms, particularly bacteria, often increases continuously between divisions, although it has not always been possible to describe the exact form of the increase with any degree of accuracy. The large random error incurred in measuring volumes of cells, particularly bacteria, makes it especially difficult if not impossible to distinguish exponential from linear growth.

Observations have also been made on the rate of increase of dry mass of individual organisms between divisions using either a scaled down version of the Cartesian diver toy or a rather more sensitive method which employs the interference microscope. With *Amoeba proteus,* this increase takes place exponentially over the whole of the cell cycle. With the fission yeast *Schizosaccharomyces pombe* growing at its optimum temperature of 32°C, the rate of increase in dry mass is linear over the cycle; when the yeast is grown at 17°C, the rate remains linear but does not extend over the entire cell cycle.

9.2.2 OBSERVATIONS ON RANDOMLY DIVIDING POPULATIONS

Data on growth of individual microbes can also be obtained from experiments using randomly dividing populations of cells. Observations on the distribution of generation times have been made using microcultures of organisms. These experiments have revealed that interdivision times of sister cells of *Escherichia coli*, *Proteus vulgaris* and *Pseudomonas aeruginosa* show a strong positive correlation, and that there is a positive association between interdivision times of second cousins (that is of microbes whose nearest common ancestor is three generations removed).

A novel approach to studies on growth of micro-organisms using randomly dividing populations was made by John Collins and Mark Richmond at the National Institute for Medical Research in London, England. These workers turned to batch cultures of *Bacillus cereus* and derived equations showing that the kinetics of growth of individual cells in the culture determine the form of the size distribution of the bacteria in the population. They measured size distributions, by making measurements on bacteria in stained and live preparations of samples from a batch culture, and concluded that the rate of increase in length of bacteria increases as the organism grows in length, and that there is no hesitation in this process before or after division. A follow-up of this work has come from the laboratory of Allen G. Marr in the University of California at Davis, U.S.A. Workers in this laboratory undertook a rigorous development of the Collins and Richmond approach to studies on growth of micro-organisms, both as regards its theoretical basis and the techniques employed. They calculated the volume distribution of bacteria by measuring the amplitude of pulses generated when the culture is passed through a Coulter transducer. Differentiation and integration of the pulses, followed by automatic pulse-height analysis, permitted a precise measurement of the volumes of bacteria over the range 0.25-20 μm^3. Using this apparatus, Marr and his colleagues showed that the specific growth rate of individual cells of *Escherichia coli* and *Azotobacter agilis* decreases at the time of division and increases in between divisions.

9.2.3   OBSERVATIONS ON SYNCHRONOUSLY DIVIDING POPULATIONS

Valuable data on the kinetics of synthesis of macro-molecules in intervals between divisions have come from experiments using *synchronously dividing cultures*. There are two main ways of obtaining synchronously dividing cultures of micro-organisms. The first of these depends upon a manipulation of the environment. A commonly used technique involves submitting a culture of micro-organisms to single or multiple changes in temperature. Usually the changes involve an alternation between the optimum temperature for growth and either a suboptimum or a superoptimum temperature. For example, synchronously dividing cultures of *Tetrahymena pyriformis* have been obtained by incubating a randomly dividing culture at 40°C and alternating this treatment with periods of incubation at 28°C. Methods based on cyclic changes in incubation temperature have been designed on the assumption that repeated stimuli applied at intervals closely corresponding to the generation time will gradually induce synchronous division. Methods of inducing synchronous division based on changes in medium composition have also been used. Auxotrophic micro-organisms, for example, can be subjected to a period of starvation in a medium lacking an essential growth factor before being placed in a nutritionally complete medium. Exposure of bacteria to compounds such as chloramphenicol that inhibit protein synthesis has also been used. Other techniques that have been reported to be successful in inducing synchronous growth in micro-organisms include treating the organisms with a sublethal dose of radiation and, with photosynthetic organisms, exposing them to alternating periods of light and dark.
    Very little is known of the biochemical bases of these methods for obtaining synchronously dividing micro-organisms. It has been suggested that the changes in incubation temperature, or subjecting the micro-organisms to starvation conditions, synchronise the synthesis of one or more compounds that trigger off the cell-division process when they reach a certain concentration within the cell. As nothing is yet known of the chemical nature of these compounds, if indeed they exist, it is clearly not possible to verify this hypothesis.

In the second type of method, organisms at a particular stage in the cell cycle are mechanically selected from a randomly dividing population. Two types of cell-selection method are now widely used and, on the whole, they are preferred to methods which involve environmental manipulation. The first of these is particularly useful for obtaining synchronously dividing cultures of bacteria, and involves filtering a portion of culture through a membrane filter and, after the filter has been inverted, eluting the cells from the filter with fresh medium. Samples of effluent suspension are incubated, and the bacteria in these samples divide synchronously for at least two or three generations. The other technique involves centrifuging a suspension of randomly dividing organisms through a density gradient of some non-metabolisable solute such as sucrose or dextran, and has been used extensively with yeasts. The gradient is centrifuged until the cell layer has moved about two-thirds of the way down the tube. The small cells at the beginning of the division cycle move more slowly down the tube and can be separated off the top layer with a syringe. When these organisms are suspended in fresh medium, they divide synchronously for three to four generations.

The time-course pattern of synthesis of various macro-molecules in the cell cycle is studied by removing portions of a synchronously dividing culture and analysing the cells for the component or enzyme activity under investigation. Studies of this type have been carried out for a variety of cell components in both prokaryotic and eukaryotic microbes, notably in the laboratories of Harlyn Halvorson at Brandeis University in Massachusetts, U.S.A., Murdoch Mitchison in the University of Edinburgh, Scotland, and of Arthur Pardee at Princeton University in New Jersey, U.S.A. There are four basic patterns of synthesis for any one cell component (*Figure 9.4*) although more complex patterns are sometimes encountered which are not easy to relate to the basic patterns.

Patterns of DNA synthesis have been closely studied, for these clearly affect the patterns of synthesis of other cell components. In fast-growing bacteria, dividing for example every 30 minutes, DNA is synthesised continuously over the entire cell cycle. However, in more slowly growing microbes, DNA synthesis is confined to a small period in the cell cycle (called the S period) with G (for growth) periods on either side of the S period. In *Sacch. cerevisiae*, dividing once every two hours, the G1, S and G2 periods each occupy about one-third of the

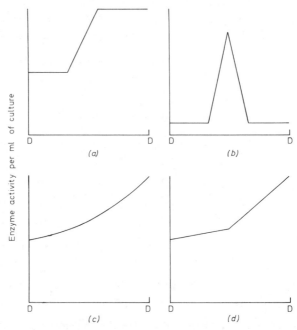

Figure 9.4 Patterns of enzyme synthesis during one cell cycle of a synchronously dividing culture of a microorganism. (a) Indicates a periodic-step pattern; (b) a periodic-peak pattern; (c) a continuous-exponential pattern; and (d) a continuous-linear pattern. D indicates a cell division

cell cycle. But in *Schizosaccharomyces pombe,* which is Mitchison's favourite organism, DNA synthesis takes place during a 10-15 minute period at the very beginning of the 150-minute cell cycle.

   Patterns of synthesis of close to 100 enzymes have now been studied during the cell cycles of a range of microbes. Synthesis of many enzymes follows the periodic-step pattern (*Figure 9.4*) but the S period for synthesis of the enzyme can be at any stage in the cell cycle. For example, in *Schizosacch. pombe,* the mid-point of the S period for synthesis of homoserine dehydrogenase occurs at about one-sixth of the way through the cycle, while that for synthesis of tryptophan synthase occurs when the cell cycle is almost complete. The periodic-peak pattern for enzyme synthesis is less commonly encountered, although it has been observed with synthesis of glycylglycine dipeptidase in *E. coli,* and of protease in *Sacch. cerevisiae.* Continuous production of an enzyme during the cell cycle occurs quite frequently, as for example with synthesis of acid and alkaline phosphatases, β-fructofuranosidase and maltase in *Schizosacch. pombe.*

9.3   GROWTH OF MICROBIAL POPULATIONS

In most organisms, cell reproduction is a consequence of growth. In multicellular organisms, this leads to an increase in the size of the individual organism, whereas in unicellular organisms it leads to an increase in the number of organisms in the population. However, in coenocytic micro-organisms, such as members of the Phycomycetaceae, cell multiplication is not a necessary consequence of growth.

Many different methods have been devised for measuring growth of microbial populations. These can be divided into those that measure the total number of organisms in a population, both living and dead, and those that measure only the number of viable organisms. With unicellular organisms, viable counts are made by plating out a portion of diluted suspension or culture on to a suitable medium, and counting the number of colonies that appear on the plate following a period of incubation. It is assumed, often without foundation, that each colony arises from one organism so that the total number of viable organisms in the original suspension or culture can be calculated from the number of colonies that appear. Total counts can be determined either by direct observation, that is by counting the number of organisms in a suitably diluted portion using a haemocytometer slide, or indirectly by measuring some property of the culture or suspension that is proportional to the total number of organisms present. The dry weight of organisms in a culture is most often used as a measure of the total numbers present. Since dry weight measurements are tedious and time consuming, growth of unicellular micro-organisms is frequently measured turbidimetrically, and the turbidity readings related to dry weight by a calibration curve. However, dry weight measurements, either direct or indirect, do not always provide an accurate measure of the number of organisms present, since the dry weight of an individual organism

N

can vary in different environments and also the dry
weight-turbidity relationship may not be constant under
all conditions. To overcome these objections, methods
have been devised for estimating the total number of
organisms in a culture by measuring the amount of some
chemical component of the organisms or of a metabolic
product which is excreted into the culture fluid. Since
the amount of DNA in each microbial cell is approximately
constant in a population of mature organisms, measurements
of the total DNA contents of the micro-organisms provide
a useful and accurate measure of growth. The amounts of
other cell constituents, such as RNA and protein, vary so
widely from one organism to another, and also with the
age of the organisms, that they cannot satisfactorily be
used as a basis for measuring growth.

In the laboratory, growth of microbes is usually studied
using separate batches of growth medium to give what are
referred to as *batch cultures*. The batch culture,
convenient though it is for the microbiologist, presents
the microbe with a very different environment from that
usually encountered in natural environments. Micro-
organisms can also be studied in the laboratory in
continuously growing cultures. The conditions which
obtain in *continuous cultures* in some way approximate
more closely those in natural environments, although in
the microbial physiology laboratory these types of culture
are usually operated for other quite different reasons.

9.3.1 BATCH CULTURES

Unicellular micro-organisms

When growth of a unicellular micro-organism in *liquid
medium* is followed using one of the methods already
described, it is usually found that the rate of growth
varies with time. A generalised growth curve for a
unicellular micro-organism is shown in *Figure 9.5*. It is
possible to distinguish four main phases of growth in
liquid batch culture.

After a portion of medium has been inoculated with a
micro-organism, a period of time normally elapses before
a constant rate of growth is established; this period is
called the *lag phase*. If the micro-organisms in the
inoculum are already adapted to growing under the conditions
obtaining in the fresh medium, then the lag phase may be

shortened or even disappear completely. Although there is
little or no increase in the number of organisms in the
culture during the lag phase, it is nevertheless a
period of intense metabolic activity during which the
organisms become accustomed to the conditions in the
medium in preparation for the period of rapid growth that
is to follow. During the lag phase there is a considerable
increase in the contents of RNA and total protein in each
organism; the DNA content however remains approximately
constant. There is also an appreciable increase in the
size of many micro-organisms during this phase of growth.

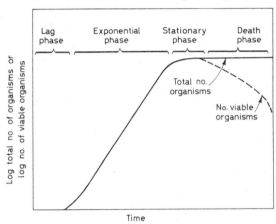

*Figure 9.5 Generalised growth curve for a unicellular
micro-organism*

The delay in the onset of growth during the lag phase
is due to *adaptation* which can be defined as a change in
an organism that increases its fitness for a particular
environment. Adaptation of micro-organisms can be genetic
or phenotypic (non-genetic). *Genetic adaptation* results
from the selection of those organisms in the inoculum that
are capable of growth and reproduction in the particular
environment. Mutant strains are continually being
produced in a population of micro-organisms but, as a
result of back-mutation and the operation of selection
mechanisms, an equilibrium is usually set up. However,
when a population encounters an environment in which only
certain of the mutant strains present can multiply, then
these mutants will rapidly take over as the dominant
members of the population. The rate at which they take

over will depend upon the numbers present in the original
population and the rate at which they multiply in the
fresh medium.

Non-genetic or phenotypic adaptation, on the other hand,
does not involve any alteration in the heritable
characteristics of the predominant organisms in a
population. In this type of adaptation, each of the
organisms acquires the ability to grow in the environment
by, for example, synthesising an inducible enzyme required
for metabolism of a nutrient present in the medium.

When a constant rate of growth has been achieved in
batch culture, the micro-organisms are said to be in the
logarithmic or exponential phase of growth. In this phase,
the organisms are growing at the maximum rate possible in
the particular medium. As each organism reaches a certain
age, it divides to produce two daughter individuals. The
two daughter cells are usually formed by binary fission,
but in yeasts (except *Schizosaccharomyces* spp.) and in a
few bacteria reproduction is by budding. The organisms
may separate immediately after division or only after a
period of time; they may also remain in chains, clusters
or packets.

A detailed analysis of the kinetics of growth of
microbes in batch culture was first made by Jacques Monod
at the Institut Pasteur in Paris in the 1940s. During
exponential growth, the biomass in the culture doubles
at a constant rate. If the initial concentration of
organisms is x, then the exponential rate of growth (μ)
is given by:

$$\mu = \frac{1}{x} \cdot \frac{dx}{dt} = \frac{d(\log_e x)}{dt} = \frac{\log_e 2}{t_d} \qquad (9.1)$$

where t_d is the time taken for the culture to double its
biomass. It follows from this equation that the *culture
doubling time* (t_d) equals 0.69 divided by the value for
μ. When cultural conditions are not varied, values for
μ and t_d are constant, but if conditions are varied,
especially the concentration of essential nutrients in
the culture medium, the values do not remain constant.

If the concentration of one essential nutrient in the medium is lowered, then the value for μ decreases correspondingly. The dependence of μ on substrate concentration (s) was shown by Monod to be of a form that can be represented by a Michaelis-Menten type function (*Figure 9.6*):

$$\mu = \mu_{max} \left(\frac{s}{K_s + s} \right) \qquad (9.2)$$

where μ_{max} is the maximum value of μ, when the value of s is no longer limiting growth, and K is a saturation constant which is numerically equal to the concentration of growth-limiting substrate at half of the maximum rate (that is when it equals $\mu_{max}/2$). Values for K_s are usually of the order of μg per ml when a carbohydrate is the growth-limiting nutrient, or μg per litre when the availability of an amino acid limits growth.

In addition, a constant relationship exists between the rate of growth (μ) and the rate of substrate consumption, as described by the equation:

$$\frac{dx}{dt} = -Y \frac{ds}{dt} \qquad (9.3)$$

where Y is known as the *yield factor* and, over a finite period of time during exponential growth, is equal to:

$$Y = \frac{\text{weight of micro-organism formed}}{\text{weight of substrate consumed}}$$

It follows that a complete description of the growth of a batch culture can be obtained if values for μ_{max}, K_s and Y are known.

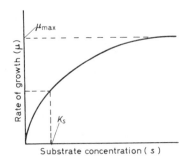

*Figure 9.6   Relationship between rate of growth of a
micro-organism and the concentration of limiting substrate.
See text for explanation*

During the exponential phase of growth, which in batch
cultures of many micro-organisms in liquid media lasts
only a short period of time, the nutrients in the medium
become depleted and waste products of metabolism
accumulate, so that the medium gradually becomes less
favourable for growth. Ultimately, the culture enters
the *stationary phase of growth* in which the number of
organisms in the culture remains constant. This phase
can last for a considerable period of time but, sooner
or later, it is followed by the *death phase* in which the
number of viable organisms, although not necessarily the
total number of organisms, declines.

On *solid medium*, unicellular micro-organisms reproduce,
in or on the surface, as a localised mass of organisms
known as a *colony*. Under these conditions there is
physiological crowding of the organisms with individuals
competing with each other for the limited amounts of
nutrients and oxygen that are available. During the
development of myxobacterial colonies, the organisms
spread over the surface of the solid medium by a gliding
movement. This gives rise to a very thin and rapidly
expanding colony. Despite the importance attached to
colonial growth of organisms in microbiology, the laws
which govern growth of organisms in colonies have not
been fully elucidated.

Filamentous micro-organisms

When a hypha of a filamentous micro-organism develops in
a suitable *liquid medium,* growth and elongation are almost
entirely restricted to the region immediately behind the
tip or apex. With septate hyphae, the segments once
formed do not elongate appreciably. With aseptate or
coenocytic hyphae, growth takes place only between the
apex of the hypha and the point at which the youngest
branch arises. Microscopic measurements of the increments
of growth at the tips of hyphae have shown that growth of
an individual hypha in a very young culture at first
increases at a rate proportional to the total length of
the hypha. A commonly realised growth rate is 200-400 μm
h^{-1} but later the growth rate declines. When filamentous
organisms are grown in static liquid culture, the
mycelium forms a mat or pellicle on the surface of the
culture; but in liquid medium that is constantly agitated,
the organisms usually grow in the form of bead-like
spherical colonies.

The hyphae of filamentous organisms grown on *solid
medium* grow in all directions but can spread only in two
dimensions to give a circular colony. The distance
between the centre and the edge of the colony increases
in proportion to the time of incubation, but the total
amount of cell substance in the colony increases in
proportion to the square of the time.

9.3.2 CONTINUOUS CULTURES

During the exponential phase of growth of a micro-organism
in batch culture, the medium becomes depleted of nutrients
and there is an accumulation of waste products of metabolism.
Either or both of these factors lead ultimately to a
decrease in the rate of growth with the result that the
culture enters the stationary phase. Clearly, if fresh
nutrients could be added to the culture and, at the same
time, waste products of metabolism removed, there is no
reason why micro-organisms should not be maintained
indefinitely in the exponential phase of growth.

Over the past two decades, an increasing number of
microbiologists have become interested in growing microbes
in continuous culture. This interest has led to the
development of a variety of different pieces of apparatus,
some quite simple, others rather more complicated. All
continuous cultures start their existence as batch
cultures, in that the medium in the growth vessel is
inoculated with microbes that proceed to grow in batch

culture. If, during the exponential phase of growth,
fresh medium is added to the culture at a rate sufficient
to maintain the culture population density at a fixed
value, lower than μ_{max}, then growth should not ultimately
cease as it does in a batch culture but continue
indefinitely. Obviously, the rate of input of fresh
medium would have to increase exponentially, with the
increase in biomass, if provision were not made for
continuous removal of culture at a rate equal to that at
which fresh medium was being added. The maintenance of
a constant volume of culture at a constant microbial
population density forms the basis of one type of
continuous culture apparatus, known as the *turbidostat*.
Microbes can also be cultured in another type of apparatus,
known as a *chemostat*. The chemostat resembles a
turbidostat in that a culture is contained in a suitably
constructed vessel into which fresh medium is pumped at
a constant rate. However, with a chemostat, the
microbial population density is not controlled directly.
Instead, the composition of the medium is compounded such
that all of the constituents essential for growth of the
microbe, except one, are in the medium at concentrations
that are in excess of the requirement for growth of the
selected organism. The growth-limiting nutrient is
present in the medium at a concentration sufficient to
support only a restricted amount of growth.

Types of continuous culture

Simple chemostats, such as that shown diagrammatically in
Figure 9.7, can have a working volume as small as 100 ml,
and are valuable tools in research on microbial
physiology. Apparatus with larger working volumes can
easily be constructed, and are useful for studies where
large quantities of microbial cell material are required.
However, with larger working volumes, problems can arise
when it comes to providing the required volume of sterile
medium.
    Several more involved types of continuous-culture
apparatus have been constructed. *Multi-stage systems,* in
which the effluent culture from one vessel is fed into a
second vessel, and from the second vessel to a third, and
so on, have been used in many laboratories in studies on
microbial product formation. A multi-stage system of
chemostats can be modified further by metering additional

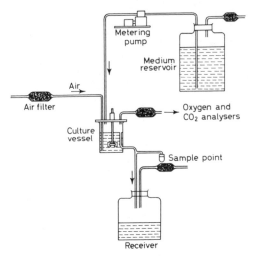

Figure 9.7 Schematic representation of a simple chemostat

fresh medium into one or more of the vessels in the chain, and by partially recycling culture from the last to the first vessel (*Figure 9.8*).

The chemostats so far described are *open chemostats,* in that the concentration of cells in the effluent from each vessel is equal to that inside the vessel. *Partially closed systems* can be constructed in which the cell concentration in the vessel is greater than that in the effluent. This deviation from the open system can be described by the *closure index (C):*

$$C = 100 \left(\frac{1 - x_b}{x} \right)$$

in which x equals the concentration of cells in the vessel, and x_b the concentration of cells in the effluent. A partially closed chemostat can be obtained by incorporating a cell filter in the effluent stream, or by decreasing the degree of turbulence in the vessel such that, in a 'quiet zone' in the culture, it is zero. When the latter method is employed, the continuous culture ceases to be a homogeneous one. Heterogeneous, partially closed chemostats have the advantage that a high localised concentration of cells (sometimes referred to as a plug) can be maintained in the culture, a situation which can

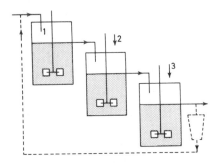

*Figure 9.8   Diagrammatic representation of types of
multi-stage chemostat system.   The continuous line
indicates the flow of culture in a simple chain of
chemostats.   The arrows entering vessels 2 and 3 indicate
the metering of additional medium into these vessels.
Also indicated is a partial recycling of culture from
vessel 3 to vessel 1*

be used to increase the rate of conversion of a substrate
into a microbial product.   The system has been used for
fermentation of brewer's malt wort to make beer.

A simple chemostat can also be modified to give a
synchronously dividing continuous culture.   This
modification has been pioneered by Peter Dawson at the
Prairie Regional Laboratory in Saskatoon, Canada, and is
known as a *continuous phased culture*.   A simple chemostat
culture contains cells at all stages of the cell cycle.
If, instead of adding fresh medium continuously to the
culture, a single addition of medium is made at the end
of a cell cycle to a culture already synchronised, then
synchronised growth can be maintained indefinitely.   By
altering the period of time in between adding a batch of
medium to the culture, the growth rate can be changed.
For convenience, the volume of culture is halved after
each addition of medium to avoid an otherwise exponential
increase in culture volume (*Figure 9.9*); the half portion
removed through an overflow device can conveniently be
used for analysis or processing.   Already, much useful
information on changes in composition of microbes during
the cell cycle has come from experiments using continuous
phased cultures.

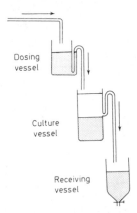

Dosing
vessel

Culture
vessel

Receiving
vessel

*Figure 9.9 The principle of operation of a continuous
phased culture*

Theory of continuous cultures

In order to appreciate fully the behaviour of micro-
organisms in continuously growing cultures, it is
necessary to examine closely the theory underlying the
technique. It is evident that, in a chemostat, the most
important factor controlling growth of the microbes is
the rate at which fresh medium is added to the culture
vessel. The ratio of the rate at which fresh medium is
added to the culture (f) to the operating volume of the
culture (V) is known as the *dilution rate* (D), i.e.
$D = f/V$. The dilution rate is equal to the number of
volumes of medium which pass through the culture vessel
in one hour. The reciprocal, $1/D$, gives the mean
residence time which is equal to the average time that an
organism spends in the culture vessel.

In the chemostat vessel organisms are growing, but they
are also being washed out of the culture vessel. The net
change in concentration of organisms (x) with time will
therefore be determined by the relative rates of growth
and washout. Thus:

$$\text{increase} = \text{growth} - \text{washout}$$

$$dx/dt = \mu x - Dx$$

$$dx/dt = x(\mu - D) \qquad (9.4)$$

From these equations it follows that, if μ is greater than D, the value for dx/dt will be positive, and the concentration of organisms in the chemostat will increase with time. On the other hand, if the value for μ is less than that for D, dx/dt will be negative, with the result that the concentration of cells in the chemostat will decrease with time and ultimately the culture will be 'washed out' from the vessel. Only when μ equals D will the value for dx/dt be zero. Under these conditions, the concentration of organisms in the culture remains constant with time, and the culture is said to be in a *steady state*.

The rate of growth of organisms in a chemostat can be adjusted, within certain limits, to any desired value. However, the specific growth rate (μ) cannot be made to exceed μ_{max} (*see* equation 9.2). Steady-state conditions can only be obtained at dilution rates below a critical value, referred to as D_c, which nearly equals μ_{max}. If the dilution rate is adjusted to a value greater than D_c, the culture will be washed out of the chemostat. It is a moot point whether there exists a minimum growth rate in continuous cultures, although it is generally conceded that this value is probably smaller than $0.05 \, \mu_{max}$.

To obtain a more complete quantitative assessment of the behaviour of microbes growing in continuous culture, we must consider the effect of dilution rate on the concentration of growth-limiting substrate (s) and of microbial cell material (x) in the culture. The growth-limiting substrate enters the culture vessel at a concentration S_r. There, it is consumed by the organisms in the culture, and emerges in the overflow at a concentration of s. The net change in substrate concentration as a result of passage through the culture vessel is:

$$\text{change} = \text{input} - \text{output} - \text{consumption}$$
$$ds/dt = DS_r - Ds - \text{growth/yield}$$
$$= DS_r - Ds - \mu x/Y$$

Re-arranging, and substituting for μ (equation 9.2):

$$\frac{ds}{dt} = D(S_r - s) - \frac{\mu_{max} x}{Y} \left(\frac{s}{K_s + s} \right) \quad (9.5)$$

Similarly, by substituting for μ in equation 9.4 it is possible to relate the increase in concentration of organisms with time to μ_{max}, s and K_s:

$$\frac{dx}{dt} = x \left[\mu_{max} \left(\frac{s}{K_s + s} \right) - D \right] \qquad (9.6)$$

It can be deduced from equations 9.5 and 9.6 that, when the chemostat is in a steady state (when values for S_r and D are constant, and D is less than D_c), unique values exist for both the concentration of organisms in the culture, and the concentration of growth-limiting nutrient. These values are designated $\overline{x}$ and $\overline{s}$. Since, at equilibrium:

$$\mu = D = \mu_{max} \left(\frac{s}{K_s + s} \right)$$

equating equation 9.5 to zero and solving for $\overline{x}$ gives:

$$\overline{x} = Y(S_r - \overline{s}) \qquad (9.7)$$

Similarly, equating equation 9.6 to zero and solving for s gives:

$$\overline{s} = K_s \left(\frac{D}{\mu_{max} - D} \right) \qquad (9.8)$$

By substituting $\overline{s}$ (equation 9.8) in equation 9.7 it follows that:

$$\overline{x} = Y \left[S_r - K_s \left(\frac{D}{\mu_{max} - D} \right) \right] \qquad (9.9)$$

For some time, it was assumed that the yield value (Y) was independent of the growth rate of a culture. However, it is now known that this is usually not so. A constant value for Y would imply that the concentration of the growth-limiting compound in the organisms did not vary with the rate at which the organisms were grown. But analyses have revealed that the contents of many of the compounds that are made to limit growth in a chemostat (for example, Mg^{2+} and PO_4^{3-}) vary with the rate at which microbes are grown. In addition, the yield value may be expected to vary following synthesis of intracellular storage compounds such as glycogen and poly-β-hydroxy-butyrate.

When the rate of growth is limited by the availability of a carbon source in a chemostat, the yield value also varies with the rate of growth. This is not because the carbon content of organisms varies with the rate at which

they are grown (in fact it does not), but because the
carbon source provides material for assimilation to give
new cell material and for dissimilation leading to ATP
production. The ratio of the amount of carbon substrate
oxidised to that assimilated clearly depends on the rate
of synthesis of new cells (that is on growth rate) and
this causes the yield to vary especially at low growth
rates.

   The variation in the yield of carbon-limited cultures
with growth rate has been used to measure the *maintenance
energy requirement* of cells (*see* page 247 for a more
detailed consideration of the importance of maintenance
energy). John Pirt at Queen Elizabeth College in the
University of London, England, used the following argument
to derive an expression from which the *maintenance
coefficient* (*m*) can be calculated. The substrate balance
in a continuously growing culture is given by the
expression:

$$\begin{array}{ccc}
\text{overall rate of} & = & \text{rate of utilisation} & + & \text{rate of utilisation} \\
\text{utilisation} & & \text{for maintenance} & & \text{for growth}
\end{array}$$

i.e.

$$\frac{ds}{dt} = \frac{ds}{dt}_{\text{maintenance}} + \frac{ds}{dt}_{\text{growth}} \qquad (9.10)$$

We saw on page 379 that the relationship between the rate
of substrate utilisation and μ, x and Y is given by the
equation:

$$\frac{ds}{dt} = -\mu x - \frac{\mu x}{Y}$$

Substituting in equation 9.10:

$$-\frac{\mu x}{Y} = -mx - \frac{\mu x}{Y_G}$$

where *m* is the maintenance coefficient and Y_G the true
growth yield. Re-arranging, this equation becomes:

$$\frac{1}{Y} = \frac{-m}{\mu} + \frac{1}{Y_G} \qquad (9.11)$$

From equation 9.11 it follows that, if *m* and Y_G are
constants, a plot of $1/Y$ against $1/\mu$ will give a straight
line with a slope equal to *m* and an intercept on the
ordinate equal to $1/Y_G$. *Figure 9.10* shows two such plots

for *Aerobacter cloacae* grown in a glucose-limited chemostat under aerobic or anaerobic conditions. Values for the maintenance coefficients from the data in *Figure 9.10* are 0.09 g glucose per gram dry weight cells per hour for cultures grown aerobically, and 0.47 g glucose per gram dry weight cells per hour for anaerobically grown cultures. It is to be expected that the value for anaerobically grown cultures would be greater because of the less efficient energy-yielding metabolism in anaerobically growing cells.

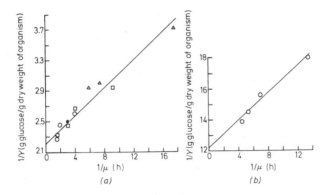

Figure 9.10 Double reciprocal plots of the yield value and the growth rate for glucose-limited cultures of Aerobacter cloacae *grown aerobically (a) or anaerobically (b). See text for further details*

USE OF CONTINUOUS CULTURES IN MICROBIAL PHYSIOLOGY. The advantages offered to the microbial physiologist by the chemostat are the ability to control independently growth rate and the concentration of nutrients in a culture. In a batch culture, the rate of growth of a microbe varies, even during the exponential phase of growth, and at the same time the concentrations of nutrients and other compounds in the culture are continually altering. Changes in the physiological behaviour of a microbe growing in a batch culture therefore cannot be attributed unequivocally to variations in growth rate or to changes in the concentration of nutrients. But, in the chemostat, the effects of each of these parameters can be studied independently of the other.

Among the most valuable data that have come from chemostat experiments concern the effects of strict nutrient limitation on the composition of microbes. Reference is made in Chapter 3 (page 137) to alterations in the contents of phosphorus-containing polymers in the cell-envelope layers of bacteria grown under conditions of phosphate limitation. Also, many studies have been made on the effects of nutrient limitation on synthesis of enzymes by microbes growing in a chemostat since, under these conditions, the organisms are largely freed from repression of enzyme synthesis by the growth-limiting nutrient (*see* page 355). One such study was made some years ago at the Microbiological Research Establishment at Porton, England. Workers at this establishment studied the effect of carbon source limitation on synthesis of the anaplerotic enzyme **isocitrate lyase** (*see* page 251), in *Pseudomonas ovalis* grown in a chemostat. When the pseudomonad is grown in a batch culture with acetate as carbon source, appreciable amounts of isocitrate lyase are synthesised. But synthesis of the enzyme is prevented by adding a small amount of succinate to the batch culture, and is only restored when all of the succinate has been used up. Since the concentrations of bacteria and of the substrates varied continuously with time in these batch cultures, it was impossible to assess quantitatively the roles of acetate and succinate in controlling synthesis of isocitrate lyase. But, with a chemostat, this can be done. When low concentrations of succinate were included in acetate-limited chemostat cultures of the pseudomonad, synthesis of isocitrate lyase was largely unaffected. Indeed, even when succinate was the sole carbon and energy source in the chemostat culture, these carbon-limited organisms still synthesised substantial amounts of isocitrate lyase. But, when the bacterium was grown continuously in an acetate-containing medium containing growth-limiting concentrations of ammonium ion, synthesis of the enzyme was severely repressed by adding a small amount of succinate. These experiments were among the first to demonstrate unequivocally catabolite repression of anaplerotic enzyme synthesis by succinate in bacteria.

Another example of the use of continuous cultures in microbial physiology came from work in my own laboratory. My colleagues and I were interested in the increased synthesis of unsaturated fatty-acyl residues in microbial lipids when the growth temperature is lowered below the optimum. While it was assumed that this effect was due

to the change in growth temperature, there was the
possibility that it was caused, in part or in whole, by
the lower growth rate or the higher dissolved oxygen
tension in cultures grown at suboptimum temperatures.
Using chemostat cultures of the yeast *Candida utilis,* in
which the dissolved oxygen tension could be controlled,
it was shown that a decrease in growth temperature led
to an increased synthesis of unsaturated fatty acids even
when the growth rate and the dissolved oxygen tension
were kept constant. But, in cultures grown at a fixed
rate and a constant temperature, an increase in the
dissolved oxygen tension in the culture also caused an
increased synthesis of unsaturated fatty acids, indicating
that the effects observed in batch cultures are not caused
solely by the lowering of the growth temperature.

9.4 SURVIVAL

Over the years, a large number of studies have been made
on factors that affect survival of micro-organisms in a
population. Much of the earlier work was concerned with
the effect of chemical and physical factors on microbial
survival and the length of time for which micro-organisms
can remain viable. Recently, a more fundamental
understanding of the physiological basis of microbial
survival has emerged.
   To the microbiologist, a viable micro-organism is one
which is capable of dividing to form one or more viable
daughter cells when provided with a favourable environment.
Many methods have been devised for measuring microbial
viability. Some of these are indirect methods, based on,
for example, the supposed inability of viable micro-
organisms to take up dyes. However, these methods are
usually rejected in favour of those that make a direct
assessment of viability. Undoubtedly the most widely
used method for assessing the viability of unicellular
micro-organisms is the *colony count* (*see* page 375).
Viability measurements using the micro-culture or
slide-culture technique are also widely used, and are
particularly useful in that they provide a ratio of the
total number of organisms to the number of viable
organisms in a population.
   Death of micro-organisms in a population is a
statistical phenomenon and, if the logarithm of the
number of surviving organisms is plotted against time,

a curve known as the *survival curve* is obtained (*Figure
9.11*). Quite often, this curve is linear (curve A;
Figure 9.11) and the slope of the line represents the
death rate of the population. Exponential survival curves
are often obtained following treatment of a population
with heat, ionising radiation or an antimicrobial compound

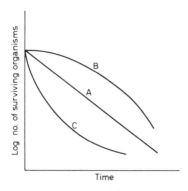

Figure 9.11 *Different types of survival curve for
micro-organisms (see text for explanation)*

such as phenol. It has been shown that, when *Aerobacter
aerogenes* is grown continuously in a medium containing a
growth-limiting concentration of Mg^{2+}, death of these
bacteria when suspended in buffer also follows an
exponential curve. Deviations from the exponential curve
(curves B and C; *Figure 9.11*) have also been reported as,
for example, with *A. aerogenes* grown continuously in media
containing growth-limiting concentrations of either
nitrogen-containing or phosphorus-containing nutrients.
Whereas the straight-line relationship is characteristic
of a unimolecular reaction, these non-exponential curves
have been taken to indicate the operation of bi- and
polymolecular reactions during killing of the micro-
organisms. Several hypotheses have been advanced to
explain non-exponential killing of micro-organisms. One
possibility is that it may be caused by clumping of the
organisms such as frequently occurs with staphylococci,
when all of the organisms in a clump need to be killed
before the loss of one colony-forming unit is observed.
A more widely applicable explanation is that the
population of organisms is heterogeneous with regard to
the ability to survive under the conditions employed.

Most of the studies on the physiological basis of
microbial survival have been made on populations that have
been subjected to starvation, for example, by suspending
them in non-nutrient buffer. The survival of micro-
organisms under starvation conditions has been shown to
depend on a number of factors. With certain bacteria,
the ability to tolerate starvation conditions is inherited.
The nutritional state of the organism too is important.
Thus *A. aerogenes,* grown in a complex non-defined medium,
is longer lived than when cultured in defined medium in
which the availability of mannitol limited growth.
Another interesting finding is that compounds that have
limited growth of organisms often accelerate death when
these organisms are suspended in non-nutrient buffer;
this effect has been termed *substrate-accelerated death.*
Starved populations of *A. aerogenes* also die at a slower
rate the denser the population, which suggests that there
may be some interaction between individuals in a
population. This population effect may be attributed in
part at least to *cryptic growth* which occurs when dead
organisms release nutrients for the multiplication of
survivors, although a population effect has also been
observed under conditions that are not complicated by
cryptic growth.

A few studies have been made on the biochemical changes
that take place in starved bacteria, mainly in *A. aerogenes.*
Dying populations show a rapid initial breakdown of
intracellular RNA with a release of phosphate and
nitrogenous bases. Intracellular protein is degraded
after a lag but intracellular polysaccharide and DNA are
scarcely degraded at all. It has also been shown that
death of *A. aerogenes* by starvation does not necessarily
destroy the osmotic barrier of the organism, for these
bacteria can respond to a mild osmotic stress long after
they have lost the ability to multiply.

It has also been shown that the magnitude of the
adenylate charge of a microbe is an important factor in
its ability to survive (*see* page 342). In cells of *E. coli*
in the exponential phase of growth, the charge has a value
of 0.8. During the stationary phase of growth, and during
starvation in a carbon-limited medium, the value for the
adenylate charge declines slowly to a value of about 0.5
and then falls more rapidly. During the slow decline in
the charge, all of the cells are capable of forming
colonies, in other words they are viable. But the steep
drop in the adenylate charge is accompanied by a rapid
decline in the viability of the organisms in the

population. These data suggest that cells die when their
adenylate charge falls below a value of about O.5, a
finding which is supported by the more limited amount of
data available for adenylate charge of populations of
other microbes that have been starved.

FURTHER READING

GROWTH

Replication and biogenesis of organelles

  REPLICATION OF GENOMES

MATSUSHITA, T. and KUBITSCHEK, H.E. (1975). DNA
  replication in bacteria. *Advances in Microbial
  Physiology,* 12, 247-327
NEWLON, C.S., PETES, T.D., HEREFORD, L.M. and FANGMAN,
  W.L. (1974). Replication of yeast chromosomal DNA.
  Nature (London), 247, 32-35
PATO, M.L. (1972). Regulation of chromosome replication
  and the bacterial cell cycle. *Annual Review of
  Microbiology,* 26, 347-368

  BIOGENESIS OF RIBOSOMES

NOMURA, M. (1970). Bacterial ribosome. *Bacteriological
  Reviews,* 34, 228-277

  REPLICATION AND BIOGENESIS OF MITOCHONDRIA

KROON, A.M. and SACCONE, G. (Eds.) (1974). *The Biogenesis
  of Mitochondria,* 574 pp. New York; Academic Press
LINNANE, A.W., HASLAM, J.M., LUKINS, H.B. and NAGLEY, P.
  (1972). The biogenesis of mitochondria in micro-
  organisms. *Annual Review of Microbiology,* 26, 163-198

  *REPLICATION AND BIOGENESIS OF THE PHOTOSYNTHETIC
  APPARATUS*

KIRK, J.T.O. (1971). Chloroplast structure and
  biogenesis. *Annual Review of Biochemistry,* 40, 161-196

GROWTH OF PLASMA MEMBRANES AND CELL WALLS

BERAN, K. (1968). Budding of yeast cells, their scars
and ageing. *Advances in Microbial Physiology, 2,*
143-171

COLE, R.M. (1965). Symposium on the fine structure and
replication of bacteria and their parts. *Bacteriological
Reviews, 29,* 326-344

FIEDLER, F. and GLASER, L. (1973). Assembly of bacterial
cell walls. *Biochimica et Biophysica Acta, 300,* 467-485

KEPES, A. and AUTISSIER, F. (1972). Topology of membrane
growth in bacteria. *Biochimica et Biophysica Acta, 265,*
443-469

SCHLEIFER, K.H., HAMMES, W.P. and KANDLER, O. (1975).
Effect of endogenous and exogenous factors on the
primary structures of bacterial peptidoglycan. *Advances
in Microbial Physiology, 13,* 245-292

Growth of individual micro-organisms

OBSERVATIONS ON INDIVIDUAL CELLS

MITCHISON, J.M. (1957). The growth of single cells. I.
Schizosaccharomyces pombe. Experimental Cell Research,
13, 244-262

OBSERVATIONS ON RANDOMLY DIVIDING POPULATIONS

COLLINS, J.F. and RICHMOND, M.H. (1962). Rate of growth
of *Bacillus cereus* between divisions. *Journal of
General Microbiology, 28,* 15-33

HARVEY, R.J. (1968). Measurement of cell volumes by
electric sensing zone instruments. In: *Methods in
Cell Physiology,* Ed. D.M. Prescott, Vol.3, pp. 1-24.
New York; Academic Press

OBSERVATIONS ON SYNCHRONOUSLY DIVIDING POPULATIONS

HALVORSON, H.O., CARTER, B.L.A. and TAURO, P. (1971).
Synthesis of enzymes during the cell cycle. *Advances
in Microbial Physiology, 6,* 47-106

MITCHISON, J.M. (1971). *The Biology of the Cell Cycle,*
313 pp. Cambridge University Press

MITCHISON, J.M. (1973). The cell cycle of a eukaryote. *Symposium of the Society for General Microbiology*, 23, 189-208

Growth of microbial populations

BROWN, C.M. and ROSE, A.H. (1969). Fatty-acid composition of *Candida utilis* as affected by growth temperature and dissolved oxygen tension. *Journal of Bacteriology*, 99, 371-378

DAWSON, P.S.S. and KURZ, W.G.W. (1969). Continuous phased culture - a technique for growing, analysing, and using microbial cells. *Biotechnology and Bioengineering*, 11, 843-851

EVANS, C.G.T., HERBERT, D. and TEMPEST, D.W. (1970). The continuous culture of micro-organisms. 2. Construction of a chemostat. In: *Methods in Microbiology*, Eds. J.R. Norris and D.W. Ribbons, Vol.2, pp. 277-327. London; Academic Press

KUBITSCHEK, H. (1970). *Introduction to Research with Continuous Culture*, 210 pp. Prentice Hall

RICICA, J. (1970). Multi-stage systems. In: *Methods in Microbiology*, Eds. J.R. Norris and D.W. Ribbons, Vol.2, pp. 329-438. London; Academic Press

TEMPEST, D.W. (1970). The continuous cultivation of micro-organisms. 1. Theory of the chemostat. In: *Methods in Microbiology*, Eds. J.R. Norris and D.W. Ribbons, Vol.2, pp. 259-276. London; Academic Press

TEMPEST, D.W. (1970). The place of continuous culture in microbiological research. *Advances in Microbial Physiology*, 4, 223-250

SURVIVAL

CHAPMAN, A.G., FALL, L. and ATKINSON, D.E. (1971). Adenylate energy charge in *Escherichia coli* during growth and starvation. *Journal of Bacteriology*, 108, 1072-1086

STRANGE, R.E. (1972). Rapid detection and assessment of sparse microbial populations. *Advances in Microbial Physiology*, 8, 105-141

10
DIFFERENTIATION

When they encounter certain environmental conditions,
especially starvation conditions, some vegetative
microbes produce morphologically different structures.
The range of structures which are formed when vegetative
microbes differentiate is vast. Very many of these
structures have been described in detail, often for
taxonomic purposes, but with a few exceptions little has
been reported on the biochemical changes that take place
when a particular differentiated structure is formed from
a vegetative cell.

Differentiation can be defined as the formation of a
morphologically different form from a vegetative microbe.
The differentiated form may be merely a vegetative cell
that has assumed a different shape. On the other hand
it may be a resistant structure or spore which is formed
inside or outside a cell, or it may be a multicellular
structure formed from a population of single vegetative
microbes. Indeed, some microbial physiologists consider
that growth of vegetative microbes is a differentiation
process, since a newly formed cell grows and then forms
a cross septum (or bud) and so becomes morphologically
different from the original cell. However, the cell
cycle differs from other differentiation processes in
that the end-product and the initial structure, namely
the vegetative cell, are morphologically the same.

The challenge offered in a study of any microbial
differentiation process is daunting. Ideally, to begin
with, the molecular architecture of the differentiated
structure should be established so as to define completely

the result of the differentiation process. When one
considers that knowledge of the molecular architecture
of even vegetative micro-organisms is confined to a
relatively small number of species, it is clear that this
first requirement in studies of differentiation is
exceedingly demanding. When the physiologist comes to
research on the time-course of the molecular events which
lead to the formation of the differentiated structure, this
being the second stage in the overall strategy of research
on differentiation in microbes, the task assumes even more
taxing proportions. Finally, a differentiation process
also involves an ordered arrangement of molecules and
structures in space, often called *vectorial metabolism,*
a phenomenon which is at present not amenable to easy
experimentation.

Some microbiologists would agree with the American
biologist James T. Bonner that differentiation in microbes
should be studied, in the first instance, in order to
understand still further the chemical activities of that
microbe. Nevertheless, many biologists who research on
differentiation in microbes do so because the microbe that
they have selected is a useful *model* a study of which will
hopefully illuminate differentiation phenomena in higher
organisms. For instance, the biochemistry of meiosis in
plant and animal cells is almost completely clothed in
mystery, but a study of ascospore formation in the yeast
Saccharomyces cerevisiae, a process which involves
formation of haploid spores from a diploid cell and which
is very amenable to experimental investigation, could
clarify some of the basic phenomena associated with
meiosis. Similarly, a study of the aggregation of amoebae
in the slime mould *Dictyostelium discoideum* promises to
furnish a valuable insight into the fundamental basis of
cellular aggregation, a phenomenon which is of seminal
importance to the embryologist. The value of microbes as
model organisms in studies on basic differentiation
phenomena largely explains why detailed information on
microbial differentiation is confined to a quite small
number of micro-organisms.

10.1 MORPHOLOGICAL DIFFERENTIATION IN VEGETATIVE ORGANISMS

The shape and size of vegetative cells of micro-organisms are controlled to some extent by the composition of the medium in which they are grown. With some microbes, very little alteration in cell shape and size is observed no matter what the composition of the growth medium, but with other organisms, which are usually referred to as *pleomorphic forms,* cell shape and size are considerably influenced by environmental factors. Studies have been made on some pleomorphic microbes with the aim of establishing the biochemical basis of the switch from one morphological form to another. Two examples are described in the following paragraphs.

10.1.1 CELL SHAPE IN *ARTHROBACTER* SPECIES

Species in the genus *Arthrobacter* and related genera are well known by bacteriologists to be pleomorphic. Fortunately, it is possible to control experimentally the morphological form of arthrobacters with some degree of precision, and this makes them useful for biochemical studies on cellular differentiation.

When *Arthrobacter crystallopoietes* is grown in a glucose-salts medium, it grows more slowly than in a chemically more complex medium but, during growth, the cells remain and continue to divide as spheres. However, if peptone, or an organic acid such as lactate, malate or succinate, or an amino acid like arginine, lysine or phenylalanine, is included in the glucose-salts medium, then the rate of growth of the bacterium is increased and the cells grow not as spheres but as rods.

Since the rigid cell wall which encases a bacterium determines the morphological form of the microbe, a switch from a spherical to a rod-shaped form suggests that changes have occurred in the composition of the cell wall, and most likely in the peptidoglycan which is the fibrillar component in the bacterial wall (*see* page 32). Analysis of walls from the spherical and rod-shaped forms of *A. crystallopoietes* confirmed these suspicions. The tetrapeptide in the peptidoglycan from this bacterium has the structure:

L-alanyl-D-isoglutaminyl-L-lysyl-D-alanine

Many of the ε-amino groups on the lysine residues are
substituted with L-alanine residues, and the peptides are
linked by cross-bridges. A comparison of wall
peptidoglycans from spherical and rod-shaped cells
revealed three main differences in composition. Firstly,
the polymer in rod-shaped cells is more extensively
cross-linked. Secondly, the amino acid glycine is
present in the link between peptides in the peptidoglycan
in spherical but not in rod-shaped cells. Thirdly, the
polysaccharide backbone in the peptidoglycan from
spherical cells is about 40 units long, while that in
rod-shaped cells is about three times longer (*Figure 10.1*).
These differences in peptidoglycan structure are
consistent with the notion that a spherical cell wall
requires more structural flexibility.

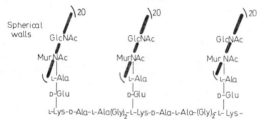

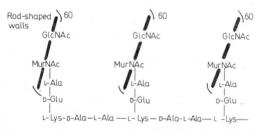

*Figure 10.1 Postulated structures of the peptidoglycans
from the spherical and rod-shaped forms of* Arthrobacter
crystallopoietes. MurNac *indicates a residue of N-acetyl
muramic acid; other abbreviations are explained on page (ix)*

These modifications to the peptidoglycan structure could
result from a repressed (or derepressed) synthesis of
enzymes which catalyse incorporation of glycine into the
cross-link, together with alterations in the activities of
other wall-synthesising enzymes and also of enzymes that
lyse the wall peptidoglycan. Evidence that the activity

of lytic enzymes is important came from the discovery
that spherical cells have a much higher N-acetylmuramidase
activity compared with rod-shaped cells. The activity of
this enzyme could explain the difference in the length of
the polysaccharide backbone in the two peptidoglycans.
However, little is known of the way in which the composition
of the growth medium affects synthesis of the enzymes.

10.1.2 YEAST-MYCELIAL DIMORPHISM

Under certain conditions, some yeasts can grow not in the
characteristic single-celled yeast form but as elongated
cells or mycelium. Conversely, some mycelial fungi can
be induced to grow chiefly or exclusively as budding yeast
cells. The biochemical basis of the switch from the yeast
to the mycelial form of growth has been examined in
several yeasts and fungi, including *Mucor rouxii* and
Paracoccidioides brasiliensis. However, the most detailed
picture available is on the yeast-mycelial dimorphism in
the yeast *Candida albicans*, largely as a result of the
work of Walter Nickerson and his colleagues at Rutgers
University in New Jersey, U.S.A.

If *Candida albicans* is grown in a nutritionally rich
medium, with glucose as a carbon and energy source, it
grows in the yeast form, that is as single discrete cells.
But if the microbe is grown in a nutritionally less
favourable medium, such as one in which the carbon and
energy source is glycogen, filamentous or mycelial growth
occurs (*Figure 10.2*). However, if a compound with a thiol
group, such as cysteine, glutathione or thioglycollate is
included in this medium, the organism continues to grow in
the yeast form. This last finding suggested that an
inability to grow in the yeast form results from a lack
of a suitable amount of reducing power in the microbe,
a deduction which was supported by the discovery that a
mutant of *C. albicans* which ordinarily grows in the
filamentous form can be induced to grow in the yeast form
by including thiol compounds in the growth medium. Later,
the terminal hydrogen acceptor in the cell-division process
was identified as the -S-S- linkages in the cell-wall
glucan-protein complexes (*see* page 30). Reduction and
hence splitting of these linkages is catalysed by a protein
disulphide reductase which is thought to be a flavoprotein. Splitting of the disulphide bridges ruptures
covalent linkages between macromolecular complexes in the

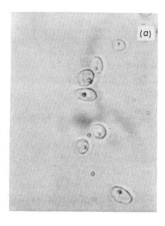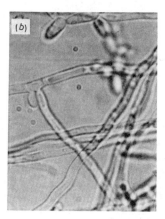

*Figure 10.2   Light micrographs of the yeast form (a) and
the filamentous forms (b) of* Candida albicans.   *(By
courtesy of Solomon Bartnicki-Garcia)*

cell wall, and increases the plasticity of the wall
thereby facilitating bud initiation and extension.

10.2   SPORE FORMATION

Formation of resistant structures or spores is the
response of many microbes when they encounter
nutritionally unfavourable conditions. The diversity in
structure among the spores formed by microbes is vast,
and physiological and biochemical studies have as yet
been made on just a few of these, notably endospore
formation in bacilli, ascospore formation in *Sacch.
cerevisiae,* conidiation in aspergilli, and cyst formation
in certain amoebae. The following account is confined to
a discussion of just two of these processes of spore
formation, ones which have received greatest attention,
namely endospore formation in bacilli, and sporulation in
Sacch. cerevisiae.

10.2.1 ENDOSPORE FORMATION IN BACILLI

Vegetative bacteria of the genus *Bacillus,* when they
encounter unfavourable conditions, produce intracellularly
a spore which is subsequently released from the cell.
A bacterial endospore is a highly efficient survival pack,
because it is as much as 10 000 times as resistant to
heat as the vegetative cell, and up to 100 times more
resistant to ultraviolet radiation. It can survive for
years, possibly centuries, in the absence of nutrients.

The physiology of endospore formation in bacilli is
currently being studied in a number of laboratories
throughout the world, and our understanding of this
differentiation process deepens as each year passes. In
laboratory studies, strains of *Bacillus subtilis* are
preferred. Vegetative cells are grown aerobically,
usually at 37°C, in a rich medium containing hydrolysed
casein, alanine, asparagine and glutamate, and salts.
Exponential-phase cells are harvested, resuspended in a
simple salts solution, and incubated aerobically at 37°C;
and this produces a yield of about 80% of spores in around
eight hours.

Molecular architecture of the bacterial endospore

The anatomy of the bacterial endospore is shown in
Figure 10.3. Until quite recently, little was known of
the molecular architecture of the endospore, largely
because of the difficulties encountered in disrupting
these structures, but the problem has now been solved,
and the anatomical parts of the endospore are revealing
their molecular secrets.

Protoplast. The spore protoplast, sometimes referred to
as the *core,* contains macromolecules, such as DNA, RNA
and enzymes, that are found in the protoplasts of
vegetative cells. Interest in these polymers in the
protoplast of the bacterial endospore centres mainly
around the basis of their resistance to heat and radiation.
The radiation resistance of the DNA is thought to be due
not to the operation of an extremely efficient repair
mechanism, as is the case with *Micrococcus radiodurans*
(*see* page 128), but to the physical state of the DNA in
the spore.

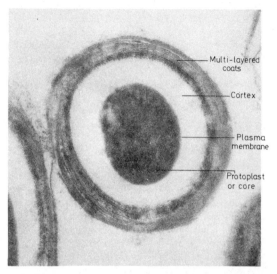

*Figure 10.3   The anatomy of the bacterial endospore.   The
electron micrograph is of a thin section through an
endospore from* Clostridium sporogenes, *and the names given
to the anatomical parts of the endospore are shown.
Magnification is ×50 000.   (The micrograph was kindly
provided by Grahame Gould)*

   Several spore enzymes have been extracted, purified and
examined, and their properties compared with their
counterparts in vegetative cells.   These studies have
revealed that the spore enzyme usually differs little
from that synthesised in the vegetative cell, except that
the spore enzyme is often somewhat smaller in size.   For
instance, the fructose diphosphate aldolase from
endospores of *Bacillus subtilis* has a molecular weight of
40 000 daltons, whereas the corresponding enzyme from
vegetative cells exists in an equilibrium of 115 000 and
70 000 daltons.   The spore enzyme has been shown to be
more heat-stable than that synthesised by the vegetative
cell, and it is believed that it is formed from the
vegetative cell enzyme by proteolytic digestion.   The
suggestion has been made that the smaller protein can be
more easily stabilised by intramolecular bonds, and so
be more resistant to heat.
   Exactly how this stabilisation is effected is far from
clear.   The heat stability of bacterial endospores has

however been correlated with the presence in the protoplast of dipicolinic acid, which is a strong chelating agent and occurs in the core complexed with calcium ions. Dipicolinic acid can account for more than 10% of the dry weight of the spore, and it is a sporulation-specific compound in that it cannot be synthesised by vegetative bacilli. The enzyme which catalyses the terminal step in the pathway leading to synthesis of dipicolinic acid, namely dipicolinic acid synthase, must also be sporulation-specific. It has been suggested that calcium dipicolinate acts as a metal buffer in the spore protoplast, and that its presence alters the conformation of enzymes in the protoplast to one which is relatively stable to heat. Sulpholactic acid (and therefore one or more enzymes that catalyse synthesis of this compound) is another sporulation-specific compound (although it is not found in all bacterial endospores). This compound, the physiological function of which is a complete mystery, may also be located in the spore protoplast.

Cortex. Around the spore protoplast there is a rigid structure known as the cortex. The rigidity of the cortex is attributable to the presence in this layer of peptidoglycan which, interestingly, differs to some extent in structure from the polymer that occurs in walls of vegetative cells. Four main differences have been discovered in the peptidoglycans synthesised by spores and vegetative cells of *B. subtilis*. The spore polymer, but not that in the wall of the vegetative cell, contains the lactam of muramic acid, is less extensively cross-linked, and contains diaminopimelic acid residues which have fewer amide groups. Also, the muramyl residues in the spore polymer are predominantly substituted with L-alanyl residues whereas in the vegetative-cell polymer all of these residues are substituted with peptides. The significance of these differences in peptidoglycan structure is unknown, but they are presumably related to some extent to the shapes of the cell and endospore.

Coats. Surrounding the cortex are layers that go to make up the spore coats, often distinguished as the inner and outer coats. These layers are composed almost entirely of polypeptides that are exceptionally rich in cysteine residues. The coat proteins can be extracted using

alkaline sodium dodecyl sulphate containing sulphydryl-
reducing agents, and they have been shown by immunochemical
and chemical means to be different from any of the
proteins synthesised in the vegetative cell, and so are
sporulation-specific compounds. The hydrophobic
properties of bacterial endospores can be ascribed to the
presence of these proteins in the outer spore coat.

Exosporium. In some bacilli, such as *Bacillus cereus*,
an outer loose envelope surrounds the endospore; this is
known as the exosporium. Few data are available on the
composition of the exosporium, although that formed by
B. cereus is made up mainly of protein and polysaccharide,
with small amounts of lipid.

Biochemical changes during endospore formation

A diagrammatic representation of the cytological changes
which take place during formation of a bacterial
endospore is shown in *Figure 10.4*. These changes, and
the biochemical events which accompany them, are well
defined, and are recognised as distinct stages,
designated O to VII. Stage O represents the vegetative
bacterium which has two chromosomes. In Stage I, which
lasts for one and a half hours ($t_0 - t_{1.5}$), the chromosomes
form a continuous thread which lies along the whole
length of the cell. Each chromosome may be attached to
the plasma membrane by a mesosome. During Stage I, there
is an onset of protein turnover in the cell, and
excretion of antibiotics and of the enzymes amylase,
protease and ribosidase.

The period from one and a half to two and a half hours
($t_{1.5} - t_{2.5}$) constitutes Stage II, and is characterised
by the formation of a spore septum (*Figure 10.4*). During
this stage, the chromosomes become separated, one being
drawn to one end of the cell and enclosed in a compartment
formed by invagination of the plasma membrane. Alanine
dehydrogenase is synthesised during Stage II.

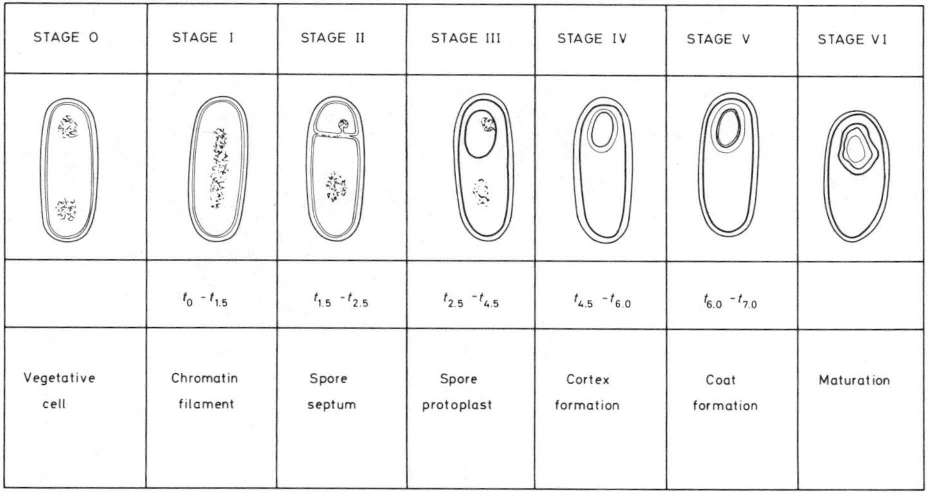

STAGE 0	STAGE I	STAGE II	STAGE III	STAGE IV	STAGE V	STAGE VI
	$t_0 - t_{1.5}$	$t_{1.5} - t_{2.5}$	$t_{2.5} - t_{4.5}$	$t_{4.5} - t_{6.0}$	$t_{6.0} - t_{7.0}$	
Vegetative cell	Chromatin filament	Spore septum	Spore protoplast	Cortex formation	Coat formation	Maturation

Figure 10.4 Diagrammatic representation of the cytological changes that take place during formation of a bacterial endospore. Also indicated are the durations of each stage, in hours, and the major cytological changes which are characteristic of each stage

A recognisable spore protoplast appears in the cell during Stage III, which extends over the period $t_{2.5}$ to $t_{4.5}$. The protoplast is formed by the plasma membrane extending from the point at which it is attached to the septum to the pole of the cell. During Stage III, the cell synthesises enzymes of the TCA and glyoxylate cycles, as well as catalase and alkaline phosphatase. Sulpholactic acid is also synthesised during this stage.

Stage IV lasts from $t_{4.5}$ to $t_{6.0}$, and it is the stage during which the spore cortex is formed. Cortex material is laid down between the double membranes of the spore protoplast and, as a result, the spore is visible in the phase-contrast microscope as a refractile area. Biochemical events which take place in this stage include synthesis of dipicolinic acid and of the enzymes adenosine deaminase and ribosidase. There is also an appreciable uptake of calcium ions.

Spore coats are laid down during Stage V, which lasts from $t_{6.0}$ to $t_{7.0}$. Coat material appears to be synthesised some distance from the cortex, and then to be deposited on this layer. Large amounts of cysteine are incorporated into the coat material. Also during Stage V, the spore becomes resistant to octanol. A final stage (Stage VI) lasts from $t_{7.0}$ to $t_{8.0}$, and involves maturation of the endospore and finally its release from the cell.

Considerable attention has been given to the significance of the biochemical events which take place in the various stages. Certain events are known to be essential for endospore formation, including synthesis of dipicolinic acid and of the cysteine-rich coat proteins. But other events appear to have absolutely no role in spore formation, as shown by the ability of mutant strains which cannot synthesise the particular component to form spores. Biochemical events which come into this category include synthesis of amylase in Stage I and of aconitase in Stage III. The likelihood is that synthesis of these enzymes is derepressed gratuitously by the conditions required to induce spore formation.

Regulation of metabolism during endospore formation

Fortunately, the genome in most species of *Bacillus* can be mapped using transformation and transduction techniques, and these studies have led to the localisation on the genome of certain *spore genes*. However, the situation is complex, since the sequence of spore genes on the chromosome does not correspond with the time of expression of the gene. Moreover, spore genes are scattered over the entire genome, and not confined to one segment.

In as much as sporulation involves expression of certain genes that are not required for vegetative growth, it is reasonable to infer that there occurs an initiation or 'triggering' event which starts the process. Despite considerable speculation, nothing is known of the nature of this triggering event, but it has been established that the process of spore formation can be halted in bacteria providing they are returned to a nutritionally rich medium, or glucose is added to the sporulation medium, before a certain period of incubation in sporulation medium has elapsed. Experiments of this type have led to the concept of *commitment* which can be defined as the physiological point of no return or, in other words, that when committed the cells carry on with sporogenesis even if they are returned to a medium that can support vegetative growth. Recent experiments, especially in Joel Mandelstam's laboratory in the University of Oxford, England, have indicated that there is no single point of commitment for the whole sporulation process. Instead the cells, after initiation in the sporulation medium, become successively committed to one sporulation event after another. Fairly early on, they are committed to synthesising alkaline phosphatase but not at the same time to refractility, whereas later they are committed to refractility but not simultaneously to synthesising dipicolinic acid.

An insight into translational control measures that operate during endospore formation has come from studies in which actinomycin D, which inhibits mRNA formation (*see* page 283), is added at different times to populations of cells that are undergoing spore formation. In this way it has been shown that there are fairly long intervals of time between synthesis of a particular mRNA and the translation of this mRNA molecule to give a protein. For instance, the messenger for alkaline phosphatase is synthesised for about an hour, and indeed most of it has been synthesised by the time any alkaline phosphatase activity is detectable in the cells. *Figure 10.5* gives further examples which suggest that synthesis of long-lived mRNA molecules may be an important process in the control of endospore formation in bacilli. It will be recalled (*see* page 72) that mRNA molecules synthesised in vegetative microbes usually have a half-life shorter than a few minutes.

It would appear, therefore, that control of endospore formation occurs at two levels. There exists control at the transcriptional level when expression of spore genes is triggered off by some as yet unknown mechanism. This is accompanied by control at the translational level when some equally unknown mechanisms permit stable mRNA molecules to be expressed in the same order as that in which the genes are expressed.

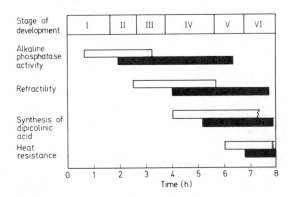

Figure 10.5 Diagram showing the times at which Bacillus subtilis *becomes committed to various events associated with endospore formation. The white rectangles represent intervals of time during which cells acquire the potential ability to carry out the process; these are the periods during which the appropriate mRNAs are synthesised. Black rectangles indicate the periods during which the mRNA molecules are translated. The stages of endospore formation are as indicated in Figure 10.4*

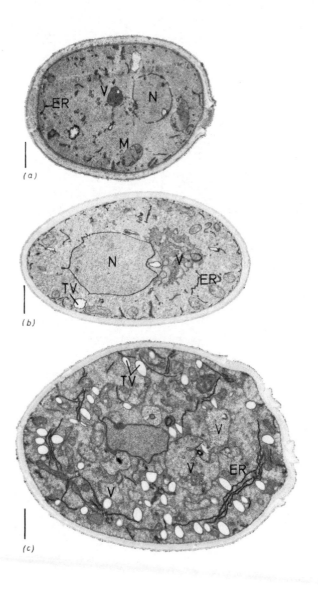

(a)

(b)

(c)

Figure 10.6   See page 413 for details

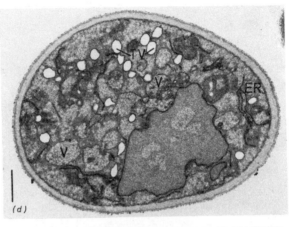

(d)

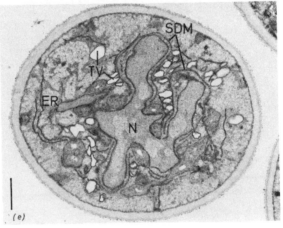

(e)

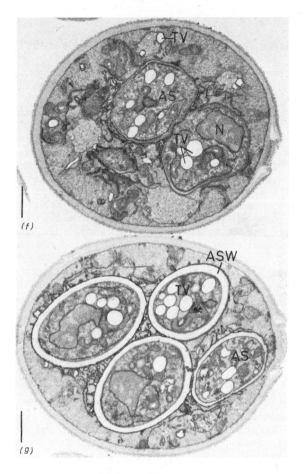

Figure 10.6 Electron micrographs of thin sections through cells of Saccharomyces cerevisiae during ascospore formation. (a) Shows a vegetative cell, and intermediate stages in the formation of a mature ascus are shown in micrographs (b) to (g). (b) Shows the appearance of a lobed vacuole; (c) the formation of numerous electron-transparent vesicles and of an extensive membrane system; (d) the formation of a lobed nucleus containing four haploid genomes; (e) the alignment of vesicles around lobes of the nucleus; (f) the formation of separate spore protoplasts and the beginning of wall growth; and (g) an immature ascus. The abbreviations used are: AS, ascospore; ASW, ascospore wall; ER, endoplasmic reticulum; M, mitochondrion; N, nucleus; SDM, spore-delimiting membranes; TV, electron-transparent vesicle; V, vesicle. The vertical bar on the micrographs represents a length of 1 μm

10.2.2 ASCOSPORE FORMATION IN *SACCHAROMYCES CEREVISIAE*

Diploid strains of *Saccharomyces cerevisiae,* when in a
nutritionally complete medium, divide mitotically by
budding. But if the cells are removed from such a
culture and resuspended in a nitrogen-free nutritionally
poor medium, the cells stop budding and, providing that
they are heterothallic, start a differentiation process
that results in the formation of up to four thick-walled
ascospores in the cell. The cell is then termed an *ascus.*
One of the most commonly used sporulation media contains
just sodium acetate (0.5%) and potassium chloride (1.0%).
Yeast ascospores are only slightly more resistant to
heat and to alcohols than the vegetative cells.

Ascospore formation by strains of *Sacch. cerevisiae* is
being studied by several groups of physiologists, mainly
because it is a useful model system with which to study
the biochemical basis of meiosis. Despite its popularity,
the amount of data available on the molecular anatomy of
the ascospore, the biochemical changes that take place
during spore formation and the way in which these changes
are regulated is small compared with that published on
endospore formation in bacilli.

Molecular architecture of the yeast ascospore

When a thin section through a mature ascus is examined
in the electron microscope, ascospores are seen as roughly
spherical structures with a smooth surface. The spore
protoplast contains a nucleus, mitochondria and vesicles,
and is surrounded by a plasma membrane *(g)* in
Figure 10.6). Outside the plasma membrane there is a
thick electron-transparent spore wall. In mature asci,
the outer edge of each ascospore always appears electron-
dense.

As yet, almost nothing is known of the molecular
architecture of the ascospore formed by *Sacch. cerevisiae.*
To some extent, this dearth of information is explained
by the technical difficulties that are encountered when
attempting to obtain a sizable quantity of ascospores
uncontaminated by ascan material and vegetative cells.
This is most easily achieved using zonal centrifugation
of a preparation of disrupted asci. The only data that
are available on the composition of ascospores are for
certain differences that have been noted between asci

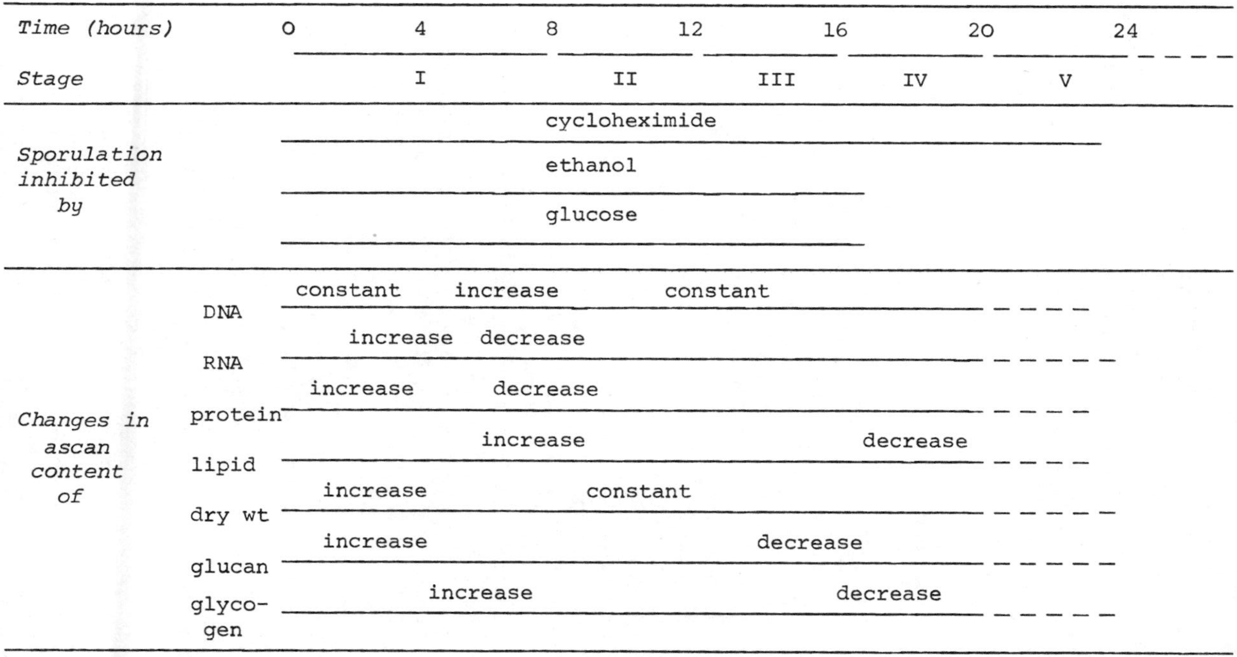

Time (hours)	0	4	8	12	16	20	24
Stage		I	II	III	IV	V	
Sporulation inhibited by		cycloheximide					
			ethanol				
			glucose				
Changes in ascan content of — DNA		constant	increase	constant			
RNA		increase	decrease	decrease			
protein		increase			decrease		
lipid		increase		constant			
dry wt		increase		decrease			
glucan			increase				
glycogen					decrease		

Figure 10.7 A diagram showing stages in the formation of mature asci by Saccharomyces cerevisiae, the effect of inhibitors on the process, and the changes in the composition of the developing ascus

and vegetative cells. In general, it seems that
ascospores are richer in glucan, mannan and trehalose
than vegetative cells, and in addition contain more lipid.
The outermost electron-dense layer, which is responsible
for the hydrophobic properties of the ascospore, is made
up of protein. Conceivably, this protein is a spore-
specific component, for there is no evidence for the
presence of a similar protein in vegetative cells.

Biochemical changes during ascospore formation

The morphological changes which take place during
ascospore formation in *Sacch. cerevisiae* are shown in
Figure 10.6, and the principal biochemical changes are
indicated in *Figure 10.7.* The time taken to form a
mature ascus containing four ascospores varies with the
strain of yeast, and with the composition of the
sporulation medium and the incubation conditions. However,
it is usually possible to detect immature asci in the
light microscope after 24 to 36 hours incubation.
The most important event during ascospore formation in
Sacch. cerevisiae is the production of four haploid
genomes from a single diploid genome. Doubling of the
DNA content of the sporulating cell occurs between t_4
and t_8 (*Figure 10.7*). The RNA content of the developing
ascus also increases from t_4 to t_6, after which there is
a decrease to about 50% at t_{24}. There are two bursts of
RNA synthesis, as measured by incorporation of radioactive
precursors, at t_{10} and t_{25}. It has not been possible to
detect differences between the RNAs in the vegetative
cell and the developing ascus, except for the appearance
in the sporulating cell of a 20S RNA which is a precursor
of 17S RNA (*see* page 72). Protein synthesis is required
for ascospore formation, as shown by the inhibitory action
of cycloheximide (*Figure 10.7*). Important changes take
place in the lipid content and composition during formation
of a mature ascus. The total lipid content of the cell
increases by a factor of about four. This increase is
mainly attributable to an increased synthesis of sterol
esters and triacylglycerols and to a lesser extent of
phospholipids. The increase in phospholipid synthesis
can be explained by the formation of an extensive system
of double membranes (*Figure 10.6*) which become aligned
around the lobes of the nucleus and between which
ascospore wall material is inserted. The increased

synthesis of sterol esters and triacylglycerols correlates
with the appearance in the sporulating cell of electron-
transparent vesicles which appear to be involved in growth
of the envelope layers of the developing ascospore. The
role which these vesicles play in envelope growth is far
from clear, but they could act as carriers of wall
material and of enzymes concerned in synthesis of wall
material, as well as of compounds that are to be
incorporated into the growing membranes.

10.3   FORMATION OF MULTICELLULAR STRUCTURES

Some micro-organisms can undergo a more extensive type of
differentiation and give rise to multicellular structures.
The morphological changes which take place during this
type of differentiation have in many instances been fully
described, but little is known about the biochemical bases
of the changes. Perhaps this is not altogether surprising
when one considers the very incomplete understanding that
microbial physiologists have of the much less extensive
differentiation that occurs, for example, during endospore
formation in bacilli. Serious attention to the biochemical
changes which take place in the formation of multicellular
structures by microbes has been restricted largely to
differentiation systems in the cellular and acellular
slime moulds. The following account is confined to the
cellular slime moulds which are a most interesting group
of microbes, in many ways in the hinterland between true
Protista and the Metaphyta and Metazoa. Of the dozen or
so recognised species of cellular slime mould, most
experimental work has been done with strains of
Dictyostelium discoideum, the life cycle of which is
shown in *Figure 10.8.*

The amoebae are found in humus where they feed on
bacteria. Under certain conditions these protozoa
aggregate and give rise to a grex which is a sausage-
shaped structure that is capable of a limited amount of
movement. Eventually, this mass of cells rights itself
and the leading one-third of the cells differentiate into
stalk cells which have a central cylinder that is stiffened
by cellulose fibres. The rest of the cells stream up to
the top of the stalk where they become capsulated to form
spores. This structure is known as the *sorocarp.* When the
spores are released, they germinate to give amoebae thereby
completing the life cycle.

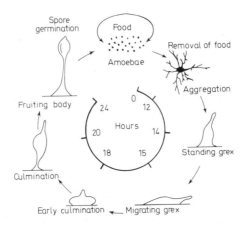

Figure 10.8   Life cycle of the cellular slime mould
Dictyostelium discoideum. *The times refer to the duration
of development on Millipore filters*

In the laboratory, the developmental process is studied
by, firstly, growing amoebae on solidified media sown with
lawns of bacteria, usually *Aerobacter aerogenes*. Development is initiated by harvesting the amoebae from the plate,
centrifuging and washing the suspension of organisms, and
then filtering them through a Millipore filter. Generally,
filters of about 45 mm diameter are used, on which there
is a population of around 10^8 amoebae. As shown in *Figure
10.8*, the developmental process is complete in about 24
hours after the filters and the amoebae have been incubated
at around 25°C in a humid atmosphere.

Several of the individual stages in the development of
the sorocarp by *D. discoideum* have been studied from a
biochemical standpoint, and three of these are now
described further. Considerable attention has been given
to the aggregation phase. It is now generally agreed that
the chemical compound which induces aggregation of amoebae
is 3',5'-cAMP (*see* page 355). Production of cAMP by the
amoebae increases by about 100-fold during the period
between vegetative growth and aggregation as a result of
a relief from catabolite repression. At the same time
that this compound is being produced, however, it is being
destroyed by a phosphodiesterase which is produced by the
aggregating slime moulds. Destruction of the attractant
is probably necessary to increase the concentration

gradient towards the centre of its production. There is evidence that cAMP is not the only attractant used by cellular slime moulds, but the nature of the other compounds involved is as yet unknown. Earlier data suggested that certain amoebae in the population begin secreting cAMP before the main body of the population, and so initiate the aggregation process. It is likely that these initiator cells are aneuploid, rather than haploid like the majority of the amoebae, and since they possess more genes can secrete more cAMP as a result of a gene-dosage effect.

Another process which has attracted attention is the spatial arrangement of cells in the developing grex. Specifically, the physiologist would like to know the mechanism by which anterior cells are caused to form stalks in the sorocarp, and posterior cells spores. If the structure of the grex is deranged by excising either portion, then the small slugs which are formed quickly adjust and the correct ratio of cells is restored. How this happens is an absolute mystery. It appears that the regulation system must somehow include a mechanism by which each cell is aware of its position in the aggregate and by which cells make their presence known to other cells. One possible explanation is that there exists a gradient of some chemical compound along the long axis of the grex, but what this compound is, if indeed it exists, is not known.

Finally, a number of physiologists have studied the development of enzymes that catalyse synthesis of oligosaccharides and polysaccharides that go to make up the mature sorocarp, and in particular of the spore-wall mucopolysaccharide and of the disaccharide trehalose which is an energy reserve in the spore. These studies have concentrated on four enzymes, namely UDP-glucose pyrophosphorylase, UDP-galactose epimerase, UDP-galactose polysaccharide transferase, and trehalose phosphate synthetase. The changes in activity of each of these enzymes in the developing sorocarp have been extensively studied, and one interesting finding which has emerged is that the mRNAs which are involved in synthesis of certain of these enzymes are, like some of those encountered in synthesis of some components of the bacterial endospore (page 410), long-lived.

The biochemistry of development in *D. discoideum* will continue to provide challenges to the developmental biologist for many years to come, and the information

which will emerge from this research promises to be most
valuable in illuminating developmental processes in
higher plants and animals.

FURTHER READING

GENERAL

ASHWORTH, J.M. and SMITH, J.E. (Eds.) (1973). Microbial
    differentiation. *Symposium of the Society for General
    Microbiology,* 23, 1-450

SMITH, J.E. and BERRY, D.R. (1974). An introduction to
    the biochemistry of fungal development. 326 pp.
    London; Academic Press

Morphological differentiation in vegetative organisms

BARTNICKI-GARCIA, S. and McMURROUGH, I. (1971).
    Biochemistry of morphogenesis in yeasts. In: *The
    Yeasts,* Eds. A.H. Rose and J.S. Harrison, Vol.2,
    pp. 441-491, London; Academic Press
CLARK, J.B. (1972). Morphogenesis in the genus
    Arthrobacter. *Critical Reviews in Microbiology,* 1,
    521-544

Spore formation

ENDOSPORE FORMATION IN BACILLI

DAWES, I. and HANSEN, J.N. (1972). Morphogenesis in
    sporulating bacilli. *Critical Reviews in Microbiology,*
    1, 479-520
MANDELSTAM, J. (1971). Recurring patterns during
    development in primitive organisms. *Symposium of the
    Society for Experimental Biology,* 25, 1-26

ASCOSPORE FORMATION IN SACCHAROMYCES CEREVISIAE

ILLINGWORTH, R.F., ROSE, A.H. and BECKETT, A. (1973).
    Changes in the lipid composition and fine structure of
    Saccharomyces cerevisiae during ascus formation.
    Journal of Bacteriology, 113, 373-386

TINGLE, M., SINGH KLAR, A.J., HENRY, S.A. and HALVORSON, H.O. (1973). Ascospore formation in yeast. *Symposium of the Society for General Microbiology*, 23, 209-243

Formation of multicellular structures

ASHWORTH, J.M. (1971). Cell development in the cellular slime mould *Dictyostelium discoideum*. *Symposium of the Society for Experimental Biology*, 25, 27-49
GARROD, D. and ASHWORTH, J.M. (1973). Development of the cellular slime mould *Dictyostelium discoideum*. *Symposium of the Society for General Microbiology*, 23, 407-435
KILLICK, K.A. and WRIGHT, B.E. (1974). Regulation of enzyme activity during differentiation in *Dictyostelium discoideum*. *Annual Review of Microbiology*, 28, 139-166
NEWELL, P.C. (1971). The development of the cellular slime mould *Dictyostelium discoideum*: a model system for the study of cellular differentiation. *Essays in Biochemistry*, 7, 87-126

INDEX

423

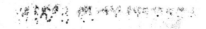